RELENTLESS SKIES

The Life of Australia's
Pioneer Aviator and Wartime Commander
Air Vice-Marshal Don Bennett CB CBE DSO FRAeS

Volume 1:
The Most Efficient Airman (1910–1942)

Ian Campbell

Copyright © Ian Campbell, 2024
Published 2024 by the Book Reality Experience, an imprint of Leschenault Press, Leschenault, Western Australia

ISBN: 9781923020450 - Paperback Edition
ISBN: 9781923020641 - Hardback Edition
ISBN: 9781923020658 - E-Book Edition

All rights reserved.

The right of Ian Campbell to be identified as author of this Work has been asserted by him in accordance with sections 77 and 78 of the Copyright, Designs and Patents Act 1988. No part of this publication may be reproduced, stored in retrieval systems, copied in any form or by any means, electronic, mechanical, photocopying, recording or otherwise transmitted without written permission from the publisher. You must not circulate this book in any format.

All source material has been fully acknowledged and used with permissions where applicable. Should you feel any material within is in breach of copyright please contact the publisher in the first instance.

The author asserts that no Artificial Intelligence methods, techniques or tools have been used within the researching or production of this biography.

Without in any way limiting the author's [and publisher's] exclusive rights under copyright, any use of this publication to "train" generative artificial intelligence (AI) technologies to generate text is expressly prohibited. The author reserves all rights to license uses of this work for generative AI training and development of machine learning language models.

Cover Design by Brittany Wilson | Brittwilsonart.com

Recommended citation: Campbell I 2024. Relentless Skies, Vol 1: The Most Efficient Airman (1910–1942). The Book Reality Experience, Leschenault, Western Australia.

In memory of

Allan J. Vial DFC OAM Chev Ld'H [Fra] OPR [Pol] COM [Pol] MRIN
Pathfinder, Navigator and Life President of the PFF Association (Qld)

Allan's post-war friendship with Air Vice-Marshal Don Bennett and his wife, Ly, led to a lifetime quest to see Don recognised for his feats as a pioneering aviator and for his contribution to the Allied war effort in the Second World War.

Allan's tireless work is the sole reason the Bennett archival legacy is preserved at the Queensland Air Museum as part of Australian and international aviation history.

Contents

Foreword i
Introduction iii
Acknowledgements ix
Maps xii
Author's Note xv
1. Decision Time at 'Kanimbla' 1
2. A Glimpse of the Future 4
3. The Last Hope of the Bennetts 7
4. *Nil Sine Labore* – the Brisbane Years 19
5. The Hinkler Effect 25
6. Entering the New World 33
7. Fighter Boy 50
8. Boats That Fly 59
9. Under Harris's Wing 68
10. Back to the Future 78
11. The Invisible Barrier Over the Horizon 83
12. The Unshackling Begins 90
13. The Vega's Leg 94
14. Ly and her Complete Air Navigator 107
15. Imperial, Empire and 'Empires' 119
16. *Mercury* and the Atlantic 136
17. *Mercury* and an Obsession with Records 171

18. War 187
19. Rescuing the Polish General Staff 196
20. Fast Tracking the Ferry 206
21. Beaten by Bottlenecks and Bowhill 223
22. A Very Short Hiatus 234
23. 77 Squadron – Enter the 'Hot' War 240
24. Bureaucracy Sows the Seeds of Bitterness 261
25. Shot Down 267
26. Escape 277
Appendix 1 Chronology 289
Appendix 2 Aircraft Don Bennett flew to July 1942 292
About the Author 293
Bibliography 294
Endnotes 310
Index
    People 342
    Places 351
    Air Force 359
    Aircraft 362
    General 366

Foreword

As a former Commander of the Royal Australian Air Force's F-111 Strike Reconnaissance Group, I am honoured to write this foreword for Ian Campbell's book, *Relentless Skies*. This well researched, very readable and engaging biography brings to life the extraordinary story of an Australian aviation pioneer and wartime commander, Air Vice-Marshal Don Bennett CB CBE DSO FRAeS. Don Bennett is perhaps best known as the Commander of the Path Finder Force of the Royal Air Force's Bomber Command during World War Two, but his earlier career in both civil aviation and the military in Australia and the United Kingdom is less well-known. This first volume, *The Most Efficient Airman (1910–1942)*, addresses this deficiency.

My first exposure to the Don Bennett story resulted from one of the annual visits to the F-111 base at RAAF Amberley by the members of the Pathfinder Force Association. This association was established by the returning Australian airmen who had flown as part of the Path Finder Force during World War Two. At Amberley, the Association members presented annually a 'Young Pathfinder' award to one member of the top strike aircrews and charged them to preserve the PFF wartime heritage. The award continues to this day.

Although Australian airmen and airwomen played a significant role in the development of aviation in its formative years, it is unfortunate that many of the early pioneers are not well known. Best recognized in Australia are those who captured the headlines with their exploits – people such as Charles Kingsford Smith and Bert Hinkler. However, there were many others who contributed greatly to the fledgling civil aviation industry as well as to the development of military aviation. Don Bennett is, I believe, one of these.

i

Although there have been two previous biographies of Don Bennett, this is the most comprehensive and the first written by an Australian. In writing this first volume, Ian has conducted very meticulous and extensive research, drawing on sources not available to previous authors such as Don's lecture notes from his pilots' course at RAAF Point Cook. This research has brought to light much greater detail about Don's pre-war career and exploits, including the precarious rescue of the Polish General Staff from France in 1940 and his escape from Norway after being shot down on the *Tirpitz* raid in 1942.

As the Curator of the Queensland Air Museum's Don Bennett Archive, Ian has developed a deep understanding of Don the person and how his character was manifested in his aviation exploits and achievements. What comes to light is that Don was truly a master aviator. Not only was he an exceptional pilot and a brilliant navigator, he was also a very competent mechanic and a skilled wireless operator. He was fascinated by how the aircraft equipment and systems worked and often made detailed notes and drawings of the aircraft he flew and the systems he operated. On reading the manuscript I truly came to know Don Bennett the person and aviator and the aircraft he flew.

I congratulate Ian on producing this scholarly work that not only relates to us the early aviation career of a true pioneer, but also allows us to understand the complex character of Don Bennett. I commend this book to all those with an interest in aviation history and I eagerly await the second volume.

Air Vice-Marshal David Dunlop CSC (RAAF Ret'd)
Caloundra, Queensland, March 2024

Introduction

This is the first volume of a two-part work on the aviation life of Air Vice-Marshal Donald CT Bennett CB CBE DSO FRAeS, an Australian pioneer aviator and wartime commander.

From when he sighted his first aeroplane in Toowoomba, Queensland, at the age of three in 1913, through to his death in 1986, aviation underwent an astounding array of developments, forged in peace and war, in civil and military aircraft. Don Bennett participated in many of those developments. He piloted dozens of aircraft, including the unique and celebrated *Mercury*, the Mayo Composite floatplane. During the war, he was the Air Officer Commanding, Path Finder Force, Bomber Command, making a significant contribution to the strategic air offensive against Nazi Germany, and in so doing becoming the youngest ever air vice-marshal of the Royal Air Force (RAF) at the age of 33.

This volume begins with his early life in Queensland, including who and what shaped that life, and his determination to pursue a career in aviation despite family expectations and pressures to consider more traditional options. It explores the foundational aspects of his character and what drove him. His time at Point Cook, then the Royal Australian Air Force's (RAAF) pilot training school, is considered in detail in the context of military aviation of the period, followed by his short stint as a fighter pilot in England. It was during his time operating RAF flying boats that he developed a reputation for navigation, met Arthur Harris, bought his own aircraft, went moonlighting with a civil airline, obtained every possible aviation qualification, participated in the 1934 Centenary Air Race, and married a Swiss girl called Ly.

He departed the RAF in 1935, considered returning to Australia, then joined Imperial Airways. Piloting flying boats and land-based aircraft, he flew the Mediterranean and more exotic routes in Africa and the Middle

East, South Asia and Southeast Asia. He was part of the Golden Age of aviation with the Empire flying boats of the late 1930s. He set world records in *Mercury*, participated in the first in-flight refuelling of a commercial aircraft, helped pioneer commercial transatlantic flights, and made a daring rescue of the Polish General Staff near Bordeaux, one day before the fall of France in 1940.

Recalled to join the war effort, Don played a lead role in establishing the Atlantic Ferry, flying American military aircraft across the North Atlantic for the first time. On rejoining the RAF, he entered what he called the 'hot' war as Officer Commanding No 77 Squadron in Bomber Command, flying a range of operations over Europe. It was on his taking command of No 10 Squadron that he participated in an ill-fated raid to bomb the German battleship *Tirpitz*. The story of this raid, the loss of his Halifax, and his escape from the Germans to neutral Sweden is recounted here in detail.

Within weeks of returning to England he was appointed by Air Marshal Sir Arthur Harris, head of Bomber Command, as commander of the newly established Path Finder Force. He was 31. That period of his life, and how his career unfolded post-war, is the subject of the second volume.

I decided early to call him 'Don', not 'Bennett', for a multitude of reasons.

My primary concern was the tendency of some biographies to be a collection of events, activities and achievements, leaving the reader little wiser about the person. Calling him 'Don' is part of an attempt to show the man behind the activities and achievements, inviting readers on a more personal journey.

Further, Don is principally known as Air Vice-Marshal Donald CT Bennett, Air Officer Commanding, Path Finder Force, Bomber Command. He was in that role, however, for less than three years of his 76-year life. The trap is seeing everything in his life as leading up to that period or examining his post-war life in light of that period; the 'Air Vice-Marshal' dominates. He never saw his time as AOC PFF as the zenith of his life and career, significant as it was to his reputation and standing; he did so much else. I hope referring to him as 'Don' throughout gives a sense of continuity and connection with the person through all stages of his career, irrespective of titles and positions.

Using a first name is also a very Australian thing, countering officiousness and class structures with what some societies would regard as over-familiarity; he is an egalitarian 'Don'. Calling people by their surnames or their titles runs against the Australian cultural grain. Don himself struggled with the class structures and social mores of Britain.

Finally, calling him 'Don' reflects my own roots in south-east Queensland. To me, he was a young kid born in Toowoomba, 'up the range' from where many of my paternal relatives lived. The family home of his teenage years was in the Brisbane suburb adjacent to where my maternal grandparents lived. He went to the same school as my father, albeit in different eras. He loved body surfing at the Gold Coast, just as I did when growing up. I know the street in downtown Brisbane where his father, George, had his real estate agency. He wasn't born an air vice-marshal or a pioneering aviator, just 'Don'.

This is a biography largely devoted to Don's aviation life, not his life as a whole. Such is the extent of his achievements in the field and the events of which he was a part, it still requires two volumes. Only periodically do I venture into his family life. His early years involved family members who played a significant role in his growing up, and so are given far greater treatment than in previous biographies. His wife, Ly, also plays a prominent part, not just because she was his soulmate, but because she played such an active role in his professional life. His two children, Noreen and Torix, are mentioned infrequently. Despite his high public profile in Britain, Don was a private person and preferred a private family life. I don't believe this biography is diminished by my respecting that.

That said, I am thoroughly convinced I came to understand Don the person very well indeed because, to a significant extent, aviation was his life. Readers will not be left wondering about Don's character by my focusing predominantly on that life.

In one of Don's handwritten notes about his life, he is drawn to a quote Sholto Watt included as the preface to his book *I'll Take the High Road*. The quote is an extract from a letter Voltaire wrote to M Bertin de Rocheret on 14 April 1732 about the difficulties of writing contemporary history.

He highlighted several problems, including officers offering entirely different accounts of the same battle and everyone thinking they deserve a mention whatever their rank. 'If the subalterns grumble at my silence,' Voltaire wrote, 'the generals and ministers complain of my outspokenness.' Don was particularly attracted to the end of the Voltaire quote, writing it down in full:

> Who so writes the history of his own time must expect to be attacked for everything he has said, and for everything he has not said: but those little drawbacks should not discourage man who loves truth and liberty ...[1]

When Don made those notes is not known, but Watt's book was published in 1960, two years after Don had published his memoirs, *Pathfinder*.[2] The memoirs were controversial, and rightly so; Don was never one to mince words. He encountered significant pushback from contemporaries who challenged his recollections, leaving him to feel the heat of the 'outspokenness' Voltaire had referred to. But, in aligning himself with Voltaire's noble ideals, he likewise would not be dissuaded from recording history as he saw it.

Voltaire raised vital issues when it comes to writing history, of which Don's memoirs and now this biography are a part. Hilary Mantel, author of the *Wolf Hall* trilogy, understood Voltaire's dilemma.

> Evidence is always partial. Facts are not truth, though they are part of it – information is not knowledge. And history is not the past – it is the method we have evolved of organising our ignorance of the past. It's the record of what's left on the record ... It is no more 'the past' than a birth certificate is a birth, or a script is a performance, or a map is a journey. It is the multiplication of the evidence of

fallible and biased witnesses, combined with incomplete accounts of actions not fully understood by the people who performed them. It's no more than the best we can do, and often it falls short of that.[3]

Voltaire's 'subalterns', 'generals' and 'ministers' are all, in Mantel's terms, 'fallible and biased witnesses' of the events of which they have been a part. Don is no different, and my research was made considerably more complex by having to wrestle with the recollections of his memoirs. Any objective reader of *Pathfinder* will pick up Don's sense of grievance at how he believed he was treated, especially during the war; there is an undercurrent of frustration and bitterness. Despite his many achievements, he tends towards self-promotion which, on occasions, is achieved at the expense of others. His renowned sarcasm is regularly on display. There are strong polemical undercurrents.

Despite the warning signals, many biographies and short pieces on Don's life quote or paraphrase his memoirs assuming his recollections were accurate and without bias. I discovered early in my research, however, that many failed to withstand scrutiny, starting with his recalling seeing the Wright brothers at Toowoomba when he was very young. These errors come as something of a surprise given his prodigious memory and his ready access to materials to ensure the accuracy of his accounts.

This produced a significant conundrum because *Pathfinder* is, necessarily, a principal biographical source. It could not just be set aside, nor should it be. My starting point was knowing it contained what Don thought was important about his aviation life up until 1958. Notes he made for writing *Pathfinder* helped confirm that. If these elements were important to him, they were important to me, highlighting key avenues for research. The style of writing itself – how he thought about things and the way he expressed himself – reveal elements of Don's character.

Ultimately, my approach accorded with the adage: 'Trust, but verify.' I would accept Don's account until clear evidence proved his recollections incorrect, or the balance of probability suggested wariness. I have regularly interacted with *Pathfinder* in the narrative, drawing Don's recollections alongside other accounts and sources. I recognise twin dangers in following this path: that some may conclude *Pathfinder* is severely diminished as an historical document because of its many inaccuracies; and that the narrative is marred by my 'sparring' with Don over his recollections. Hopefully not. Errors or no, memoirs are always valuable, if not essential, in coming to understand the person. Don's are no different.

In late 1999, *The Australian* newspaper marked the passing of the 20th century by publishing a collection of names of prominent Australians under the overarching title 'This Living Century'. These individuals were grouped under different fields of endeavour, one of which was 'The Warriors', a selection of ten military commanders who had held senior rank and engaged in significant combat operations. First on the list was General Sir John Monash, followed by General Sir Thomas Blamey. Third was Air Vice-Marshal Donald Bennett. Those making the selection acknowledged the air vice-marshal was probably the most controversial choice because he served most of his time in the RAF, not the RAAF. However, as they said, 'he was to play a major role in one of the most formidable campaigns of World War II', the Allied air war against Germany.[4]

Don Bennett has never been a household name in Australia like Monash and Blamey. There isn't a street or suburb in Canberra named after him, like so many other prominent military figures. After leaving Australia in 1931 on gaining his wings, he never returned to live here, carving out both military and civil careers overseas. I hope this biography contributes to Don's recognition in Australia as both a pioneer aviator and a wartime leader of note.

For those already familiar with elements of Don's life, this goes further than previous biographies by providing new information and new insights and, in doing so, corrects some errors about his life that have persisted over the years. And in that context, another hope is that those of other countries and cultures who may have formed stereotypical views of Don based on his being Australian will come to see him on his own terms, and with greater nuance.

That said, it is, as Mantel says, no more than the best I can do and, in recognition of my own fallibilities, may well fall short of that. Hopefully, readers will find something that broadens their understanding of Don, the contributions he made and the times of which he was part.

Acknowledgements

In the past three years of research and writing, I have had the privilege of encountering an extraordinary array of people whose expertise and knowledge has underpinned every part of this book.

Archivists Vivien Harris at Brisbane Grammar School and Denise Miller at Toowoomba Grammar School helped with the school years of Don and his brothers. Gary McKay kindly emailed me his work on the Real Estate Institute of Queensland. Pat McCallum filled out my understanding of 'Kanimbla', the Bennett family property. Dr Bill Metcalf not only provided his research, especially regarding Dr TP Lucas, Don's grandfather, but took me for a walk around South Brisbane to see places of historical importance to the Bennett family story.

Steve Campbell-Wright sent me his book on the history of RAAF Point Cook. John Evans was of considerable help regarding Don's time at No 210 (Flying Boat) Squadron, Pembroke Dock, Wales. He also provided invaluable information on Don's colleague at RAF Calshot, 'Crackers' Carey. Carmel John Attard, who lives on Malta, helped with Don's DH.53 Humming Bird.

Julia Wallis mined the Jimmy Woods archive at the State Library of Western Australia to find material for the chapter on the Centenary Air Race. There were some magnificent finds. That story can now be recounted with fresh detail.

Regarding Don's time at Imperial Airways, my heartfelt thanks to Paul Sheehan. Nobody knows more about the aircraft of Imperial Airways than Paul. He patiently built my understanding of those extraordinary years in Don's life, layer by layer. He would furnish original documents and photographs, I would do my research, and we would discuss my findings and conclusions. We even made some fresh discoveries together; the chapter on Don's rescuing of the Polish General Staff is testimony to that.

Through Paul, we obtained some excellent material from Alice Vivancos, Musée de l'Hydraviation, Biscarrosse, France.

Tony Pilmer, Librarian and Archivist at the Royal Aeronautical Society, was crucial to my understanding Don's fellowship.

The Atlantic Ferry is a complex story that took much longer to research than I had anticipated. I am grateful to Sandra Seaward at the North Atlantic Aviation Museum, and Robert Pelley, both in Gander, Newfoundland, for their input. Isabella Sun at the Directorate of History and Heritage, Department of National Defence, Canada, went above and beyond in finding a range of archival materials. I thank Carl Christie for his definitive work on the Atlantic Ferry – *Ocean Bridge*. His book anchored all my other research.

My thanks to Paul Markham of the 77 Squadron Association for kindly and enthusiastically laying the groundwork for my research on Don's time as commanding officer of the squadron. I was only able to build a comprehensive picture of Don's participation in the ill-fated attack on the *Tirpitz* due to Nigel Smith's book – *Tirpitz: The Halifax Raids*. A particular value of this book lay in his recording the accounts of aircrew, including on Don's Halifax. David Gray, at the University of Dalarna in Sweden, filled out my understanding of Don's time at the Falun internment camp, and produced an exquisite gem of an anecdote going to the core of Don's character.

Those stalwarts of Australia's aviation history fraternity – Ron Cuskelly and the late Geoff Goodall – helped considerably by sharing their wealth of knowledge, either directly or through their websites.

In my pursuit of photos of aircraft Don flew, I was greatly assisted by Ron Dubar and Bill Pippin (1000AircraftPhotos.com), David Carter and Peter de Jong (AirHistory.net), Linzee Duncan (archieRAF.co.uk), Dick Flute (UKairfieldguide.net), Julie O'Donoghue at the Foynes Flying Boat & Maritime Museum, Stuart McKay of the de Havilland Moth Club UK, James Osborne of Mediahuis Ireland, and Phil Vabre of the Civil Aviation Historical Society in Australia.

The leadership at the Queensland Air Museum, which houses the Don Bennett Archive, has been consistently supportive throughout. They have continued to encourage my work curating the archive, which is being done for the first time. Their support has been crucial in my use of the archive to write what is the first comprehensive Australian biography of Don.

In tandem has been the support and guidance of Gary Vial, the son of Allan Vial. Allan, a friend of Don's, was the driving force behind assembling the archive.

Kerri Setch and Di Bricknell performed their magic on the photographs and maps respectively.

There was no question after his superlative work on my previous book, *Thinks He's a Bird*, about Pathfinder pilot Flight Lieutenant Keith Watson, that I would turn again to Andy Wright for his editorial expertise. Nothing escapes his eagle eye, whether it be a matter of aviation detail or an obscure issue of grammar. He has remained tolerant and of good humour throughout.

My thanks to Ian Hooper and his team at Leschenault Press for their expertise, responsiveness and encouragement. They relieved me of a range of publishing issues, allowing me to focus on the manuscript.

Last of all, my wife, Kathy, who made all the usual sacrifices, delivered with good cheer and the subtle wisdom that comes from knowing me better than anyone else.

Maps

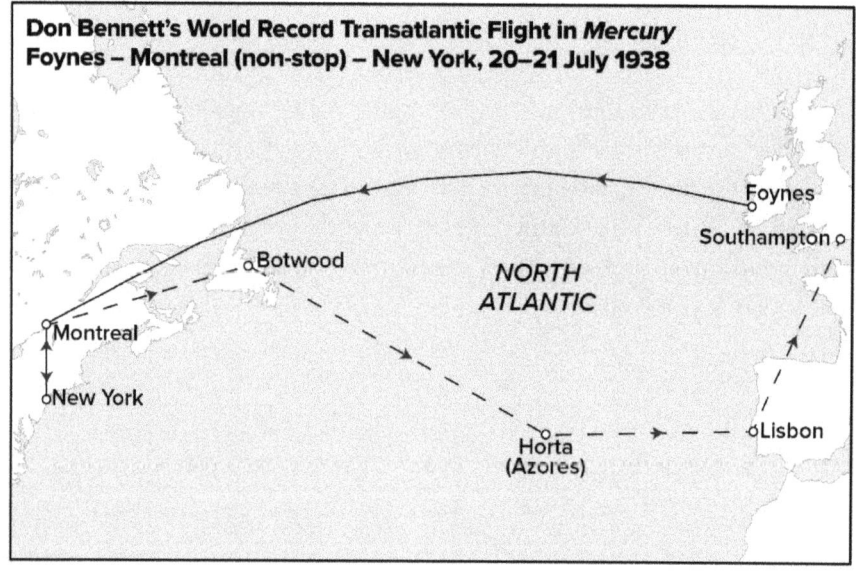

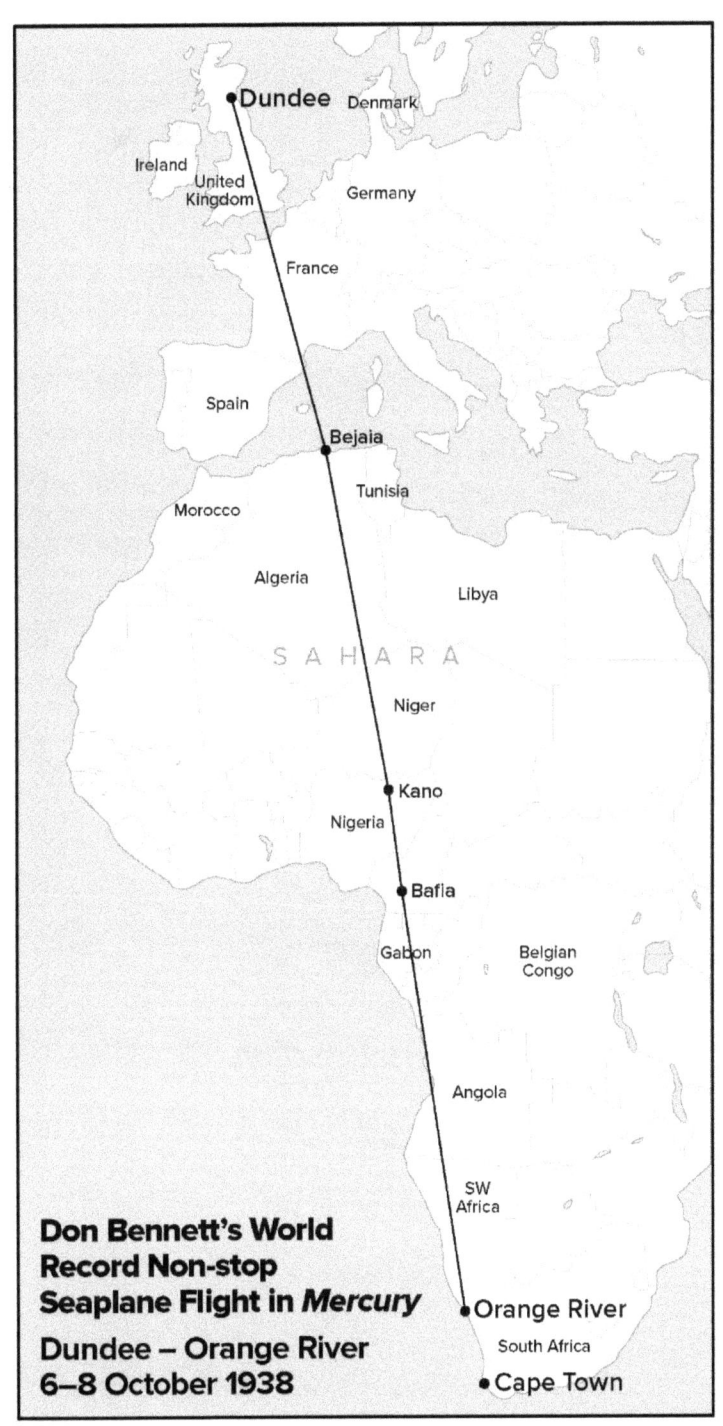

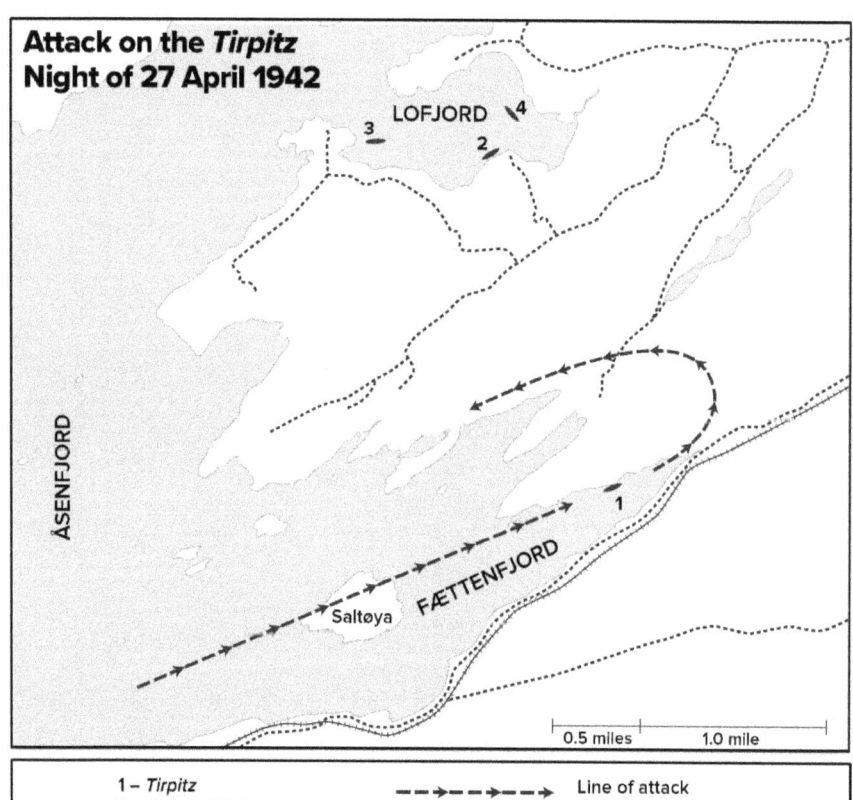

Author's Note

Measurements

In keeping with the times, all measurements – distances, speeds and weights – are in imperial not metric.

Distance:
   1 mile = 0.87 nautical miles (nm) = 1.61 kilometres
   1 nautical mile (nm) = 1.15 miles = 1.85 kilometres
   1 foot = 0.305 metres (m)

Speed:
   1 mph = 0.87 knots = 1.61 kph
   1 knot = 1.15 mph = 1.85 kph

Weights:
   1 pound (lb) = 0.45 kilograms (kg)

Terminology

In the choice between 'Path Finder Force' and 'Pathfinder Force', I decided on the former, despite the latter being common. My reasoning was straightforward: Don referred to it in his memoirs as the 'Path Finder Force'; it was the title on the No 8 Group crest; and it lends itself more easily to the acronym 'PFF'. That said, 'Pathfinder Force' was often used during the war, including by Air Chief Marshal Sir Arthur Harris, head of Bomber Command.

I have, however, adopted the familiar 'Pathfinders' to identify those who were a part of the Force.

Aircraft

I am aware aviation enthusiasts will be reading this biography not just for a greater understanding of Don, but to learn more of the aircraft he flew. Many will want to know not just the type, but the actual individual aircraft.

I have done my best in this regard, but it was not possible in every instance, nor appropriate. At all times, such details have been included to serve the biographical nature of the narrative. Appendix 2 offers a full list of aircraft types Don flew between July 1930 and July 1942.

1
Decision Time at 'Kanimbla'

Don Bennett had painted himself into a corner. He had been messing around for years, unable to find a direction that would motivate him to knuckle down and apply himself. His limitless energies had been expended on things unrelated to his school studies. Little in the grammar school syllabus attracted and held the interest of his precocious intellect. He had got away with loafing, as he himself called it, for a long time. Suddenly and unceremoniously, his father, George, had finally called him on it.

Now he was out on the family's cattle station 'Kanimbla' just south of Condamine, about 200 miles west of Brisbane. It was late 1927, early 1928; he had just turned 17. Like all the Bennett boys, Don was attracted to the station as something of a rugged counterpoint to life first in Toowoomba and later when he was living in Brisbane.

> My environment at 'Kanimbla' was of horseflesh, of 180 square miles of sandy soil often parched and spoilt by prickly pear. It was sleeping on the ground with a saddle for a pillow, hearing the howl of the dingo, and the raucous screams of the cockatoo. In short, it taught me the toughness of an outdoor life.[1]

Over the years, on the periodic trips out to 'Kanimbla', he had learned to be the self-described 'ordinary jackaroo', just like his older brothers, the station hand doing the hard work with axe and barbed wire, boundary riding to check the fences, and mustering and branding cattle. It sure beat being in school.

Ostensibly, he was now there to consider forging a career on the land, centred, at least initially, on managing the station on his father's behalf.

Away from the family home back in Brisbane he had, at last, spent time thinking deeply about his future.

Reaching the decision about what to do with his life was the product of his thoroughly analytical mind progressively eliminating the careers he definitely did *not* want to pursue. Those options involved too much academic effort in subjects of no real interest in order to live up to family traditions and expectations, as his brothers had done. His parents had expected him to follow a medical career like his maternal grandfather, Dr TP Lucas, and one of his older brothers, Aubrey. He had drifted along with those expectations. Given his intelligence, medicine was well within his reach academically had it been of interest.

Failing the choice of medicine as a career, he could work for his father in the family real estate business in downtown Brisbane, as his brother Clyde had done prior to going into the law. Law had also been the choice of his other brother, Arnold, two years his senior. He considered both those options.[2]

Or he could manage 'Kanimbla'. His father would have settled for that but, much as he loved the place, a few weeks traversing it had convinced him it was a career 'dead end – absolutely' and 'utterly devoid of prospects'.[3]

Once his mind was made up, he set out from 'Kanimbla' in early 1928 with what became his trademark ferocious, single-minded, take-no-prisoners determination: he would have a career in aviation. There was no time to waste. The transformation from his prior indecision, indolence and drift was swift and thoroughly remarkable, made all the more so by his knowing little about the world of aviation.

Reflecting on this seminal turning point decades later, he admitted to having often wondered how much the famous early aviators had influenced his decision.[4] As will be seen, many of his heroes did not make their famous flights until after his return to Brisbane from 'Kanimbla' and, as such, had no influence on it, but their arrival in the skies soon after undoubtedly reinforced and enlivened that decision.

How he conducted his subsequent career suggested the skies, and those who flew them, offered the prospect of engaging his love of mechanics and getting his hands dirty, exploring constantly emerging technologies, and deploying the only school subjects he had aptitude for and interest in, namely physics, chemistry and mathematics. It all seemed wrapped up in some inherent drive to demonstrate he could forge a path in the world that would make him worthy of the family name.

Whereas boundaries on the ground – family life and expectations, the confinements of grammar school traditions, accepted career pathways for the best and brightest – all suggested he find fulfilment through conformity, the world above the ground spoke of limitless horizons where boundaries only existed for the purpose of being crossed. The sky was the one place capable of accommodating that combination of an enormous, restless intellect, an inexhaustible energy, and the desire to master all things mechanical. He was not anti-social, but he did prefer his own company and setting his own path without having to comply with the less palatable demands of others. There would be less need for that typically blunt reaction he offered to anyone trying to hem him in due to their failure to understand the import of what he was doing or who felt it was acceptable to waste his time. Free of all strictures – aside from the immutable laws of physics and mathematics – the relentless skies were where he would prove himself.

2

A Glimpse of the Future

Don's recollection that he saw the Wright brothers give a flying display at Toowoomba as a very young child is partly true and partly a false memory.[1] He was, in fact, nearly three when he saw his first flying machine, but it was not the Wright brothers. In 1981, his brother Arnold recorded what happened:

> My first memory of an aircraft goes right back to about 1913 ... when I was five years old. The aeroplane, a flimsy thing of wooden struts, wires and fabric, was, I believe the first ever to come to our town, Toowoomba, then only a country town with a population of about 20,000. We children were playing at my uncle's home at Ascot on the western side of the town, when 'the thing' appeared. It seemed to swoop down just over the trees. Somebody shouted it was going to land at the racecourse. One of my elder brothers caught me and my little brother [Don] by the hand and dragged us racing in that direction. Everybody seemed to be running towards the racecourse. Fortunately, the plane continued to circle at what would now be a prohibited level and we were able to arrive at the racecourse in time to see it landing.
>
> The aircraft, a bi-plane [*sic*], seemed to us to be a most wonderful creation, and everybody was excited about it, and laughing. Man was vying with the birds, though there were still those who said, 'If God had intended us to fly, he would have given us wings.'[2]

Nearly 70 years after the event, Arnold's recollection was unerringly accurate. The day was Saturday 14 June 1913, and the aviator one Mr

A Glimpse of the Future

Arthur W Jones of Shropshire. He had emigrated to Australia around 1909, settling in the town of Warwick, 50 miles south of Toowoomba, where he worked at the Warwick Motor Car Company.[3] He built a Blériot monoplane in his spare time. After some experimental flights, he returned to England, spending several weeks at the Aviation School at Hendon and purchasing a single-engine, two-seat Caudron G biplane. On returning to Warwick in March 1913, he commenced a career as a professional aviator, joining a few other early aviation pioneers, such as AB 'Wizard' Stone, doing exhibition shows in Australian country towns and cities.[4]

These exhibitions followed a typical pattern. The pilot placed an advertisement in the local paper announcing the date of his arrival to do his show, usually at the local racecourse. Admission was one shilling to get into the ground and an additional shilling to do a close-up inspection of the flying machine or occupy one of the seats in the stands with the best views. The pilot might perform exhibitions or 'stunts', or take people on joyrides.

Jones duly landed at the Clifford Park Racecourse at Toowoomba on 14 June, where an estimated 1,000 people had paid for admission.[5] Many more, including the Bennett boys, gathered outside the grounds. The whole place was abuzz. The first manoeuvre was nothing more than accelerating across the racecourse, ascending to 15 or 20 feet and landing again. He then conducted his 'show trip', rising to between 150 and 200 feet and flying a circuit.

> Every eye strained to follow his dangerous career. When he came towards the grandstand, the pent-up feelings of the onlookers burst out with loud plaudits of cheering and clapping, for the exciting suspense created by his perilous position now gave way to a sense of relief at the hero's completely successful management of his machine. The people were delighted to see him come through his ordeal safe and sound.[6]

Having made a smooth landing, he was swamped by those paying the extra shilling to get up close 'to examine the wonderful contrivance which had afforded them so pleasing an exhibition'.[7]

The day before, *The Toowoomba Chronicle* had done an extended piece on Jones. The First World War was still over a year away, but the tone of the article was decidedly military:

> Aviation, with the practical use of flying-machines, as a means of defence, is a matter of the deepest concern to all nations at the present time ... There is no doubt in the minds of the military strategists that these inventions with the terrible powers they can command, are destined to play an important part in the next great war ... It is very important therefore, that our ruling authorities should feel themselves bound to give the matter of aviation their serious and immediate consideration ... A few powerful and reliable aeroplanes might inflict more damage and cause a hostile foe more reason to pause in an intended attack, than other military measures we might adopt, that would cost infinitely more money.[8]

In that context, the reporter outlined who he regarded as the type of person most suited to piloting such flying machines:

> Aviators are born, not made. Certain qualifications are essential to make a successful aviator that are inherent in his nature. He must have an iron nerve, cool head, quick perception and indomitable courage, with many other qualities seldom found combined in an individual.[9]

The correspondent may have had Jones in mind, but as a description of the character of the forthcoming aviator, Don Bennett, who would rush to the racecourse the following day to glimpse his future, it could not be surpassed. Perhaps he was born to it.

3

The Last Hope of the Bennetts

Three years before Mr Arthur Jones brought his Caudron to Toowoomba, Donald Clifford Tyndall Bennett was born on 14 September 1910 at 'Fairthorpe', the family home at 117 West Street, just over a mile from the racecourse.

He was the youngest of four boys who had arrived in a two-by-two arrangement, all in Toowoomba. Clyde and Aubrey had been born in May 1900 and December 1901 respectively. Seven years passed before Arnold made his presence felt in November 1908, with Don bringing up the rear just under two years later.

They were in Toowoomba because their father, George Thomas, had established a real estate business in Ruthven Street – G.T. Bennett and Co. – with his brother Fred. Typical of a country town, the business took on the roles of 'auctioneer, land, estate, stock, and station agent',[1] selling and renting properties in Toowoomba while handling rural properties in the surrounding Darling Downs.

During those years, George had purchased 'Kanimbla', a sizeable rural property on the southern outskirts of Condamine, 120 miles to the west, for running cattle. 'Kanimbla' comprised a number of leases, totalling 30,731 acres, described as having '… Open Box, Brigalow, Belah, Sandalwood … splendidly watered by several creeks and lagoons; 6-roomed house, man's room and storeroom, dairy and meat house, cart shed and motor garage … also good cattle yards.'[2] However, it was also the land of the prickly pear, and figures for the number of cattle[3] suggest that although George's heart lay in the land, 'Kanimbla' was an insufficiently productive concern, even at that size, to sustain the family.

Consequently, he had appointed a farm manager, Steve Patch, in 1915 to oversee the work because the family, while visiting regularly from Toowoomba, and later Brisbane, would not be living there permanently.[4]

Don's great-grandfather, Isaac Bennett, a tailor and draper,[5] had emigrated from England to Queensland with his wife and seven children in the mid-1800s. Their fifth child, George Henry, married Sarah Makepeace in 1871. She was a Brisbane girl, from Moggill, cementing the family connection with the city. A year later, George and Sarah's first, George Thomas, was born in Rockhampton.

What George Henry did is something of a mystery. In saying he came from 'a family of cattlemen',[6] Don suggests his paternal grandfather lived and worked on the land. That he moved the family to the country in southern Queensland likely explains Don's father's love of the land, too. George Thomas may even have traversed the area around Condamine in his younger years, setting his heart on having a cattle property of his own and ultimately being able to afford 'Kanimbla'.

Although Don grew up in the suburbs, once in England he appears to have emphasised this 'cattlemen' aspect of his heritage, along with detailing his time out at 'Kanimbla' as a jackaroo. Perhaps he enjoyed feeding the perception he was a bit 'rough and ready' from the Australian outback rather than the city grammar school boy. Or it may have been a distraction from what transpired during his grammar school days as a teenager.

Methodism was the Bennett religious DNA through several generations. Aside from the strongly held beliefs and practices, it was the catalyst for bringing Don's parents together.

After a number of adventures, the young George Thomas Bennett found himself in Brisbane in the early 1890s, attending the Kangaroo Point Methodist Church, directly across the river from downtown Brisbane. Here he fell in love with Celia, the daughter of an infamous and eccentric local doctor, Thomas Pennington Lucas. The good doctor proved to be a

The Last Hope of the Bennetts

hard nut to crack when it came to agreeing to let his daughter marry George, but they finally tied the knot at the church on 4 March 1896. Sometime before May 1900, when Clyde was born, George and Celia moved from Brisbane to Toowoomba to start the real estate business, remaining there until 1919.

As George built the business, the Bennetts decided to educate their children at Toowoomba Grammar School (TGS), founded in 1875 and run along traditional English lines. Once Clyde and Aubrey began their schooling, George and Celia must have realised quite quickly they had some very bright boys on their hands.

In the *Toowoomba Grammar School Magazine and Old Boys' Register*, the boys' achievements feature regularly. In November 1914, Clyde, then Form IV Prefect, described the form's contribution to school life over the previous six months, concluding with this evocative vignette of empire:

> We were sorry to say good-bye to Mr. Willmott, after having had him with us only a few weeks. He has gone to serve his king and country, and we hope to receive Christmas cards from him from Berlin (if not this Xmas, the following one).[7]

Both boys consistently won awards in Mathematics. Aubrey took a prize in essay writing and later in languages. He would go on to win the W.H. Groom Prize for best pass in the Junior examinations. In 1917, they found themselves on opposite sides at the Debating Society engaging on the subject: 'That a finer career is open to one who chooses a military life than to one who chooses the Navy.'[8]

They were both sporting types, particularly running, with Clyde also in the TGS Rugby First XV. He headed for Law School, with Aubrey sailing into Sydney Medical School two years later having won the Form VI 'General' prize. Don said they were both 'dux' of the school.

Arnold and Don began school life at the Toowoomba Prep School, called 'Gill's' after its founder, EA Gill. How well they did academically is unknown, but their achievements later in life suggest George and Celia's second brace was just as intellectually capable as their first, if not more so.

With four exceptionally bright boys, Celia came to refer to Don as 'the last hope of the Bennetts'.[9] Early on, she probably meant it as a pure statement of pride given all her sons seemed destined to be high achievers, each offering great hopes for the Bennett name. Later, after the family moved to Brisbane, she may have uttered it with more than a hint of irony as Don was spending his teenage years giving every impression of becoming a 'no hoper'.

The contrast between Don's immediate Bennett and Lucas forebears could not have been stronger in terms of class and interests:

> Dr Thomas Pennington Lucas was in some ways typical of his era and social class. He was a misogynistic, ambitious, hard-working, scientifically and religiously driven Victorian gentleman who became well-known across a variety of spheres. He was also atypical in being driven by the conviction that he could save humanity through his fundamentalist religious views, in his arguments against surgery and the germ theory of disease, and particularly through his vehement commitment to the use of fermented papaw as a panacea. In some respects, Lucas was generations behind time – but in other respects he would fit into today as a New Age, holistic healer and religious fanatic. He was an eccentric Queensland writer, scientist, inventor, medical doctor and utopian dreamer.[10]

He was also Celia's father and Don's maternal grandfather.

Dr Lucas,[11] the inventor of the well-known 'Lucas' Papaw Ointment', was born in Scotland, then raised in England. His father, Samuel, a Wesleyan Methodist minister, was given to wrestling with marrying his religious convictions and science, particularly due to the 19th century advances in geology and evolution. While perpetuating his father's strongly held religious beliefs, Thomas studied medicine and homeopathy, opening a surgery in Lambeth, London, along with a chemist shop next door for making his own medicines.

His first wife, Mary, bore him six children, including Celia in 1874, before dying the following year. Thomas developed tuberculosis shortly after. On recovering, he took a job as the surgeon on the *Essex*, departing in December 1876 for Melbourne. Celia, not yet three, and her two siblings (three had died young) were left in England with friends and relatives, effectively orphans.

Thomas opened a medical practice in Melbourne in 1877. Having re-married, another Mary, he brought the children out from England in 1879 and added to their number. He moved the family to Brisbane in 1886, establishing a new medical practice.

Mary died in 1888, leaving Thomas with five children, including Celia, now 14. Shortly after, Thomas re-married, this time to Susan, the middle-aged daughter of a local Congregationalist minister, who the family later described as 'a real tartar and a source of great unhappiness'.[12]

Amid all the family trauma, Thomas simply carried on with his medical practice while indulging other interests such as natural history and, as his father had done, the intersection of science and religion. He also began doing scientific experiments on papaws, leading to 'Lucas' Papaw Ointment', which he claimed cured everything from constipation to cancer.

He was already on the wrong side of the medical establishment when the Bubonic Plague broke out in Brisbane in 1900. Lucas mounted a sustained campaign that it was no such thing, convinced it was influenza or 'dengue catarrh'. Everyone and everything was in the firing line: government, 'paid officials', the medical establishment and the class system.

On the nose with the medical establishment in Australia, he took his ointment to England, where he was likewise rebuffed. 'The British medical establishment, like that in Australia, regarded Lucas as a crackpot.'[13]

After a number of legal travails back in Australia around his business activities, he died in November 1917. The 'Lucas' Papaw Ointment' business would thrive under coming generations.

Given Thomas died when Don was seven, it is hard to know how well Don knew his grandfather or felt the dominance of his presence in family life. Evidently enough to weaponize one of his grandfather's sayings – 'The only crime an Englishman cannot forgive is to be right'[14] – when he, too, encountered boundaries and blockages in England after moving there in 1931.

Thomas's influence on Don was most on view through Celia. The trauma of her own upbringing, due in large part to her father's absenteeism from parental duties, appears to have instilled in her a clear determination to do the opposite with her own children. She was a wonderful mother and her children loved her very much in return.

Interestingly, rather than reject religious faith due to her difficult upbringing and her father's dystopian fundamentalism, Celia embraced it, making it integral to how she lived and raised her own family. During Don's teenage years when the family were living in Brisbane, Bennett family life revolved around the Auchenflower Methodist Church to a significant degree. At one point, Don progressed from Sunday School student to teacher.[15]

Celia also held strongly to her English roots despite having left the Mother Country at a very young age. Don referred to it as an 'intense love'.[16] This may have given Don an overly optimistic or rose-coloured picture of England or led him to believe he understood the English better than it turned out to be.

Perhaps the strongest element Don inherited of his grandfather's approach to life was the stupendous work ethic, provided, of course, he thought it was something worth working at. This came not just from Don's maternal side. His father, George, also had a very strong work ethic. One contributor may have been a perception he had married above his station in life, a need to prove to Dr Lucas he was worthy of Celia's hand in marriage.

Another was his being firmly in the mould of the Queensland Wesleyan Methodists, characterised by a pietistic Evangelicalism, one element of which was 'the relationship between piety and prosperity, or more specifically between poverty and impiety'.[17] This included a strong self-reliance streak more commonly expressed as, 'God helps those who help themselves.' Nothing would, indeed should, fall in your lap; the abundant blessings of God were realised in your life by dint of hard work. It was integral to a religious life of dutifulness, devoutness, adhering to religious practices (including temperance and protesting against the 'desecration of the Sabbath'), following biblical commands, and showing reverence for God. In this regard, George and Celia spoke with one voice. In the

Brisbane region, Wesleyan Methodists were strong among the small business community and the Bennetts played a prominent part.

Attempting to understand something of Don's own religious outlook is crucial to understanding how he lived his adult life. Having an extensive Wesleyan Methodist pedigree on both sides of the family, English and Australian, certainly set the tone.

Don retained some religious belief throughout his life, living out elements of the Methodist moral code, including the strong work ethic and teetotalism. The latter became a source of difficulty at times, particularly in the Air Force where alcohol played such a significant role in social life. He also upheld other Methodist shibboleths including refraining from smoking and swearing.

Given his introverted nature and desire for a private home life, it is difficult to tell exactly how central a role Don's religious faith played in his life. (One thing is for certain, in any clash between science and religion, a preoccupation of his maternal grandfather and great-grandfather, science would win every time: it is inconceivable Don thought prayer an integral part of detailed flight planning.) There are clues, however. He recalls an encounter with a Church of England Church Army lady whom he picked up as a hitchhiker on the outbreak of the Second World War:

> [In her view] the only way of preventing warfare was Christianity. Now, I have myself been brought up by a most religious mother, and I have tried to be Christian, but I have also tried to be a practical man, and I am afraid I turned on the good Church Army lady very harshly indeed.[18]

In his memoirs, Don returned to two hymns his mother taught him which he regarded as summarising his mother's religious teaching: 'Dare to be a Daniel' and 'If I were a beautiful [twinkling] star'.[19] Whether they did or not, the chorus of the first hymn speaks volumes about how Don interpreted, and in adult life applied, her teaching:

> Dare to be a Daniel
> Dare to stand alone
> Dare to have a purpose firm
> Dare to make it known

As he saw it, to stand alone, to declare the rightness of your position and hold firm, placed you in the tradition of the brave biblical heroes who stood up to authority, made their case from the moral high ground, and persisted with it irrespective of the consequences. It was an archetype of a moral nobility imbued with honesty, integrity and honour.

Having this sense of rightness, that he somehow always knew with headstrong certainty what he stood for, what he was doing and where he was going, is a common picture of Don. But the contents of a remarkable notebook, dating from his teenage years in Brisbane, show an earnest young fellow searching for a framework by which to live his life as a man. While religious elements definitely play a part, he ranges much more widely.

The first entry is Rudyard Kipling's poem *If*. The poem is written from a father to his son. It lays down a series of character traits that, if mastered, will qualify the son for manhood. It is essentially a treatise on how to behave, how to get along with others, and how to respond when things do not go well in your life. It is strewn with elements of the adult Don: 'If you can trust yourself when all men doubt you', 'If you can bear to hear the truth you've spoken, twisted by knaves to make a trap for fools', 'If you can force your heart and nerve and sinew to serve your turn long after they are gone, and so hold on when there is nothing in you except the Will which says to them: "Hold on"', 'If you can talk with crowds and keep your virtue, or walk with Kings – nor lose the common touch'. Finally, and tellingly speaking to that Wesleyan work ethic:

> If you can fill the unforgiving minute
> With sixty seconds' worth of distance run,
> Yours is the Earth and everything that's in it,
> And–which is more–**YOU'LL BE A MAN, MY SON!**[20]

Don writes the final words in capitals, in large bold print. Perhaps the poem was from his mother or father, a set of criteria to aim for, or practise, so he could be sure to walk through the appropriate door into adulthood.

The following page of his notebook is headed 'LIFE AND GOD', in large print, highlighted in blue and red:

> Fear God and keep his commandments for this is the duty of man, viz:-
>
>> Worship only God,
>> Observe the Sabbath.
>> Don't be profane.
>
> Love one another as Christ loved, hence; honour father and mother; Do not Steal, Kill, Commit adultery, Bear false witness, Covet other's things.
>
> For God shall bring every work into judgement with every secret thing whether it be good, or whether it be evil.

Evidence suggests he kept all the commands except two. 'Observing the Sabbath' would be difficult given his chosen career. 'Do not kill' would go by the way due to bombing the enemy while on operations in the Second World War or strategising for others to do so. Perhaps three if you include 'Love one another'. Air Vice-Marshal Sir Ralph Cochrane, a Royal Air Force contemporary, undoubtedly believed Don did not show him a lot of love. Mind you, Don was adamant he did not receive much either!

Perhaps the most crucial section in the notebook, running across two pages, are his 'MAXIMS OF LIFE', an eclectic amalgam of quotes accumulated over time but, as the consistency of handwriting shows, all written in one sitting. He offers no source for any except the last:

> Life is an arrow – therefore you must know,
> What mark to aim at, how to use the bow.
> Then draw to the head and let it go.
>
> All success depends on the possession and pushing
> of an idea that betters the condition of humanity.
>
> Think out your work, then work out your think.

Nobody knows what he can do until he tries.

Everything comes to him who hustles while he waits.

Better to climb and fall than never to strive at all.

Success don't consist in never making blunders, but in never making the same one twice.

When you don't know whether to fight or not, always fight.

The great thing in life is not where we stand, but in what direction we are moving.

A Gentleman is one who knows his promises, obligations and duties to his fellow men, even to who cannot enforce him to do so. [To which he adds …] One with a God-like capacity to think and feel for others irrespective of a rank.

If you would be a MAN, speak what you think today … and tomorrow, speak what you think tomorrow, even though it may contradict everything you said today.

The Cheerful Optimist makes the progress of Humanity.

Look for your profits in the 'earts of friends, for 'ating pays no dividends.

Lack of Definite Instructions causes much inefficiency.

Few will stand in the way of a man with a purpose which he is determined to carry into effect. (Arn) [Arnold, his brother]

Written across the bottom of the two pages is:

'WILLING WORKERS WORK WONDERS'

At the top of the second page, inscribed with a different pen and undoubtedly added after the above entries is: 'Better be not at all than be not noble. Aub.' Aubrey would deliver this saying to Don at a crucial time in early 1928, showing the other entries were written prior to that date.

They are, as the heading suggests, not contributions to an overall coherent philosophy of life, but a collection of adages and aphorisms that attracted his attention sufficiently to become lodged in his memory. That said, anyone with knowledge of the adult Don would see just how determinative or influential his thinking was at this early stage on how he lived his life.

There is no doubt he consulted the notebook in adult life because he noted down thoughts on scraps of paper in his typical scrawl and tucked them in it. None can be dated, unfortunately, though the handwriting is 'mature Don'. He draws on biblical verses from the books of Romans and Proverbs, such as:

> Do not fight other people's conceptions of the Ideals of Christianity and concentrate on living up to your own. R[omans] 14.
>
> [Proverbs] Ch 10: 4. But the hand of the diligent maketh rich.
> 8. A prating fool shall fall.

He notes down 'Proverbs 6:16-19' in large handwriting on another piece of paper. The verses list seven things the Lord finds 'detestable' including 'hands that shed innocent blood'. Although he provides no explanation nor context, the verses have struck him hard enough that he has not only noted down the reference but put it with his lifetime notebook.

One scrap of paper makes direct reference to the second of his maxims with:

> Humanity requires:-
>
> (1) Food
> (2) Clothing
> (3) Comforts
> (4) Pleasure
> (5) Education
>
> Transport is the servant of all these and in itself is a fundamental requirement.

Without overstating the case, the evidence suggests the notebook entries continued to serve as a periodic point of reference, elements of a kind of rudimentary personal moral code that dovetailed with his Methodist upbringing. Whether he jettisoned anything of what he had written down as he matured is not known. Some elements he would find difficult to keep; others he would reframe or expand upon to encompass life's experiences or later thinking. The various sections of the notebook and its inserts offer us a tantalising glimpse of his inner private life and the ideas he found impactful, especially in those crucial early years when he was laying down a framework for living while searching for direction.

4
Nil Sine Labore – the Brisbane Years

In April 1919, George handed over the Toowoomba business to his brother Fred and headed back to Brisbane where he set up G.T. Bennett and Co. in the Union Bank building on Queen Street. The business grew over time. George ended up serving a term as president of the Real Estate Institute of Queensland with many more years on its board of management.[1]

Celia returned to Brisbane with Arnold and Don at the end of 1919 after Aubrey matriculated and headed for Sydney. Don was nine. He lived in 'Greenbank', the family home at 6 Kellett Street, Auchenflower, an inner-city suburb, throughout his teenage years until he joined the Air Force.

He attended Windsor State School before going to Brisbane Grammar School (BGS) in 1924, two years behind Arnold. He gained his Junior matriculation in 1926 with Passes in English, Arithmetic, Algebra, Geometry, History and Chemistry, and a Merit for Physics, but somewhere in those years the wheels fell off. School life seemed to bore him. Later life suggests this may have been because he combined a massive intellect with a formidable memory, enabling him to synthesise vast amounts of information quickly and master subjects easily, so he had no need to try. It did not help having to do subjects he regarded of no use or interest, while not responding well to grammar school regimentation and discipline either. His slothful approach to schooling was not necessarily a repudiation of the BGS motto *nil sine labore* (nothing without work), just that BGS did not offer much worth working at, and what it did offer of value required no work at all.

The problems emanating from what Don was confronting, or not confronting, at school were compounded by his parents. While he later claimed his brothers counteracted any spoiling he might have been afforded as the 'baby of the family', he doesn't dispute it occurred. The evidence suggests Celia certainly gave Don plenty of latitude in those Brisbane years. Was he indulged or simply too strong-willed to be controlled? Maybe the former had led to the latter. Commenting on Don receiving a Companion of the Bath in June 1944, Celia volunteered, 'I think he was born a man.' It prefaced a series of comments about Don's years in Brisbane, where he continued to avoid study ('Nobody will make me swot') while indulging a range of passions such as cars, sport, and outdoor pursuits with Arnold, including camping in the open spaces around Brisbane, and canoeing, either on the river or down at what later became the famous Gold Coast.[2] Like their father, the boys also enjoyed time at the beach swimming and bodysurfing. It is hard to tell whether Celia really thought Don was maturing earlier than other children or just that he was independently minded and self-sufficient from a young age. Whichever it was, he delivered his rebelliousness in a cheeky, obstinate, dismissive manner and mostly got away with it.

Despite regularly expressing misgivings at Don's lack of application at school, Celia hints there was more than a touch of *laissez faire* in the parental approach: 'They [Arnold and Don] were boys, and I expected them to be boys. But I was in a simmer over them sometimes.'[3] 'Simmer' did not exactly speak of a tough disciplinary regimen for young Don. The seeming lack of discipline was something of a parental contradiction given both parents expected their boys to excel academically, and that demanded commitment to schoolwork.

George did not help. For one thing, he had passed on to all his boys his love of the land and the outdoors. It appears they all enjoyed the regular, sometimes lengthy, periods out at 'Kanimbla', where the contrast with city life and school could not have been greater. George also did nothing to dissuade Don from indulging his emerging love of cars and all things mechanical, a love George himself shared. Don learned to steer a Model T Ford while sitting on his father's lap and by age 11 was behind the wheel proper of his father's Overland. He says his father 'was always generous, letting us boys have cars, and we often had private races with our friends'.[4]

Celia recalled Don driving the family car fast down Kellett Street, much to the dismay of neighbours. (To stand on this street opposite the Bennett house and take in the steep gradient, ending at the bottom of the slope at a T-intersection, is to appreciate why Don gave the neighbours conniptions.) Don's response was a simple 'I've got to practise',[5] while dismissing any suggestion he might be a danger to anyone. It was an early sign he was prepared to take risks, but not without assessing them. It also spoke of a particular character trait: once he found a subject or activity of interest, he had to keep practising it until he knew he had mastered it completely. There was no middle ground, you either ignored it as irrelevant or you gave it everything you had.

This applied as much to the engines of cars as driving them. George was interviewed about Don for a broadcast in October 1944. 'According to his father … [Don] was a very easy-going youth, his outstanding talent being a genius for taking the family car to bits – and putting it back together again.'[6]

Being around cars, understanding how they worked and how to improve them, was infinitely more interesting, challenging and useful than sitting behind a school desk. The very first reference to Don in the Australian press dates from 5 December 1928. The Daily Commercial News & Shipping List record of Applications for Letters of Patent received from 1 to 7 November, includes:

> 16.549. D.C.T. Bennett. Transmission mechanism for use on motor vehicles.[7]

As it turned out, it was not just cars catching his interest but aeroplanes too. Arnold mentions two episodes, the first of which does not appear in Don's memoirs:

> My first flight was as a school boy in the early twenties during Easter holidays at Southport, on what is now Queensland's famous Gold Coast. My young brother and I managed to extract ten shillings each from our father for a joy ride. Looking back, it must have been pretty

hectic, though we thought nothing of it at the time. The pilot in his Cirrus or Gypsy [*sic*] Moth was operating from Main Beach, north of the surfing area. The flight took us out over the ocean, and then back across the boat passage above which we orbited in what must have been extremely steep turns. I remember leaning away from the wing which seemed to be pointing straight down to the water. Then the pilot landed the aircraft on the sloping beach in a strong crosswind ... The pilot had to avoid the many spectators, too, ...[8]

Then this:

My youngest brother (then a teenager) had really become smitten and, for some time, had been helping a friend to build an aeroplane. They built it, but the friend, taking off from a farm paddock, hit the boundary fence and that was the end of their work.[9]

Don recalls it as a Farman biplane that he had helped his neighbour rebuild.[10] This story will be relayed shortly in its context.

In October 1927, just after Don's 17th birthday, George received Don's latest school report and decided he had given his son sufficient time to mend his slothful ways and confront that perpetual drift. He was pulling Don out of Brisbane Grammar School. There was no point in putting good money after bad either at BGS or for a university course should Don somehow pull himself together academically and matriculate. He would force his youngest to make some decisions.

The school did its very best to put a positive spin on Don's academic performance with this entry in the BGS General Register on his departure: 'Of fair ability and steady industry.'[11] It would surely come to rate as one of the understatements of the century. Given his endless gyrations to avoid schoolwork, there is something of a distant metaphor in the school awarding him the title of Champion Gymnast for the year of his departure.[12]

The school also noted Don was off to work in his father's business, which he did. It only took him a few days in the real estate office in Queen Street to realise he was not cut out for the work. He decided to head west on the 200-mile trip to 'Kanimbla' to contemplate a life on the land. He was there for three months, crossing off careers in medicine, the law, business with his father, and managing 'Kanimbla' to reach that momentous decision in early 1928 to put it all behind him and head for the skies.[13]

At this point, as he set his course, his love of the mechanical, his insatiable curiosity to know how things worked, and an equally strong desire to master, then improve, anything capturing his interest cannot be overstated. His personal notebooks from various stages of his career reveal this to be an obsessive pre-occupation, first with mechanics and then with electronics: for example, the detailed and intricate drawings and sketches of items such as the coupling mechanism for the famous 'pick-a-back' plane, *Mercury*, in which he set world records. With his flair for art, the drawings are sometimes dedicated purely to form – the lines, the aesthetics – sometimes to function, and often both. It was in the mechanics and electronics, and later in navigation, that physics and trigonometry became supremely useful. (The very same notebooks, along with other papers showed him to be a poor speller his entire life, a lasting reminder that attempts by BGS to engage him in the Humanities had thoroughly failed.)

Don's quest was not for the perfect (though that is often how he would be portrayed) so much as an expression of his utter conviction there was always room for improvement. Such improvement was not sought just for its own sake but was driven by a desire to find challenges and explore possibilities while, at the same time, assessing both the risks of doing so and *not* doing so, and planning accordingly. Perhaps, subconsciously, this is what attracted him to the skies as opposed to a career in the burgeoning car industry, starting as a mechanic. In the skies he knew there were more possibilities, greater risks and higher stakes, which simply made the challenges more compelling.

There was, of course, the freedom afforded by forging a life entirely outside family tradition and expectation, but despite the apparent headstrong self-confidence at this point in his life, there was a degree of insecurity, perhaps even fragility. He admitted to having an inferiority complex in the face of his brothers' intellectual achievements. Without his Senior matriculation, he would never reach university and then the professional lives they were leading. While he resisted family tradition, he certainly felt the weight of family expectation. Given his intellect, he knew full well that in stepping outside the accepted boundaries, it would not be good enough to prove to himself he had made the right choices and was fulfilling his potential, he would have to prove it to them too; the early rebellion would ultimately need to be justified. In his own mind, the bar would always be set near impossibly high on his own performance, suggesting he would meet family expectations long before he met his own.

5

The Hinkler Effect

Don arrived back in Brisbane from 'Kanimbla' in early 1928 to break the news to his parents that a career on the land was a non-starter and he was off to be an aviator. It did not play well.[1]

They saw it first through the 'jobs' prism. Three months earlier, they had resigned themselves to Don not having a professional career in medicine or the law. Don had then quickly shown no interest in joining his father in the real estate business. Now he was ruling out managing 'Kanimbla'. All the sensible, logical options had been jettisoned by their 17-year-old in favour of some adolescent notion of a 'career' in aviation. They must have been exasperated.[2]

His parents interpreted 'career' within the context of the emerging airline, the Queensland and Northern Territory Aerial Services Ltd (known as Qantas). Founded in the outback Queensland town of Winton in November 1920 by First World War veterans Paul McGinnis and Hudson Fysh, it moved to Longreach in 1921. Routes started to open up – Charleville, Cloncurry, Mount Isa, Camooweal. Having begun with the Avro 504K biplane, the company started building its own de Havilland DH.50 aircraft in Longreach in 1926. In 1927, it built a hangar at Eagle Farm aerodrome in Brisbane and opened a flying school.

According to Don, the family's connection with Qantas was his brother Aubrey, now a doctor in Townsville. Aubrey had a 'close association with many of the Qantas pilots in Western Queensland (he flew with them before the Flying Doctor days)'.[3] This had begun as early as 1925 when Aubrey became a doctor at Camooweal on the border of Queensland and the Northern Territory.[4] Camooweal was one of the airline's early landing strips. Aubrey's stories of the Qantas boys had led George and Celia to conclude pilots were nothing but 'aerial bus drivers' engaged in inherently dangerous work.

While Don had grown up spending most of his time with Arnold, he had formed a particular attachment to Aubrey, so his parents now prevailed on Aubrey to write to Don to dissuade him. He did indeed point out the dangers in the letter, but dropped in a quote that had the opposite effect to the one his parents intended. As Don saw it:

> ... there was only one thing in the letter which mattered, and it confirmed my ambition completely. He used the phrase ... 'Better be not at all than be not noble', and he admitted to forsake one's ambition at the dictate of fear would indeed be ignoble.[5]

He added the quote to the other maxims in his notebook. It is from Alfred Lord Tennyson's poem *The Princess* (II, line 79), accurately rendered as, 'Better not be at all than not be noble.' This is where it gets problematic because the poem addresses a Victorian-era debate about the higher education of women and women's equality, using the device of a mediaeval setting where the prince is pursuing a princess who is not interested in marriage for the above reasons. The quote is from a speech by Princess Ida where the use of the word 'noble' bears no relation to what Aubrey seemed to be indicating it meant (forsaking 'one's ambition at the dictate of fear').

It didn't matter because Don was definitely not into the 'mechanics' of poetry and was going to interpret it however he wanted. The most important thing was that by interpreting it as he did, he fended off the one weapon his parents had deployed to dissuade him.

How he interpreted the quote is of considerable significance. It evidently meant a great deal because he used it at the start of his memoirs, reflecting a lifetime quest to live what he understood to be 'the noble life'. And that quest saw the pursuit of, and presumably achievement of, one's ambitions through the prism of confronting fear. By confronting one's fears, one attained a certain nobility of character. Likewise, failing to do so not only meant such nobility remained out of reach, but so did one's ambitions.

The situation in which this played out is also enlightening. The fears on view here regarding his chosen career in aviation are not his but his parents'. In effect, he sees that he's being asked to forego his own ambitions based on the fear of others. He will, from here on in, reject all efforts at curtailing him where opposition is based primarily on fear not fact. Indeed, he will be attracted to ventures, especially if they break established boundaries or require considerable courage, if an appropriate analysis of the facts countermands those fears.

In 1981, when reflecting on his parents' opposition, he couched it in these terms:

> And so I wanted to go into aviation much against the wishes of my parents. They thought something to do with the risk element in those days, which was quite wrong, of course. It was just as risky to drive down the streets of Brisbane …[6]

Aside from the wonderful irony that his mother thought *he* was the one creating the risk for others by the way he drove in Brisbane, it speaks again of an essential 'driver' in his make-up: deciding to engage in any activity was not about avoiding risk but making a very careful assessment of it. You didn't address fear, yours or others, by giving in to it, but by assessing the perceived risk associated with such fear and making a judgement as to whether it could, and should, be overcome. Failing to do so, at least in his eyes, was a failure of character and therefore ignoble.

Whatever slim hopes George and Celia might have entertained that Don would come to his senses had probably already been dashed by the arrival in Brisbane of Bert Hinkler following his world-record flight from England.

If ever there was a role model for the teenage Don, Hinkler was the one.[7] A native of Bundaberg in Queensland, Bert had experimented with aircraft from an early age. By 1914, he was with Sopwith in England before serving in the Royal Naval Air Service during the First World War. After the war, he was a mechanic and then test pilot for AV Roe. In 1920, his first attempt at flying from England to Australia failed.

On 7 February 1928, he set out for Australia from Croydon, England, in his Avro Avian (G-EBOV), with a *Times* Atlas for navigation, and landed in Darwin on 22 February. He had completed the 11,005-mile trip in 15 days and two-and-a-quarter hours, almost halving the record time set by Ross and Keith Smith in a Vickers Vimy in 1919. What is more, he had done it solo. He was a sensation everywhere he went in Australia, generating blanket press coverage nationally and overseas. *Flight* magazine compared him to Charles Lindbergh.

After arriving in his hometown on 27 February, Hinkler flew down to Brisbane on 6 March to yet another hero's welcome:

> Bert glided into Eagle Farm racecourse, to the cheers of 12,000 people, many of them schoolchildren given a half-day holiday. The tall and commanding Governor, Sir John Goodwin, towered over Bert as he led him to the judge's box where the crowd's hero gave a short speech imploring Queenslanders everywhere to embrace aviation because it was the tool to annihilate distances. With the wings of the Avian folded back she was towed behind a lorry with a mounted police escort as Bert led a motorcade through the streets ... Buildings were decorated with placards and bunting, and children waved Australian flags and Union Jacks all the way from the racecourse to the city of Brisbane 10 kilometres away.[8]

Don had returned to working for his father in downtown Brisbane. He recalls seeing Hinkler arrive. As luck would have it, the Hinkler procession wound its way along Queen Street right past the G.T. Bennett and Co offices in the Union Bank building. It was a slow crawl due to the crowds.

Here was the aviator and his Avro Avian tantalisingly close, Don's proposed future brought spectacularly into the present. Hinkler was the complete package: pilot, mechanic and navigator. It was all rather simple really: to fly from A to B anywhere in the world you needed to be a skilled pilot, an expert mechanic who knew how and why your aircraft could go the distance (and could fix problems when they arose), and a reasonably competent navigator so you could accurately find your way to your destination. Master those three components and you could go anywhere. If you managed it as a solo act you could set your own goals, perhaps even break records, as Hinkler had just done. The task of learning, improving and innovating in all three areas, and the sense of achievement it could bring, was limitless.

Hinkler issued a clarion call for his fellow Australians to follow in his footsteps because aviation was the future.

The public euphoria at Hinkler's feat had barely settled when Charles Kingsford Smith, Charles Ulm and crew flew into Eagle Farm in the Fokker trimotor *Southern Cross* on 9 June 1928, completing the first trans-Pacific flight. The crowds were back, as was the procession through town.

George and Celia now got on board with Don's dream of a future in the skies, or just gave up and admitted the proverbial pigeon had flown the coop. The likely reality was that keeping the latch fastened on their headstrong youngest son was never a starter anyway, whatever their protestations.

Somewhere amid all this, Don had determined his pathway into aviation was the Royal Australian Air Force (RAAF).[9] His parents had been clear they did not favour the Qantas route, so where could he learn to fly, something offering the prospect of a steady job and even potentially a career? The RAAF was the obvious choice. After turning 18 in September 1928, he was eligible for the 1929 intake. Knowing how competitive the selection process was for the very few positions on offer, he commenced preparing for potential selection.

On 1 July 1928, he joined the Citizen Forces of the Australian Army, now known as the Army Reserve.[10] He was appointed to the 13th Field Battery of the 5th Field Regiment, based at Kelvin Grove, not far from his old school. Suddenly, discipline and regimentation did not appear to be a problem! By 4 December, he had been promoted to bombardier, then lance sergeant just over three months later.

Knowing his failure to complete his Senior matriculation might count against him, he took some night classes to brush up on his science. He recalled these being at the University of Queensland, but there is no record of his enrolment. The campus was also used by the Central Technical College, later to become the Queensland University of Technology. It would make more sense he attended that institution.

All this was done while working for his father in Queen Street, doing office administration and sales, learning book-keeping, conveyancing, and associated legal matters.

Sometime in mid-1929, Don was interviewed by the RAAF and, to his 'intense delight', was accepted. Having achieved his immediate goal, his account of what followed bristles with indignation and frustration, inaugurating a love-hate relationship with the Air Force that lasted his entire military career.[11]

First up, he passed his medical in every respect except for it being noted he still had his tonsils. He was horrified this might rule him out. He had them removed the same day 'and reported the fact to the Air Board'. They still delayed his entry by six months.[12] He regrouped, spending the time reading everything he could about aircraft and flying.

In due course, he received the letter requiring him to report for duty in Melbourne. On 1 November 1929, he suspended his Army training at Kelvin Grove. He was preparing to leave when a further letter arrived declaring him medically unfit and asking him to reapply in a year's time. 'I was … very deeply affected by this bombshell.'[13]

Offered no explanation, he travelled to Melbourne immediately to put his case to the Air Board. He passed an on-the-spot medical, only to be told he would not be reinstated in the incoming course but had a further six-month wait. None of it made sense until he learned subsequently that a politician had used his influence to get his nephew into the course, meaning someone had to lose their place. Don was still incensed recalling the episode three decades later: 'To the best of my knowledge there was never any question of "medical unfitness" in my case – then or now!'[14]

In future years, questions would be asked why Don did not make a lifetime career in the Air Force. There were suggestions on his return to the Royal Air Force in England in 1941, after five years as a civil airline pilot, that he did not deserve his senior commands for that very reason. But right here, before he even set a foot on the base at Point Cook, he was having to negotiate through decisions by the Air Force bureaucracy he regarded as absurd, illogical, inefficient or, in some instances, unduly influenced. Unfortunately, while he came to appreciate much of what the Air Force gave him, he would arrive at Point Cook with a mindset attuned to finding fault. As a free spirit who had made a virtue of doing his own thing at his own pace, which was fast and direct, he was to learn that bureaucratic processes in the Service rarely meant getting things done efficiently by navigating the shortest distance between two points.

This initial episode with the RAAF may have been a bruising encounter but, as Don was negotiating his way to a call-up, there was much else to remind him that aviation was a field of endeavour extending way beyond the Air Force.

Geoffrey Wikner,[15] variously described as a neighbour and friend of Don's, had purchased a Farman Sport biplane (G-AUHM, later VH-UHM) in Sydney for £160 in July 1929. It was in poor condition. On its arrival in Brisbane, he housed it in a shed on Nudgee Road at Hendra, about seven miles from the Bennett family home, where he commenced restoring and modifying it. Don became part of the restoration effort, including helping to overhaul its engine. It was his very first taste of aircraft mechanics.

The Farman took to the air on 16 November with First World War pilot Les Kewell in the cockpit.[16] Over the next few months, they tried their hand at barnstorming around the local region. Don was rewarded for his efforts with at least one flight as a passenger. Then, on 29 March 1930, the Farman crashed into a fence on landing at Dayboro, writing it off.[17] Don was not on board, but it may have put a significant dent in his argument that aviation was no riskier than driving in Brisbane.

This problem was accentuated weeks later when Arnold and Don headed out to Eagle Farm on 24 May for the arrival of Amy Johnson from Darwin in her Gipsy Moth 'Jason' (G-AAAH) after flying solo from England.

> Just as the admiring thousands were expectantly awaiting her to alight, they were astounded to see her plane, the Jason, hit the ground with alarming speed on the Nundah side of the Qantas hangar. Speeding on it struck the boundary fence, bounced over, and crashed in an adjoining cornfield.[18]

In an apposite turn of phrase, Arnold commented, 'Our imaginations were set on fire.'[19] Amy was uninjured but, coming so soon after the Farman crash and just weeks before Don departed for Point Cook, it would have been unsurprising if Don's parents were entertaining serious concerns their teenager had got mixed up with the wrong crowd.

On 9 July 1930, *The Age* newspaper in Melbourne published the following:

FLYING TRAINING COURSE

LIST OF TRAINEES

The following candidates have been selected for the next flying training course at Point Cook, to commence on 15 July, and to last eleven months:

A.B. Stowe (W.A.), C.H. Smith (S.A.), R.A.R. Rae (N.S.W.), J.R. Paget (W.A.), J.C. Miles (N.S.W.), A.J.G. Maitland (W.A.), N.B. Littlejohn (N.S.W.), A.D. Grace (S.A.), J.G. Glen (V.), A.C. Drew (V.), A.J. Draper (Q), A. McD. Bowman (T.), H.R. Berg (Q.), D.C.T. Bennett (Q.), Lieut. E.K.H. Klose (Staff Corps).[20]

6

Entering the New World

Two months shy of his 20th birthday, in mid-July 1930, Don Bennett left home in Brisbane for the first time, heading for No 1 Flying Training School, RAAF Point Cook, 18 miles from downtown Melbourne on the edge of Port Phillip Bay.

First port of call was Royal Australian Air Force (RAAF) Headquarters in Melbourne where his anticipated career was immediately put in jeopardy. The recruits were advised the paucity of places for pilots in the RAAF meant the majority could only continue if they agreed to a short-service commission with the Royal Air Force (RAF) in England after gaining their wings. Don agreed immediately, describing the decision as 'the only safe course of action'.[1] He portrayed it as assessing alternatives but, in reality, if he wanted flying training and then a job, he had no choice. Consequently, he admitted deciding after a 'few moments brief thought'.[2]

Don arrived at Point Cook just over nine years after the inauguration of the RAAF on 31 March 1921. The Australian Flying Corps and its Central Flying School (CFS) had been created in September 1912, under Army command, with works commenced on establishing a base at Point Cook in 1914. The site was chosen because it could serve both land and sea-based aircraft. Training began in August that year, just a couple of weeks after the First World War broke out.

The presence of land and sea-based aircraft reflected a wartime debate in Britain and Australia about the military capabilities and uses of aircraft and whether this fledgling air arm should come under the control of the Army or the Navy. The Royal Flying Corps and the Royal Naval Air Service had evolved in parallel in Britain, amalgamating into the Royal Air Force in April 1918. Australia followed suit with the decision in April 1919 to create a separate Service. Initially called the Australian Air Corps, it became the Australian Air Force on 31 March 1921, with the 'Royal' added on 13 August.

This new Service aimed to 'unify the aerial activities' of both Army and Navy and be 'small but efficient'.[3] Don would have wholeheartedly approved of such an aspirational goal.

The CFS now became No 1 Flying Training School, with Point Cook No 1 Station RAAF. Throughout the 1920s, Point Cook played a significant role in developing military and civil aviation in Australia. Its personnel engaged in aerial mapping and the development of air routes. The first commercial air flight, between Sydney and Melbourne, landed at Point Cook on 16 April 1920. Four years later, the first aerial circumnavigation of Australia was conducted from there, completed in 44 days. In August 1928, Charles Kingsford Smith and Charles Ulm departed from Point Cook on the first non-stop east-to-west cross-continent flight. Steve Campbell-Wright summed up those years:

> Point Cook ... was frequently in the newspapers as well as the minds of the general public, being the base for many of Australia's significant pioneering aviation feats. [It] had grown into its new role and had effectively become the practical form of Australia's growing airmindedness.[4]

Despite this, by decade's end, the RAAF was already suffering from a lack of funding, reflecting the prevailing British view that, with the empire facing no threats, a bare minimum of military expenditure was needed. This attitude existed in parallel with an ongoing debate in Australia as to whether the RAAF really needed to exist in its own right.[5] The Chief of the Air Staff, Air Commodore Richard Williams, remarked in 1926:

> The very fact that the Army and Navy still contend that the Air Force is nothing more or less than the auxiliary to the older services is sufficient proof that they fail to realise the characteristics and possibilities of aircraft applied to war.[6]

This debate would continue into the early 1930s. In his 1928 review, RAF Air Marshal Sir John Salmond concluded 'the force [was] established on a firm basis, and it [had] been developed along sound lines',[7] but it needed to upgrade its aircraft, review its organisational structure, and address what he perceived as deficiencies in training. It served to highlight what Air Commodore Williams already knew. A modernisation program was begun, despite no change of military mindset and the onset of the Great Depression in 1929. It included upgrading the fleet of aircraft and new capital works, including at Point Cook. It was into this environment Don arrived less than two years after the Salmond report.

If Williams bristled at anything Salmond had highlighted, it was the quality of the training. Speaking to the Melbourne Legacy Club on 9 October 1928, he was recorded as saying:

> The Air Force in Britain was quite prepared to accept all the Australian-trained officers sent across to the Royal Air Force for experience. In all but two instances, the Australian officers who had been sent to England to undergo the special course of training there had 'topped' the lists, and a greater percentage of Australian officers had passed the examinations set by the Air Ministry for promotion.[8]

This defence of the RAAF's training vis-à-vis the RAF's was not without foundation, despite some in Britain believing Colonials could not turn out graduates of equal quality. From the outset at Point Cook, the focus was on training both pilots and mechanics. The 'sound mechanical training'[9] was matched by high engineering standards.[10] The flying cadets themselves were given hands-on mechanical experience. Further, Dr Richard Hoskins had been appointed as a specialised 'Science Instructor' at Point Cook to provide a thorough theoretical grounding for the practical aspects of the flying course. He was there 'to teach theory of aerodynamics, navigation, wireless, aerial range-finding, optics, internal combustion engines, bombing and gunnery …'[11]

By the time Don arrived, the syllabus was tried and tested. In his memoirs, he supported Air Commodore Williams's assessment by commenting that while the practical training was 'quite as good as the average', he came to regard the theoretical training at Point Cook as 'far superior' to comparable flying training courses. In a 1981 interview, he provided a different, perhaps more personally revealing, angle:

> [Point Cook] covered more subjects. The Royal Air Force equivalent, Cranwell, may have been better at writing essays and things like that, but they didn't, for example, put you through the workshops whereas Point Cook did. You had engineering training as well as flying training. You did everything … The width of knowledge was far greater.[12]

In drawing a direct comparison with Cranwell, Don's point of differentiation – what he considered made the Point Cook course superior – was the engineering training, both theoretical and practical. It is no surprise given engineering played to his interests and strengths, but undoubtedly many RAF contemporaries regarded the 'non-engineering' components of Cranwell's syllabus equally important, if not more.

Don recorded the lectures in two Simplex Note Books, two-ring binders to which he could add blank sheets when lectures or areas of interest demanded it. Inside the cover of the first volume, he wrote 'D.C. Bennett. R.A.A.F.' in large print, trying to be neat, perhaps a note of self-congratulation, or even recording a defining moment of realisation that he really had entered this new world of aviation.

In Volume 1 he would fill the blank pages with Air Pilotage; Airmanship; Rigging I, II & Practical; Engines I, II & Practical; and Photography.

Volume 2 was for Organisation & Administration I, Administration II, III & IV; Law; History & Strategy; Theory of Flight; Air Fighting; Meteorology; Electricity & Wireless Telegraphy; Armament; Hygiene; Army Cooperation; and Naval Cooperation.

For a fellow who had spent most of his teenage years avoiding schoolwork like the plague, now given a clear purpose and subjects of interest, he was studious to a fault.

No doubt he regarded the lectures as a mix of the good, the bad and the ugly. Aircraft and airmanship had to give way periodically to the bureaucratic staples of administration, organisation, law and regulations, only getting marginally more interesting, or embarrassing, with the obligatory hygiene lectures on recognising the symptoms of syphilis and gonorrhoea. Also, you needed to keep fit, avoid smoking to excess, and evacuate your bowels regularly. In the air, you should always protect your eyes with goggles, equalise the pressure on your ear drums with regular swallowing, keep your mouth shut to prevent 'phyorea?' [pyorrhoea], keep warmly clothed, and use Vaseline to prevent chaffing of skin. From the poor-spelling teetotaller came the following gem:

> Alcohol even in small doses definitely effects [*sic*] mental balance, judgement & physical stability of the human system generally.[13]

A cursory look at his notes yields a simple but revealing contrast: the 'Organisation', 'Administration' and 'Law' subjects are dealt with collectively in a perfunctory 23 pages but, when it comes to aircraft, there are 68 pages on 'Engines' alone. The notes are crisp and detailed, distilling large amounts of information. There are graphs, tables and mathematical formulae, but the highlight is the diagrams of different engine parts, sometimes coloured. Quite simply, he loved the mechanics: the design, the detail, how it worked and what was required to make it perform at its very best.

It is the same for 'Electricity and Wireless Telegraphy', another 51 pages of tight script, graphs, formulae and diagrams given over to circuits, valves, switches, coils, and oscillations relating to transmitters, amplifiers and receivers. There are many more drawings with a higher level of annotation because this field was almost entirely new to him, unlike engines where he already had experience on cars and the initiation to aircraft engineering with Geoff Wikner when rebuilding the Farman. In many places, the writing is much smaller, as are the diagrams. The overall effect draws a comparison with the computer industry: his notebook is the equivalent of a computer chip where it is becoming necessary to fit more and more information into a small space. Within a few years, this process of condensation would be seen in tables he drew where the vast volume of information was captured in words and images so small they can literally only be viewed through a magnifying glass. It is like being sucked into some 'Don Bennett information vortex', equally illuminating and intimidating. When he climbed into a cockpit, he would not be content with the basics, he needed to know every component of his flying machine inside out.

He would be this way throughout his life, immersed in the detail of the mechanical and electrical, following technological developments and reading about the latest piece of equipment. As the aircraft industry grew, with the evolution of specialist roles on multi-crew aircraft and equally specialised ground crews, Don endeared himself to those colleagues and subordinates who appreciated how much he knew about their jobs and his readiness to engage irrespective of their rank. It was one way he was seen to 'lead from the front'. He was admired by, and inspired, many Pathfinders during the war because he never asked them to do what he wasn't prepared to do himself. This was true, and one of the reasons was the simple fact he could – he had the knowledge, the qualifications and the experience.

For those less capable of, or less interested in, absorbing the detail he routinely immersed himself in, his 'nerdishness' could be a source of alienation. He did not help his cause by ultimately acquiring a reputation for positing his knowledge in a frank and direct manner. This partly reflected the Australian tendency towards plain speaking, with its attendant lack of appreciation for social niceties and conventions, but this is probably overemphasised. There was also his inherent introverted shyness, paired with a fundamental desire to get to the point to avoid wasting time. But, in the main, it arose from a soaring frustration and plummeting tolerance when faced with arguments and objections based on ignorance, arrogance, laziness, unexamined fear, group think, or a plain lack of common sense.

'Ignorance', to some degree, meant knowing less than he did or not seeing things the way he did; patience never became one of his virtues. He was open to persuasion if you gained his respect, but it was a mountain too many failed to climb. To dismiss him on the basis of this 'character flaw', however, is to miss the main point.

Don saw ignorance in the context of risk. His constant, heightened risk awareness was paramount. Risks were unavoidable if aviation was to expand. But you had to know them, assess them, then avoid or mitigate them. The greater the ignorance, the higher the risk. And worse, if the knowledge was available – the facts, the statistics, the technological advancements, the real experience of flying the conditions, etc. – but people chose not to avail themselves of it, such wilful ignorance was the worst form of risk-taking, particularly if lives were at stake.

In Don's mind, the pace of change in aviation meant unquestioned tradition, unyielding hierarchies, ossified bureaucracy, outmoded theories, and outdated experience were all significant and unjustifiable impediments to innovation and technological advances. Aviation had no place for the status quo; there was always a better way and he would be relentless on himself and others in seeking it out. If that involved taking informed, carefully assessed risks, accompanied by the requisite degree of planning, so be it.

The impact on his leadership style, which emerged over time, would be an obsessive control of the detail in circumstances where he perceived the risks demanded it, cementing his reputation for some as a micro-manager. This was especially so where subordinates failed to measure up to his standards or knowledge base and could not be delegated full authority to perform their roles without close supervision. It should be acknowledged some did, indeed, win his full trust.

Other sections of Don's lecture notes at Point Cook are informative, if not entertaining.

The 'Armament' section considers first the subject of guns, in particular the flight of bullets, deflection, types of ammunition and the use of gunsights, concluding with a section on the Vickers machine gun consisting solely of ways to deal with it when it stopped firing!

Of greatest interest, however, is Don's very first introduction to the theory and practise of bombing. This was not in terms of overall military strategy but the art of dropping bombs from an aircraft. The immediate impression is that his skills in trigonometry are to the fore, with several diagrams and formulae acting as a concise summary. There follows a detailed section on the installation and use of the Course Setting Bomb Sight Mk.IIB.

Finally, a section on bombing accuracy is broken into two parts. The first addresses bombing accuracy using a Probability Chart to show the distribution of one thousand bombs dropped on a target. The pattern yields first the MPI [mean point of impact] and then 'the 50% Zone', the circular area drawn around the MPI in which half the bombs have fallen, 'measured by radius at 10,000 feet'. It is, of course, totally theoretical, with everything depending on the MPI being somewhere proximate to the actual target during a real-life bombing exercise. As the RAF would discover in the forthcoming world war, simply identifying targets was no easy feat, let alone hitting them consistently with a high concentration of bombs.

The second part on 'Bombing Errors' is an object lesson in the difficulties of bomb aiming, particularly in a 1920s aircraft using an early model bombsight. Errors can creep in with respect to air speed, trail angle, lateral levelling, installation, wind direction, wind speed and height setting. Don captures all these in a single diagram.

The possibility of encountering enemy aircraft was covered in the subject 'Air Fighting', the lectures being delivered by Flight Lieutenant Fred Scherger, the Chief Flying Instructor (CFI).[14]

Don's notes suggest the CFI was much more interested in flying than in lecturing on the subject. There are brief tips on tactics for a fighter plane encountering, singly or in formation, enemy planes of different types – single-seat [S.S.], twin-seat [T.S.] or multi-seat. Given Scherger had no combat experience, some quotes have an anecdotal or plain off-the-cuff quality. Airships are dismissed with, 'No harder to bring down than T.S.', while the 'S.S. Fighter v T.S. Fighter' section begins with the informative, 'Rear gun appears worse than it really is. T.S. has usually work to do whereas S.S. has only to fight.' Consequently, the first recommended method of attack is: 'Dive on rear gunner.'

Other material from the CFI is just as entertaining:

Air Fighting differs from other fighting in the following points: (i) No shock attack. (ii) There is no fear of mob panic but (iii) there is not the support of close comradeship. (iv) The fighter is more personally master of his own destiny. (v) Leaders cannot make use of personality as they cannot speak to others. (vi) Time factor is limited.

It is impossible to tell whether Scherger delivered these insights as a series of points or it was merely a brave attempt by Don to make the most sense of what he was hearing. Scherger's points indeed make more sense if one interprets 'other fighting' as trench warfare, pointing to a dominant mindset within the Air Force which is explained shortly. Nevertheless, the 'Air Fighting' lectures may have been the exception to Don's view that Point Cook offered superior theoretical training. It mattered not because there was ultimately no practical training in the subject.

Don's two pages covering 'History & Strategy' provide a concise time capsule of military thinking. The notes begin:

History & Strategy

Shall include discussions on: The importance of Modern British History to officers; The history of the development of the Art of Flying & its application to War; & History of Air Force. The Employment of Air Forces when acting separately or in cooperation, etc.

Books: 'Short History of Royal Air Force'; 'RAF War Manual'; Maurice on 'Strategy'; 'War in the Air' Vol I Raleigh; Vol II Jones.

Strategy: (Mahan's def. of Maritime Strategy) has for its object – to found and increase in peace as in war the power of a country. Maurice: It is the art of employing a nation's war power in the most effective way. War is a continuation of National Policy by resort to force as the final means of settling a dispute.

The Principles of War: (i) Maintenance of the Aim (ii) Offensive (iii) Surprise (iv) Concentration (v) Economy of Force (vi) Security (vii) Mobility (viii) Cooperation

Note: Are these principles or methods? Adequate Bases are necessary for mobility. Bases (whether Naval, Army or AF) – must be in necessary position; must have resources; and Defence.

Superiority includes a moral, physical & material superiority.

After half a page on 'Australian Air Defence History', traversing seven dates of importance, and a couple of sentences on liaison between the RAF and RAAF, the whole section finishes with:

> Aim of War & aim of Air Force in War – To bring another nation, as quickly and economically as possible to conform to your will or purpose. The Services are instruments of the nation's will. Aim of Air Force is to break down enemy resistance.[15]

The final statement is fascinating in light of Bomber Command's *modus operandi* during the Second World War, but when it comes to military strategy in 1930, the *actual* role of the Air Force only becomes clear in the lectures on cooperation with the other Services.

By mid-1930, while the debate continued as to whether the Air Force should exist as a separate entity, the lecture notes on 'Army Cooperation' clearly show the Army treating the Air Force merely as the auxiliary it thought it should be. The notes are extensive, covering its organisation from War Office down to platoon/section level, and ranging across infantry, artillery, engineers, tanks and cavalry. It appears Don was taught more about how the Army was organised than the Air Force.

Each Army division would have an 'Army Coop' (A/C) squadron consisting of 12 aircraft and pilots with ground support, coming under Army command during operations. The lectures spelt out the tasks, overall role and objectives of these squadrons, all of which could be summed by the core idea of giving the Army the ability to see 'over the hill' and provide updated tactical information:

> Tasks of an A/C Squadron (a) artillery recon (b) medium reconnaissance (medium distance, camera always carried) for 8–50 miles (c) close recon to 8 miles (d) photographic recon up to 50 miles.
>
> Other unusual: (a) supply dropping (b) conveying Army Staff Officers (c) conveying orders (d) bombing ground troops (e)

Entering the New World

machine gunning ground troops (f) night recon (g) dropping smoke screens.

A/C Operations: Timely information about the enemy is the foundation of all war operations …

Air reconnaissance: is one of the methods of securing information re the topography and resources of a country and the disposition of the enemy.

Objectives: (1) To gather accurate and useful information (2) To pass on quickly to Army.[16]

There are several pages of notes on reconnaissance issues relating to close cooperation with elements of the Army. A drawing of a battlefield with an infantry march towards the enemy looks thoroughly 'Western Front' and is accompanied by an explanation of the squadron reconnaissance role including, 'Defence may be earth-works or highly organised trench systems.'[17] The notes about working with the artillery go so far as detailing specific battlefield signals.

The conceptual use of the Air Force in warfare is perhaps best summed up by: 'Reconnaissance may be strategical [*sic*] before forces are in contact, or tactical after contact.' The implication is the Air Force is not really a 'force' in its own right: while its aircraft may drop a few bombs or fire a few machine guns, its principal role, where it is actually seen to be useful, is reconnaissance.[18] Twelve years after the Armistice, it was a classic case of some senior military figures continuing to fight the previous war.

Compared with the vast amount of detail on Army Cooperation, there is just one page of notes on Naval Cooperation. The focus is again on reconnaissance. Pilots are to identify types of ships – battleships, battlecruisers, cruisers, destroyers and aircraft carriers – and notify position, course, speed and composition. It finishes with 'stay in vicinity'.

This theoretical positioning of the Air Force as predominantly the Army's and Navy's 'eyes in the air' (and mostly the Army's) was reflected in Don's practical flying training.

At 9:10 am on Monday 28 July 1930, Sergeant Preston took 'Cadet Bennett' on his first flight, 30 minutes in a de Havilland DH.60X Cirrus Moth (Serial A7-17).[19] The Cirrus Moth was the predecessor to the more powerful DH.60G Gipsy Moth, already well established in Britain by the time Don started flying. As such, it was obsolete, reflecting the RAAF's funding constraints.

Over the next month, he had a total of 16 flights of about 30 minutes each. Then, on 21 August, Sergeant Preston gave him a practise run to settle any nerves before Flight Lieutenant Scherger climbed aboard to conduct Don's 'Test for Solo'. Scherger was satisfied after 20 minutes and Don immediately took off for another 20 minutes by himself. 'For a youngster of reasonable enthusiasm and susceptibility the sheer delight of solo flight – alone for the first time – is hard to equal.'[20]

For the next six weeks, Sergeant Preston instructed and Don then flew solo to practise what he had been taught. He later described his instructor as 'a solid average type, by no means a genius, but with a simple direct approach to the job of teaching flying which suited me admirably'.[21] He had one short night flight in early September.

His first cross-country came on 17 October, a 70-mile solo flight. The first leg took him west to Anakie before he turned north-east, passing over Balliang and heading to Deer Park in the western suburbs of Melbourne, where he changed course again and made the home run south for Point Cook. At 75 minutes, it was his longest flight to date and the first real test of his navigational skills. The man who would write the international standard work on air navigation just six years later was able to note in his logbook: 'Test Cat A7 passed'.[22]

Just a week later, he had his first exposure to the camera gun and bombing. In mid-November he was practising aerobatics, forced landings, steep turns, half rolls, and crosswind landings. Squadron Leader George Jones, the commanding officer of the Flying Training School, put Don through his paces for 35 minutes on 20 November; he passed that too.

After practising in four different Moths over two days, Don finished flying the aircraft on 28 November, with 19 hours dual, 31 hours solo and an 'Above Average' rating. It had been exactly four months since his first flight. This was a leisurely pace compared with trainee pilots at the elementary flying training schools during the upcoming world war, some of whom completed their time in Tiger Moths in less than eight weeks and accumulating more hours than Don in the process.[23]

On 10 December, he had a 'taster' for the second phase of his training – learning to fly the Westland Wapiti Mk.IA. Sergeant Preston did the flying. Don was not there to take notes but to sit in the front cockpit and try his hand at some Vickers gun practice.

The theoretical work on the aircraft themselves was matched by regular 'practical ground instruction' – 'practical rigging and fitting and other trades were covered fairly thoroughly'[24] – though Don indicated this meant a lot of washing down the aircraft with soap and water.

Training on the Wapiti began in earnest in mid-January 1931. Like the Moth, the Wapiti was a two-seat, single-engine biplane, but was considerably larger, in particular the step up from the 100-hp Cirrus to the 480-hp Bristol Jupiter radial piston engine. Designated a light bomber, it was used for a range of training purposes. Also like the Moth, the Wapiti Mk.IA had already been superseded, another reflection of funding constraints. He would get a few flights in the Mk.IIA as the RAAF upgraded its fleet.

Squadron Leader Jones conducted Don's test on 2 February. Declared 'Fit for Solo', Don went up in the Wapiti by himself.

Much of the next four months was given to mastering landings, especially forced landings, and the various turns – climbing, steep and stall. He passed the altitude test and the cloud test. There was more time in the front cockpit firing the Vickers gun and a couple of occasions in the rear cockpit firing the Lewis gun at some targets in Port Phillip Bay.

There was precious little bombing practice. Don had just 35 minutes on 7 May to try his hand before a short test six days later, which he passed. It was followed by half an hour receiving 'Diving on Target Instruction' and that was it. He recalled a mixture of dropping bombs on land targets and 'on one occasion [dropping] live 112 lb bombs on a target in the bay'.[25] When using land targets, the base 'advised people living in the vicinity to leave their doors and windows open and to remove fragile items from the shelves'.[26]

The cadets did some formation flying, but no night flying. Despite Point Cook having commenced parachute jumps in 1926, Don did not do that either. He would have to wait until his arrival at No 29 (Fighter) Squadron at North Weald in England for that pleasure.

While receiving theoretical and practical training in wireless telegraphy, Don was also required to master the Popham Panel, a visual ground-to-air signalling system developed during the First World War where the numbers 1 to 9 were arranged in various patterns around a large 'T' to send coded messages to pilots and observers flying low enough overhead to read them. It had been developed when initial wireless communications were air-to-ground only and continued to be taught until two-way communications became reliable and ubiquitous. There is no suggestion Don ever put it to good use.[27]

Aerial photography was something he really enjoyed, though again the opportunities were limited. His reference to doing 'Photography (Overlaps)' on a solo flight on 12 May was later explained:

> ... we actually flew photographic runs sufficient to make up reasonable mosaics of the area which we had covered, and subsequently we developed the film ... We did stereoscopic analyses of these photos, and made out interpretation reports.[28]

Mosaics were a wartime invention where overlapping aerial photographs were used to create battlefield maps, often showing trench fortifications.[29]

Navigation was taught under the subject 'Air Pilotage', which Don defined, in part, as 'the art of conducting a M/C [machine] from place to place ...'.[30] It involved a significant amount of theory. There were the fundamentals about traversing the earth's surface – distance, velocity, gravitation, magnetic fields, winds and friction. The cadets learned to read maps and how to draw up a grid system so they could track their movement from a point of origin. They received instruction on how to use, and care for, compasses. It all fed into the lectures on interception and determining a radius of action, where Don deployed his favourite mathematics subjects – algebra and trigonometry. If the length of his notes is any indication, it was not a subject generating nearly as much interest as mechanics.

Entering the New World

Air Pilotage was one subject occupying significant flying time. In an exercise called 'Pin-points', the grid-system technique was used to locate specific points of reference on a map.[31] He had to complete a compass test. There was a short flight west to Little River engaging in road reconnaissance and locating an advanced landing ground where he made a brief touch down before returning to base.

There were two navigation tests in early May. The first lasted more than two hours, by far his longest flight to date, but he doesn't mention the outcome. Six days later, he was aloft for nearly two hours again, this time recording an emphatic double-underlined 'PASSED'!

There are other hints Don may have been taking time to master navigation. Back on 19 February, he failed to complete a solo cross-country test taking in Werribee, Newport and Bacchus Marsh. Five days later, an exercise involving road reconnaissance, designed to take him west to Little River, then back over Point Cook to Newport on the outskirts of Melbourne before turning north and heading for Sunbury, ended in less than an hour, labelled once again as 'incomplete'. It is entirely possible Melbourne's notoriously fickle weather played a part on the first occasion, with thunderstorms predicted. Not so on the second when it was fine all day.

The one cross-country flight Don recalled specifically was the overnight trip to Deniliquin in New South Wales on 22–23 May:

> It was quite an adventure, and what with one or two pupils going astray and one or two overdoing the low-flying business, it was amazing how such a simple flight could produce so many episodes – and on one aircraft, so much telephone wire![32]

It was the last of the cross-country exercises, duly noted as 'Passed'.

The spirited camaraderie hinted at here is reflected in Don's recollections of social life on the base itself, about which he gives a surprising amount of detail. Coming from a family of sporting types, and having been a gymnast at school, it is no surprise he enjoyed the sports activities. He described a game of six-a-side 'all-in rugger', using a medicine ball, which descended into something of a brawl, resulting in a significant number of casualties at the sick quarters and the game being banned.

He showed similar enthusiasm for a well-orchestrated initiation ceremony for new cadets involving sitting naked on blocks of ice, facing a red-hot poker while blindfolded, and the shaving of various parts of the anatomy using sheep shears, all finished off with a version of tarring and feathering. Welcome to the Air Force.

In describing other activities, he admitted, as a teetotaller, he was not particularly enamoured of drunks, who he found to be 'unpleasant company'. The overall impression is that despite being an introvert and exceptionally bright, he offered no sense of being distant or aloof from his fellow cadets or that he did not fit in.

His only real criticism of Point Cook concerned perceived inefficiencies in how the base operated, particularly due to its layout and having to obey station standing orders whose usefulness he questioned. To Don, the punishments for breaking standing orders were often as idiotic and wasteful as the standing orders themselves. He evidently regarded the whole business as a farce, providing a sarcastic summation: 'Thus it was that I learnt the first law of the Air Force, "Thou shalt not get caught".'[33] Standing orders were a manifestation of bureaucracy, however well intended, that he never reconciled with.

After a flurry of flights across three days, flying training was over on 12 June 1931 and Don was awarded his wings. 'I had been beaten by Clary Smith in the examinations, but fortunately, to my delight, I came out top in the actual flying.'[34] The latter came with the coveted 'Above Average' rating. *The Herald* in Melbourne, reported on 20 June:

> The following Royal Australian Air Force air cadets, who have successfully graduated the No 9 flying training course "A" at Point Cook, have been awarded the pilot's badge:
>
> Pass with Special Distinction – Air Cadets C.H. Smith and D.C. Bennett.
>
> Distinguished Pass – Air Cadets N.B. Littlejohn, A.C. Drew, J.C. Miles, J.R. Paget, and R.A.R. Rae.

Entering the New World

Pass – Air Cadets J.C. Glen, A.McD. Bowman and A.D. Grace.

It further mentioned that Smith, Bennett, Littlejohn and Drew would depart for England imminently on the SS *Narkunda*.[35]

Regarding his studies, Don had been duly rewarded for the hard work over the course of the year, applying himself in ways he had never done at school. That, in itself, was something to brag about to his parents. On his return to Brisbane, however, the bragging took a different form. He quickly made contact with Lester Brain.

In May 1924, Lester Brain had joined Qantas as a pilot up at Longreach. Hudson Fysh, the co-founder of the airline, recalled, '[He] was a cadet pilot from Point Cook, our first non-wartime pilot and one of the younger "go ahead" school who set out to make aviation their career rather than an uncontrolled bit of fun.'[36] After flying experience around northern Australia, Brain had been appointed as chief instructor when Qantas opened a flying school in Brisbane at Eagle Farm Aerodrome in March 1927 with a Cirrus Moth. Qantas also built a hangar and opened a branch with Brain, as Fysh put it, 'branch manager, chief instructor and jack-of-all-trades'.[37]

After the training school was handed over to the Queensland Aero Club in May 1929, which moved to Archerfield Aerodrome (also in Brisbane) in 1930,[38] Brain continued as the chief instructor, with Qantas retaining aircraft ownership and doing the maintenance. Brain had returned to Point Cook in 1926 for a refresher course, so he was aware of the quality of its graduates.

Don first knew of Brain because of his brother Aubrey's connection with the Qantas pilots as early as 1925 when a doctor at Camooweal.[39] On the morning of 23 June 1931, he headed out to Archerfield. At 7:00 am, Brain gave him a five-minute test in a Qantas Moth,[40] the shortest flying test of his career. With Brain satisfied, Don spent 90 minutes giving 10 passengers joyrides. None are named, but the conclusion is inescapable that at least some of the entourage were family, most likely his father George and brother Arnold.

With that, Don departed Brisbane for England, arriving in London on 7 August. He was five weeks short of his 21st birthday.

7

Fighter Boy

The newly fledged pilot officer arrived in England half a century after his mother Celia had left as a young child. He could now decide for himself if her 'intense love' of the Mother Country was justified.

With the paperwork complete at the Air Ministry, Don and his fellow Aussies were dispatched to RAF Uxbridge on the western outskirts of London for their induction into the Royal Air Force (RAF). They were fitted out in their RAF uniforms; regulations permitted them to continue wearing their Royal Australian Air Force dark blue if they wished. Some preferred the colour because it distinguished them as Australians; a bit of a flag-waving exercise that would continue with Aussie aircrew serving in RAF units during the Second World War. Don hinted some also enjoyed the fact it irritated their hosts.

In short order, they were bundled off to No 5 Flying Training School at RAF Sealand in North Wales to test their pilot credentials. It had been two months since Don had flown in Brisbane. On 20 August, Flying Officer Morris gave him 25 minutes of instruction in a DH.60M Moth so he could get some feel for British flying conditions. After two solo efforts, Flight Lieutenant FK Damant introduced him to the Armstrong Whitworth Siskin III.

'The AW Siskin III was the UK's first 'all metal' (actually a fabric-covered aluminium alloy structure) fighter and it first entered RAF service in May 1924 ... It was well-liked and considered an excellent aerobatic machine ...'[1] Don very quickly formed a different opinion. It was 'undoubtedly one of the worst aeroplanes ever built ... Most models of the Siskin were not stable in any plane.'[2] His view had not changed decades later, but he acknowledged a silver lining:

> The Siskin was a bit of an atrocity … but it was good for you; you had to fly it all the time, both in the air and landing and taking off.[3]

In effect, the Siskin's performance characteristics allowed little latitude for error, meaning if you managed to keep it in the air, you learnt a lot about flying.

It was perhaps appropriate Flight Lieutenant Damant introduced Don to spinning on his very first flight, finishing off the session with 'advanced forced landings'. After several more flights, Don went solo four days later. That achieved, the instructors set a blistering pace: 14 flights over three days. Squadron Leader James Williamson, the Chief Flying Instructor, then performed the test and it was over. Don was posted to No 29 (Fighter) Squadron at RAF North Weald, north of London.

His ten months of fighter pilot training at North Weald was almost entirely on the Siskin IIIA. The aircraft itself was just one of a multitude of challenges that included learning to fly in English weather, navigating over the English countryside, the breadth of training, and the dangers posed by the bravado of young, inexperienced pilots in high-powered machines.

The recently updated training programme for fighter pilots spent little time on the basics; it was simply assumed you could handle the Siskin or, if you couldn't, you learned fast. The training emphasis fell on six areas: radio telegraphy (R/T), gunnery, formation flying, aerobatics, night flying and navigation.

Absolute priority was given over to the practicalities of R/T. Don had studied the foundational subject of 'Electricity and Wireless Telegraphy'[4] at Point Cook, but there is no indication he had any practical experience in the air before arriving at North Weald.

By the end of the First World War, much progress had been made on wireless air-to-ground communication. Having started with Morse code, both the English and the Americans had progressed to radio transmitters in aircraft capable of transmitting voice, leading to two-way communication. With the basic technology established, the post-war focus moved to increasing its power and range.

In September 1931, there were still problems with reliability and range when Don embarked on his crash course in communicating from on high.

In an open cockpit with rushing wind, behind a 14-cylinder air-cooled radial piston engine, he flew around North Weald anywhere from 2,000 to 5,000 feet straining to hear through a headset fitted with rudimentary telephone receivers:[5]

> It was extremely difficult to decide whether one could hear a voice from the ground station or whether it was wishful imagination ... It was something of an historical occasion when I managed to work both ways at a range of forty-three miles.[6]

The primitive nature of the experience pointed to just how tethered aircraft still were to transmitting points on the ground when attempting to communicate with clarity in real time. Developing this technology had huge implications for reconnaissance, let alone anything aircraft might do in an offensive capacity, or even just reporting position, weather or mechanical difficulties.

Over time, Don was introduced to a range of R/T exercises, combining it with cloud flying, map reading, sector patrols, aerobatics, an interception exercise and, on several occasions, flight formation. They even achieved air-to-air communications in one exercise.

Formation flying was also conducted regularly. There were sector patrols and battle flight climbs up to 16,000 feet, one of which Don notes as being achieved in 14-and-a-half minutes. The new year brought fighter-affiliation attack exercises as a formation, usually targeting a Wapiti.

These came after considerable work had been done on gunnery. Practising with camera guns, they learned to attack ground targets and then engaged in air-to-air combat. In late May, the squadron moved to RAF Sutton Bridge in Lincolnshire for the annual live-firing exercises on the shores of The Wash. In a competitive atmosphere, there was a combination of firing at ground targets and shooting at drogues:

> ... to have the violent chatter of two Vickers guns just in front of one's face was one of the most war-like experiences I have ever met.[7]

Don commented about the unreliability of the Vickers machine guns and having to clear them while flying, proving the astuteness of the Point Cook lecturer who had instructed them on how to do just that. The 'violent chatter' produced mixed results: out of 200 rounds fired at ground targets, Don had a hit rate of 25% but, when it came to the drogue, he managed only 47 hits out of 900 rounds (5.2%).[8] Definitely room for improvement and perhaps a sign that as an air gunner he made a very good pilot.

The aerobatic staples of loops, rolls, slow rolls, half rolls, barrel rolls, slow flying, and spins were a regular feature as the months passed; underpinning it all was a continual fine-tuning of skills, building experience and learning the performance characteristics, and limits, of the Siskin.

There was far less night flying. Don did a baker's dozen, half of those testing night-flying equipment, and a 15-minute test. It is on the subject of night flying that Don commenced another area of criticism of the Air Force in his memoirs and in later-life interviews: fellow officers. Whether justified or not, it was an unfortunate trait, because it could well be argued that in offering public character assessments of others, he perhaps unwittingly performed one on himself.

At North Weald, the first officer in his crosshairs was the commanding officer, Squadron Leader Henry 'Paddy' O'Neill. An Irishman, O'Neill had been a flying officer in the Royal Flying Corps in the First World War. Post-war, he had served in India, then attended Cranwell in 1930 before his posting to 29 Squadron, first as a flight commander, then as CO.[9] According to Don, O'Neill had left a general impression around the squadron that he was averse to flying the Siskin. This manifest itself most obviously in O'Neill's night-flying performances where he got the primitive flarepath set up during daylight and completed his bit 'into moderate dusk conditions' before leaving everyone else to carry on flying into the night.

Don offered some faint praise of O'Neill, that he was a 'most conscientious CO' and 'kind-hearted and helpful', but it is a thin, barely disguised veneer for the underlying stinging criticism. O'Neill was Don's archetype of the RAF officer for whom he came to show disdain; he appeared not to want to fly regularly, let alone in the same conditions as his men. At best, in Don's mind, this left him short of the requisite experience to provide effective leadership and, at worst, evidenced a fear that opened him to ridicule around the squadron. To Don, O'Neill's flying history, general experience, and overall duties and responsibilities as CO were irrelevant; he was failing to lead from the front, an unforgivable sin for a squadron leader.

In the end, the issue is not principally about whether Don was justified in his criticism of O'Neill, but the manner and tone in which he did it. This may be more of an expression of how he felt in 1958 when he wrote his memoirs than in 1931 when a neophyte at North Weald.

The last significant area of training was navigation. Initially, there were frequent flights just to master map reading, leading to longer flights such as the hour-long, 100-mile return trip up to Ipswich. 'Pin-pointing' exercises were plentiful but relatively few 'cross-countries'. Perhaps the most important of these, given what was to occur, was a two-and-a-half hour flight on 26 February 1932 that saw Don fly over the eastern outskirts of London and down to Hawkinge, near Folkestone in Kent, where he turned west-south-west and headed for Tangmere, then on to Upavon, before passing over the western and northern suburbs of London on the return leg to North Weald.

Some six weeks later, on 11 April, due to a 'lapse' by his flight commander (explained shortly), Don decided to take Widows, a fellow pilot, on an unauthorised 'pin-point' exercise to a spot some 60 miles south of London. The weather was poor all the way south, deteriorating rapidly on the return journey. Don led the two aircraft straight into a massive cold front on the southern outskirts of London because fuel limits did not allow an extended detour around it. They kept in formation but the icing on the wings was becoming a significant problem:

> What was serious was that my engine was losing power. I, like most if not all other pilots, did not fully understand carburettor icing. Eventually, quite suddenly, my engine stopped altogether. Widows overshot me and was gone, and I had the nasty thought that I was on the southern outskirts of London and I could not therefore bale out. Moreover, the modern invention of a turn indicator which was serviceable at the beginning of the flight (a rare event on a Siskin) now went unserviceable, and with a partly iced-up air speed

indicator, heavy ice on the wings, no engine, an uncertain cloud base and probably a suburban terrain, life was far from happy.[10]

The end result was emerging below the cloud base with just enough height to clear a small wood before landing in a field and flipping the Siskin on its back. 'I came to a standstill, inverted, with my head dangling six inches from the ground.'

Fortuitously, Widows had emerged below the cloud base at the same spot.

At the formal inquiry, Don tried to defend his choice of 'pin-point' location by arguing the cold front had not been forecast and, therefore, had it not materialised, they would have made it back. He put it down as 'an expensive, though valuable lesson to me for the rest of my flying career'.

Taken on its own terms, the accident might have passed as a simple act of youthful exuberance and poor judgement, but Don uses it in his memoirs to take aim once again at a fellow officer. This time the target was his flight commander, Flight Lieutenant John Merer.

> John Merer was a very nice man and all that. Never before or since, however, have I ever met such a bureaucrat. Translating the orders of every other bureaucrat at higher levels as generously as he could, he produced every order in writing, he insisted upon signature *ad nauseam* and reduced our flying time, both in frequency and duration. He allowed no discretion and no action other than that which he had ordered in detail and in writing ... All this was intended to produce safety, the only ambition then existent in the RAF, and the only basis on which promotions were judged.[11]

Merer had served with No 99 Squadron in the latter stages of the First World War before occupying a number of roles in the 1920s, including as a pilot at No 207 Squadron and adjutant at the RAF (Cadet) College. In March 1932, he succeeded Flight Lieutenant AL Duke as B Flight commander at North Weald.

To Don, Duke and Merer fell into two entirely separate camps: the first focused on training the pilots while the second, in his eyes anyway, focused on bureaucracy. The underlying issue was the more important one: it seems Duke allowed his charges some latitude as they built skill and experience; Merer was a strict 'fly the plan' officer. Don's argument – one he maintained his entire flying life – was that the pilot, the one charged with the responsibility for the aircraft, must be given the discretion to make the decisions. The pilot has ultimate responsibility for assessing the risks, no one else. If the pilot is trained to make decisions purely within a narrow, regimented framework, the ability to make critical decisions when most needed decreases. Pilot autonomy is therefore paramount and training should be directed to that end.

What Don seemed unwilling to acknowledge was the context: for most young, inexperienced pilots, the capacity to assess risk and stay within boundaries is limited. It was a situation made worse by the squadron having to train on Siskins. Merer may well have been officious but, given the circumstances, it could be argued it was sensible for him to keep a tight rein on his inexperienced pilots for their own safety. Further, Don showed little appreciation that pilots could benefit from accepted practice built over time through collective experience.

As with O'Neill, Don raises a legitimate point for debate, but his criticism of Merer reveals an undercurrent pattern in his general approach to authority – praise for superiors who give him freedom, but criticism for those who attempt to constrain him, disagree with his approach, challenge his decisions, attempt to limit his actions, or stand in the way of his independent-mindedness, particularly if bureaucracy is their weapon of choice. From here on, right to his final years, it is hard to get away from a sense his superiors and contemporaries will always fall into either the 'Duke' or 'Merer' camp.

And in that regard, it is hard not to draw a comparison between Don's reaction to Merer and how he responded to his parents when they tried to impose limits on his behaviour, such as how he drove the family car. He takes it as some sort of personal affront, their touching some primal nerve by seeming to suggest he doesn't know what he is doing or isn't smart enough to make the right choices.

In this instance, he carpet-bombed his own argument in spectacular fashion in his memoirs by an oblique attempt at blaming Merer for his crash:

It is somewhat ironical, therefore, and even perhaps educational, that as an indirect result of these restrictions, I was guilty of the only aircraft accident of my flying career. It happened, believe it or not, that John Merer one day sent up Widows and myself for formation practice and forgot to put his usual restrictions on the order. Having signed his instructions, we rushed out of the office, and grinning from ear to ear we mutually agreed that we would go somewhere – and go we did.[12]

In other words, he would not have chosen to fly to the 'pin-point' south of London in bad weather, with Widows, in Siskins no less, and then crashed, if Merer had completed his flying orders to the usual degree of bureaucratic rigidity. Ironical, indeed.

By the middle of 1932, Don was hankering to move on. The training was coming to an end and he had learnt all that was on offer. Marking time was anathema; intellectual boredom and the absence of new challenges were his mortal enemies. He looked for possibilities, something new to master.

He gave no appearance of wanting to continue flying fighters. Aside from the crash, the Siskin had given him no real problems. He had nearly 184 hours solo on it. In June, the squadron started taking delivery of the new Bristol Bulldog II. He had three flights, and even put a photo of it in his logbook but it did not encourage him to stay. Fighters were just a part of the broader aviation canvas he wanted to 'paint' on. Being recognised as a 'fighter boy' ultimately meant nothing.

His friend, Ralph Cleland, had done the Flying Boat Pilot's Course at RAF Calshot. Apparently, the course was good and he could learn to fly an entirely new aircraft, the Supermarine Southampton.

There was just one thing left to do – the parachute jump. He had been on a parachute course at Henlow six months earlier in the last week of November, but for some reason had not gone beyond the theory. His lecture notes, a succinct two pages, covered all the necessaries: parachute types (pilot and observer); the different parts (canopy, rigging lines, harness and pack); care and maintenance, finished with a curt 'forget nothing';

completing the logbooks (each parachute had its own logbook, and your own, with a live jump underlined in red); and the steps to packing a chute.[13]

On 2 July 1932, the day after his last flight at North Weald, he returned to Henlow where, without the benefit of a refresher course, he launched himself into space at 1,000 feet from the wing of a Vickers Virginia, the successor to the Vickers Vimy. Back on the ground, a photo was taken of him grinning smugly which he duly pasted in his logbook. It would be ten years before he repeated the exercise, on that occasion from a burning Halifax bomber.

Two days later, he was off to RAF Calshot. The move would catalyse a remarkable blossoming of his aviation career, establishing and shaping foundational elements of his reputation. He entered Calshot primarily to broaden his skills as a pilot, emerging three years later as a captain and 'the master navigator'.

As Don departed from his short-lived tenure as a fighter boy, the last word went to Squadron Leader O'Neill. Don had left Point Cook with an 'Above Average' rating. Whether it was the crash or Flight Lieutenant Merer's report, the commanding officer decided 'Average' was sufficient. Don's assessment of his superiors at 29 Squadron was probably no different. On the upside, he departed North Weald a 21-year-old and having been promoted to flying officer.

8

Boats That Fly

Just three days after descending by parachute from the Vickers Virginia, Flying Officer Don Bennett ascended for the first time in a flying boat at RAF Calshot. The coupling of flying boat and navigation training proved a potent combination for the restless and ambitious pilot because it laid the foundation for his taking on the world by air, literally.

RAF Calshot was on a spit jutting out into the water where ferries on the Southampton–Isle of Wight run would emerge from Southampton Water into the Solent and turn south for Cowes. Given its strategic naval position, Henry VIII had built a small circular castle there in 1540. From 1913, the Calshot Naval Air Station was progressively built around the castle, officially designated RAF Calshot in 1922. Initially used to test seaplanes, during the First World War the base had performed aircrew training, and anti-submarine and convoy protection roles.[1] The School of Naval Co-operation and Aerial Navigation moved to the base post-war. No 201 Squadron was re-formed there in 1929 with Supermarine Southampton flying boats; the Seaplane Training Squadron followed in October 1931:

> It was the main seaplane/flying boat development and training unit in the UK, with the landing area sheltered by the mainland to the west, north and east, and the Isle of Wight, a few miles away to the south on the other side of the Solent.[2]

The term 'flying boat' seems passing strange because it is quite evident on looking at one that it was not a boat that flew but an aircraft that landed on water – a 'sea plane'. The reason for its name lay in its design.

The very first attempts at getting aircraft onto water followed the fairly predictable course of replacing wheel assemblies with floats and leaving the fuselage sitting above the water, hence the term 'floatplane'.

Floatplanes had limitations, particularly handling the rougher ocean surfaces, leading to designers experimenting with turning the fuselage itself into a hull. The floats were moved to the wing tips to prevent the wings touching the water. This essential design was stronger and more stable. It was the American aviator Glenn Curtiss who coined the term 'flying boat' with his Models E and F. The effect on aviation was immediate. As more aerodynamic designs materialised, flying boats could operate in a wider range of meteorological conditions, carry more freight or passengers, go further by taking on greater fuel loads, and land in more places. At a time when there were few purpose-built runways, the 'ready-built' safe harbours were relatively plentiful. It was the perfect aircraft for the expanding the civil aviation industry.[3]

In strict terms, 'seaplane' was the generic term covering 'floatplanes', 'flying boats' and 'amphibians' (combining wheels and floats), though the British came to refer to 'seaplane' and 'floatplane' interchangeably vis-à-vis 'flying boats'. The distinction between types was an important one in the context of Don's career because Imperial Airways' celebrated Short Mayo Composite in the late 1930s involved coupling a flying boat (*Maia*) and a floatplane (*Mercury*), with Don piloting the latter and setting world records.

The Supermarine Southampton flying boat was just the sort of challenge Don relished, being about as far from land-based fighter aircraft as you could get. From the drawing board of the Supermarine design team headed up by RJ Mitchell (who subsequently designed the Spitfire), it had entered service in August 1925 at RAF Calshot as part of No 480 (Coastal Reconnaissance) Flight, replacing the Felixstowe F series, which had been a mainstay during the latter half of the First World War and into the 1920s.

The Southampton Mk.I was a twin-engine biplane with a double wooden hull. The Mk.II, the main production version, was delivered in 1929 with a metal hull, a significant advance because the wooden one absorbed water over time, adding to the aircraft's weight and reducing range considerably. It was a popular aircraft in the inter-war years due to its reliability and sea-keeping qualities, even in relatively poor weather. With no enclosed cabin, however, the crew were left open to the elements, albeit with small wind shields.

The course began with lectures. Under the heading, 'Southampton Boat Seaplane', Don gave a page to outlining the basic specifications, followed by a masterpiece of an overall description:

> The Southampton is a 5-seater fleet reconnaissance boat seaplane of the tractor type, fitted with 2 Lion Engines series 5. Dual control is fitted. The hull may be wood or duralumin and the tail unit incorporates the tail plane with incidence gear and three fins with balanced rudder. A planing bottom with 2 steps is fitted to the hull and reduces the area of immersion of the boat at high speeds, thus decreasing the resistance of the water. The steps eliminate longitudinal oscillations. The position of the main step on the planing bottom is so located that it is exactly below the centre of gravity when the boat is about to get off the water. The chine line keeps down the bow wave. Wing tip floats are fitted to the outer plane. Petrol system is by gravity from 2 tanks situated under the top centre M/P [mainplane]. Armament – 3 Lewis guns, bomb gear. A special handling chassis is designed to facilitate launching and beaching of M/C [machine].[4]

What cemented Don's relationship with flying boats was the instructor who took him aloft on that very first flight on 5 July 1932 – Flight Lieutenant Laddie Clift. Don described him as an 'old salt ... in fact, everything about him was deep green from the salt sea spray'.[5] It was a term of endearment because Don discovered immediately that Laddie knew it all, knew how to teach it all, and led by example. Clift commanded respect:

> My time or should I say my training under the direction of Laddie Clift was intensely satisfying. He taught me every function of being the captain of a Southampton flying boat, particularly on the seamanship accountability. He had us double checking, going over the nature of seamanship and the vast accountability of flying a flying boat until one could prove to him that they understood that knowledge in one's mind ... His methods were more direct, the style of tuition essential, teaching us to know one's place and position as a captain of a flying boat at all times. He was very precise in the do's and don't of seamanship, attitude, making sure you were up to the task.[6]

The critical element of Don's portrayal was Clift's laser-like focus on training and equipping his charges to be competent *captains*. This effectively shifted the point of focus for Don away from merely upskilling as a pilot; Clift, and the Flying Boat Pilot's Course generally, was about learning to command an aircraft. It meant taking responsibility for all elements of flying their boat – preparation, planning, piloting, navigating, technical, flying conditions (both air and sea), crew, etc. Competency comprised not just the technical aspects of airmanship and seamanship, but understanding the qualities of captaincy, including how to lead, having the right attitude, taking responsibility and being accountable.

It made perfect sense, but Don was 'crewing up' for the first time. For the young lad who had grown up essentially doing his own thing, being self-sufficient in his own abilities – one perhaps more suited to a 'fighter boy' mindset – the shift in thinking added a degree of challenge.

Flying Officer Alban Carey, a contemporary of Don's and known to all as 'Crackers',[7] offered additional insight into why Don responded so positively to Clift's leadership:

> Laddie Clift was indeed a hard taskmaster. He was always looking to get the very best from his trainees ... Clift was the master of preparation; never came across anyone like him, second to none ...

> [He] was a very technical man with the ability to understand how to teach those who were prepared to listen and train hard.[8]

Clift set high expectations and pushed them hard, but they knew if they responded he would get the best out of them. He was technically competent, always a plus for Don. His focus on preparation showed he understood risk.

Further, Clift was a leader of precision, not waffle. He knew what he was talking about, checked you knew what he was talking about, and measured your progress. He believed in learning 'on the job', seeing flying as an infinitely more instructive training ground than sitting behind a desk in a classroom. Leadership should be conducted in that space.

Don found the entire intellectual and practical challenge of the course hugely attractive. A key contributor was that marrying airmanship and seamanship brought navigation to the fore. Laddie Clift's later assessment of Don shows once again some key aspects of Don's character:

> He performed everything I had taught him with perfection. His attitude was right, and for that reason, I felt his time had come to be a captain of a Southampton flying boat. He … in my mind would become a perfect sea boat captain.[9]

Once Don had found something challenging and of interest, he would apply himself until he had mastered it completely.

There were precious few highlights. Early on, Clift had him flying up and down Southampton Water and the Solent in a Mk.II (S1249), just getting used to the aircraft and the conditions. After the Siskin, it is no surprise he found the Southampton heavy. The course pupils flew in different weather conditions, practised take-offs and landings, did their 'wind finding' exercises, learned some navigation and reconnaissance techniques, and developed their photography skills. For the first few weeks, Don rarely flew higher than 1,000 feet. Late in life, he reflected on the essential challenges of piloting these early flying boats:

> I think a flying boat automatically involves more problems. The mere flexibility of the flying boat, the fact that it can operate

anywhere, anyhow, anytime all over the world because there's always water, automatically introduces greater judgement. You have to judge whether you can land in a crosswind in a narrow river or you can land along a swell at sea because you can't land across it … too steep.

The judgement required by a flying boat pilot is far more extensive than any land plane pilot. And to that extent it's also character-building; the fact that you do cope with bad conditions and emergencies and engine failures and things like that which in those days were very common. That, in itself, made the flying boat fraternity quite different to the rest of the Air Force.[10]

Ten days in, Clift allowed Don to go up 'solo' with just Leading Aircraftman Powell on board. Just two weeks after his first flight, he was confronted with a forced landing when the port engine seized.

For a week in early August, Don flew the Fairey IIIF floatplane, a single-engine biplane used for reconnaissance. He could now draw a direct comparison with the flying boat. He was flying it solo on the first day.

Back on the Southampton, they undertook a series of navigation exercises beyond the surrounds of RAF Calshot and in a range of conditions. The trainees might be the pilot on one exercise, the navigator on the next, or just observing. Being the School of Naval Co-operation and Aerial Navigation, the exercises were designed around maritime reconnaissance. The early routes kept them within sight of land, with prominent headlands as turning points: Selsey near Bognor Regis, St Catherine's Lighthouse and the Needles on the southern and western tips of the Isle of Wight respectively, and Durlston near Swanage. They flew a little further along the coast to Beachy Head near Eastbourne.

Their first big adventure across the English Channel to Guernsey came unstuck in thick fog, causing them to divert to Exmouth. After waiting four hours, they took off, but fell short of Calshot in poor weather, alighting on the Beaulieu River where they were forced to practise 'navigation by soundings'.

There were no navigational radio aids when on the water in fog, so finding your way to a 'stretch of sea that you knew was almost certainly clear' for take-off saw one crew member 'on the side swinging the lead and calling the soundings' while another noted the various depths on a maritime chart until you knew where you were and could set a course for take-off which wouldn't see you run aground or plough into a hill.

As Don said, everything was 'primitive', resulting in what you learnt being 'very useful'.[11] While favourably disposed to all technological developments in aviation, he was nonetheless strongly attracted to learning from first principles and engaging with 'common sense'. It was a platform for seeing the similarities between aircraft of all types rather than their differences.[12]

Then came the specific reconnaissance exercises. They did a wide sector patrol to report all ships they sighted. They were sent to intercept another flying boat. HMS *Iron Duke*, the old battleship and veteran of the Battle of Jutland, was the subject of a course and speed-judging exercise, and the following day they were ordered to shadow HMS *Valiant*, another Jutland veteran. There was a search patrol for a battleship and a 'troopship', which turned out to be the ocean liner RMS *Mauretania*. Then, in late October, they undertook their longest trip to date to intercept the ferry SS *Worthing* in the Channel between Newhaven and Dieppe.

All this was a build up to the big finish for the course, the ten-day cruise up to Scotland where everything was put to the test. The initial crew was Don and Pilot Officer Menzies, (doing turnabout in the pilot's and navigator's seats), with Farrow, Maver and Palfrey. Flying Officer Ken Pickles, who Don called 'Percy', was there as instructor, examiner and monitor of progress.[13]

The first day, 8 November 1932, was a short hop to Mount Batten, with Don piloting. Day two was a seminal event for Don because he was to navigate the longest leg of the entire trip. They headed for Penzance, then turned north for Scotland, requiring them to venture further from land than at any time to date. Crossing the western approaches of the Bristol Channel, they reached St David's Head in Wales and headed over the Irish Sea where the visibility deteriorated. Carrying on, they passed over the Calf of Man and changed course for Stranraer, arriving after a challenging flight of five hours and 40 minutes. Don assessed his first significant long-distance test as a maritime air navigator: 'Navigation worked out very well.'

For the next five days, they mooched around the Hebrides and western parts of the Scottish mainland. They did map reading, photography exercises and anchoring trials, taking in coastal spots like Oban, Iona and the Crinan Canal, along with Ardnamurchan, Loch Alsh and other inland waterways. Don had to bring the Southampton down on Loch Lomond after an oil pipe broke. They fixed it and got airborne again. Much of the time the visibility was good, but the Queenslander did remark on how cold it was.

On 15 November, they did a practice formation flight across the North Channel to Belfast and down onto Strangford Lough where they anchored for lunch, Menzies piloting and Don navigating. The following day, they swapped seats for a return formation flight to Belfast, turning north for Larne Lough where they did an air salute for HRH Prince of Wales, who was aboard the ferry MV *Ulster Prince*.

On the return journey to Calshot, Don was piloting on the leg from Pembroke Dock when they struck poor weather around Land's End and didn't clear it until they were near Calshot. This challenge was a fitting end to the cruise.

This first cruise was formative for Don's thinking. Away from Calshot, the cruise gave them freedom, discretion and responsibility. Being thrown into new environments, out of their comfort zones and over much greater distances, tested their theoretical knowledge and practical skills. In reflecting on the many training cruises he ultimately participated in, Don recorded: 'It was excellent for initiative, as we were often left very much to ourselves, and if in trouble, we had to take the appropriate action entirely on our own initiative.'[14] This spoke of self-reliance, self-sufficiency and an ability to problem-solve. It was tailor-made for someone who sought to be good at everything.

The overall Flying Boat Pilot's Course, and the cruise particularly, offered something of a training archetype for Don, which began with getting out from behind the desk and going flying. Don't learn about the job, learn on it. There was nothing like being in low visibility over the Irish Sea with no adjacent land to find out if what you had been taught actually worked, whether you were temperamentally suited to the task, and where you might improve.

Throughout his career, Don regarded his steadily rising flying hours as a mark of credibility, a basis for seeking respect, and a way to measure himself against others. Such was the pace of change in aviation that even a short time away from flying was, in his eyes, a sure sign of impending, if not actual, redundancy.

The cruise offered up echoes of the Brisbane boy too. There was something of the outdoor life of his youth with the adventure and the camaraderie, 'roughing it' as Australians would say. The Aussie 'can do' and 'make do' spirit was evident, and getting his hands dirty with the mechanical remained a legitimate activity.

The course was now over. Don achieved '1st Pilot', while topping the course in navigation, and they went on leave. '[It was] in the navigational subjects it so happened that I did well, and this was to affect my subsequent postings.'[15] As much as anything else, the course, in marrying airmanship and seamanship, provided a bridge to the relentless skies out over the open oceans, opening up a world of possibilities.

Shortly after, Don was posted to No 210 (Flying Boat) Squadron at Pembroke Dock, commencing in the first week of January 1933. There he would make the most significant connection of his entire aviation career.

9

Under Harris's Wing

The future Air Chief Marshal Sir Arthur 'Bomber' Harris, Air Officer Commanding (AOC), Bomber Command, was a mere wing commander when, in late March 1933, he was appointed station commander of RAF Pembroke Dock, on the shores of Milford Haven, the home of No 210 (Flying Boat) Squadron.

He already had a long Royal Air Force career, having flown in combat in the First World War, including at night against the Zeppelins, and ending it commanding No 191 Squadron, a night-flying training squadron, then No 44 Squadron.

During his time in charge of No 45 Squadron in Mesopotamia in the early 1920s, in keeping with his passion for innovation and the technological advance of military aircraft, Harris and his flight commanders, Robert Saundby and Ralph Cochrane, had experimented with new techniques for bombing, including installing bomb racks and crude bombsights in their Vickers Vernons (a large biplane troop carrier), accompanied by copious amounts of practice. The squadron also commenced night flying, even deploying crude marker bombs to illuminate targets.[1]

His commitment to night flying continued at No 58 Squadron, a bomber squadron at Worthy Down, where he pursued the development of blind-flying instruments such as horizon and turn indicators. Harris followed this with time as the deputy Senior Air Staff Officer to the Air Officer Commanding, Middle East Command, in Egypt.[2]

Just prior to his appointment at Pembroke Dock, he had spent six months completing the Flying Boat Pilot's Course (No 16 FBPC) at RAF Calshot, completing it with an 'Above Average' rating. Don was already halfway through his course (No 15 FBPC).³ Neither referred to knowing, or even being aware of, the other at Calshot. This changed at Pembroke Dock.

Although different personalities and from different generations, they had quite a lot in common, which may be why their coming together at Pembroke Dock produced something of a *consensus ad idem*. Like Don, Harris had rebelled against parental expectations. Born in England in 1892, his father, an Indian civil servant, had wanted him to join the Army but, at the age of 17, he escaped to Rhodesia – 'the colonies'. He came to regard himself as Rhodesian, not English. In Rhodesia, Harris drove some of the very first cars, enjoyed the sunshine and the outdoors, and 'took to general farming, tobacco, maize, and cattle'.⁴

His early break from parental expectations to establish a life for himself reflected his character:

> He became self-reliant at an early age, developed the equanimity and stoicism of one much older than his years and, as he progressed to maturity, displayed an unselfish nature and a tremendous sense of integrity and justice. His courage as a youngster was also great, but it was always tempered by plain good sense.⁵

During his time in the Royal Flying Corps and then the RAF, Harris had developed a fervent distaste for moribund military thinking (especially around lack of foresight, vision, and the failure to evolve with new military technology); disliked the Army for its various efforts at suppressing an emerging and independent RAF; and found petty bureaucracy and unwieldy structures that prevented effective decision-making anathema. His self-deprecating humour and acerbic, coruscating wit were complemented by a reputation for speaking his mind. It was a career that had been, and would continue to be, founded on a bedrock of being 'a highly experienced airman … so well versed in the technicalities associated with modern aircraft'.⁶ All of these aspects of Harris's background, outlook and character would have resonated with Don.

Where they differed was that Harris had been with the RAF from its inception, amassing nearly two decades of experience in operational and staff roles. He knew the players and the processes, and therefore the best way to deploy his practical, no-nonsense bluntness. As such, he was someone who, as Don put it, 'certainly made things move'.[7]

On completing the course at Calshot, Don was posted to 210 Squadron at the start of 1933, a couple of months ahead of Harris who was still at Calshot completing his. It was a natural next step given he loved the flying boats and had a particular aptitude for navigation.

The first two months for Don were simply an extension of the Calshot experience. He refers to being '1st Navigator' on some flights, specifically noting the outcome of one as 'Average error <2%'. Did this reveal a degree of satisfaction or was it a clear sign there was room for improvement?

There were photographic exercises, taking the usual photos and obliques, building mosaics, and testing the F24 camera. Two of those exercises were with Flying Officer Clayton DC Boyce, known as 'CDC' or 'Bruin' for his teddy bear shape.[8] Like Harris, Boyce would come to play an important part in Don's future RAF career. He was three years older than Don with extensive flying boat experience. Commencing in 1928 at No 480 (Coastal Reconnaissance) Flight at Calshot,[9] he had spent nearly three years with No 205 (Flying Boat) Squadron in Singapore from February 1929.

Boyce had been at 210 Squadron almost a year when Don arrived. In January, he was one of a few officers who flew with Don to get him acquainted with the conditions at Milford Haven, the stretch of water they flew from at Pembroke Dock. After a couple more flights together later in 1933, a decade would pass before they flew together again, at which point Don was the AOC, Path Finder Force, Bomber Command, and Boyce his Senior Air Staff Officer. That created a dynamic all of its own.

Something new was anti-submarine patrol exercises, including practising dropping smoke bombs on a substitute for a submarine – a pinnace – before undertaking a more serious naval cooperation exercise with HMS *Iron Duke*.

In the third week of March, Harris arrived to take command:

> I had three aims, to do something to break what we called at that time 'the flying boat trade union', to clear the black magic out of the trade, and to help solve difficulties they were having with night flying.[10]

On 30 March, Don flew Harris to Headquarters Coastal Area at Lee-on-the-Solent, then on to Calshot for an overnight stay. The following day, he flew Harris back to Pembroke Dock via Bridgwater, prompting a curious logbook entry: 'Prospected new land crossing – more direct.' It is impossible to see this as occurring other than at Harris's instigation given it took an hour longer to fly the route than the trip from Pembroke Dock the previous day. It certainly afforded Harris an early opportunity to assess Don's piloting and navigational abilities.

Harris had arrived at Pembroke Dock determined to challenge the prevailing thinking that flying boats really should not go up after the sun went down:

> The flying boat branch of the service had succeeded in surrounding itself with an esoteric atmosphere, based largely on spurious nautical lore, and wished it to be understood that there were so many difficulties in the way of taking up a flying boat by night that for ordinary human beings to attempt this feat was suicidal. It did not take me long to discover that the only difference between night flying in a flying boat and in an ordinary aeroplane was that it was in every way much simpler and safer in a flying boat.[11]

Having developed something of a love for flying boats, Don might have disagreed with this attack on the fraternity who were teaching him so much, such as Laddie Clift, but would have wholeheartedly embraced Harris's direction to fly at night. He, himself, had done no night flying in flying boats prior to this, either at Calshot or Pembroke Dock, so it represented a new challenge.

Harris was indeed in a hurry. In the first week of April, at least two flying boat crews, including Don, were dispatched to the relatively safe waters off Calshot to commence night flying under the oversight of No

201 Squadron. Over the coming fortnight, Don flew for over six hours at night, taking to the air around 30 times, and qualifying '1st pilot by night'.

It is instructive that Harris concluded this section on night flying in his memoirs by making his very first mention of Don:

> Don Bennett, afterwards AOC of Bomber Command's Pathfinder Force, was one of my Flight Commanders in the boat squadron … He was, and still is, the most efficient airman I have ever met.[12]

The last sentence is arguably the most famous and ubiquitous description of Don. What is intriguing is Harris's prompting to use it in the context of Pembroke Dock, and night flying in particular. What did he mean by 'efficient'? What did he see in Don so early?

Harris provided no clarification, making definitive conclusions impossible, but there are directions to look in. The first was the night flying: he wanted a job done and a point proven, and Don simply got on and did it in the most efficient manner possible, indicated by rapidly qualifying as '1st pilot by night'.

Secondly, within days of taking command, Harris had flown with Don for a total of seven hours. What did he observe on those flights? Aside from demonstrating Don's navigational expertise, it provided the perfect opportunity to observe his mastery of the cockpit.

Finally, Harris noted Don was one of his flight commanders at Pembroke Dock. Don was silent on the subject,[13] but Harris, in mentioning the fact, evidently based at least part of his assessment of Don's 'efficiency' on seeing him through that lens.

It is highly likely Harris saw something of himself in Don. They thought alike: possibilities rather than problems; breaking through bureaucracy and hidebound traditions that prevented development; and a belief that just doing it trumped protracted discussions about possibly doing it, especially if the outcome of those discussions was a product of negative group think. Harris's biographer, Henry Probert, quotes from Flying Officer Peter Tomlinson, Harris's personal assistant during his time as AOC of 5 Group, Bomber Command:

He had a reputation for being a 'no nonsense' man and I could see why … His questions were very much to the point … I knew at once I was being talked to by a pilot still interested and active as such …[14]

Probert also describes Harris as having a 'realistic, down-to-earth approach'.[15] All echoes of Don, and all founded on a shared passion for flying and pursuing technological developments in aircraft.

While speculative, one cannot escape the fact that, in the context of 'first impressions last', it was at Pembroke Dock where Don laid solid and lasting foundations with Harris. Interviewed about his 210 Squadron days in 1981, Harris made some comments about night flying and added, 'Don Bennett was an outstanding pilot and held a nap hand of "tickets" eventually.'[16]

It would be Wing Commander William 'Andy' Anderson, one of Don's original staff of five in the Path Finder Force, who offered perhaps the clearest insight into Don's 'efficiency' with this observation among his pithy reflections on his boss:

And if anything had to be done, he would lift up the phone, or press a knob to operate his incredibly complicated loudspeaker system and it would be done at once. He never put off for half an hour what he could do then and there …

To sum up. A brilliant brain, a fanatical worker. A good talker, with more wit than humour; and yet essentially not a talker, but a man of action.[17]

This is complemented by a piece of varnished plywood board Don had on display in his post-war office. Burnt into it was his characteristically blunt take on a well-known Longfellow poem about the legacy one leaves in life. It said simply: 'Footprints on the sands of time are not made by sitting down.'[18]

It all pointed to character traits undergirding the great strengths Don exhibited, but likewise the source of some of his limitations.

It is productive simply seeing Don through the prism of some basic definitions of 'efficient', both of which are pertinent:

1. (of a system or machine) achieving maximum productivity with minimum wasted effort or expense.
2. (of a person) working in a well-organised and competent way.[19]

The first definition has also been expressed as 'the ability to achieve the end goal with the least waste'. Don was definitely goal-focused, crystal clear what those goals were, absorbed detail faster than almost anyone else, evaluated options quickly, decided just as quickly, acted on the decision, and ruthlessly pursued improvements to pathways to reach set goals, perhaps with a machine-like regard for the feelings of others. As such, he likely contrasted with many elements of the RAF around him, something Harris, being of similar ilk, found attractive. To Harris, Don could cut through and get things done.

Don also had a pathological distaste for waste, whether that be pieces of paper that still had space to write on, or money expended on bureaucratic processes achieving no outcome or, ultimately in the context of the war, lives. And, in his mind, a key contributor to wastage of any kind was incompetence.

Following the initial night-flying exercises at 210 Squadron, Don was dispatched to the Central Flying School, RAF Wittering, for a three-week course on instrument flying in an Avro 504N. The time was spent learning to fly blind 'under the hood', including take-offs, spinning and flying compass courses, along with a range of instrument-flying exercises.[20]

Don signalled the highlight of his final two months at Pembroke Dock via a cryptic reference in his memoirs:

> On one occasion I was flying as second pilot to a more senior officer in the Squadron, and on the way back to Calshot his ability to fly on

instruments, even in fairly good weather which prevailed, was not all it should be! When I saw the compass spinning and the air speed climbing, I rushed for the cockpit to find him half rolled over – and with all flying wires screaming and wings flapping, he was heading for the sea. I grabbed the controls in the rear cockpit, and righted the aircraft without any resistance from the first pilot. It was an interesting moment, and a very surprising one for an old hand.[21]

The lead up to this event had begun nine days prior. Don had arrived back at Pembroke Dock from Wittering and been thrown straight into three days of night flying, starting with Boyce on the night of 22 May. The following night, he flew with Flight Lieutenant Hodder and was only back at base a short time before taking off again just after 2:00 am on the 24th for his very first night operation – a fishery protection search dropping parachute flares south-east of Swansea. Don recalled they 'very nearly managed to make arrests of French fishermen' in British territorial waters.[22]

Just over five hours later at 10:00 am, Don was flying with Canadian pilot, Flight Lieutenant Fred Mawdesley,[23] over to Calshot. Something was afoot because on arrival he found Squadron Leader AF (Frank) Lang, Officer Commanding Flying Flight at 210 Squadron, was already there. The following morning, Don flew Lang across to Headquarters Coastal Area, then back to Calshot late that afternoon before heading to Pembroke Dock, alighting after midnight.

Within days, he was back at Calshot and taking to the air with Lang on a daytime long navigation exercise out to the Scilly Isles. Don had now flown Lang for over nine hours, by day and night, in the space of five days. Something was in the offing.

Just 48 hours later, Lang flew Don and the crew of S1124 from Calshot to Dover. If Don had been unsure why this flurry of activity had occurred since returning from Wittering, he now knew. The squadron was participating in experiments being conducted by the Air Defence Experimental Establishment (ADEE) in the English Channel. ADEE was a forerunner of the Telecommunications Research Establishment (TRE), a group that worked on radar systems during the Second World War. It would play a pivotal role during Don's time commanding the Path Finder Force.

In these pre-radar days, the ADEE was experimenting with an early-warning detection system involving sound waves. A series of large spherical acoustic mirrors with microphones at their point of focus, known as 'listening ears', had been erected along the coast, designed to detect incoming enemy aircraft by picking up the sound of their engines. They were the brainchild of Dr WS Tucker.

Their lifespan turned out to be very short because radar was on the way. They also encountered difficulties distinguishing between aircraft and ships, and their usefulness for early warning was decreasing rapidly as faster aircraft were produced.

That was yet to come. Don was now to spend much of June helping ADEE with its acoustic experiments by doing a series of flights up and down the Channel. All were to be conducted at night, ensuring the 'listening ears' were not given a helping hand by 'seeing eyes'.

To get things underway on 1 June, Lang, having attended the ADEE conference during the day, decided to pilot S1124 with Don and crew on the first exercise from Dover before continuing the flight, not back to Dover but to Calshot. They took-off at 8:45 pm for what became a five-hour-and-50-minute flight, beginning with the ADEE exercise. At nearly 39, Lang was not a young man. Then, as Don recalled, the flight back to Calshot became life threatening. He arrived back at the cockpit to find Lang slumped over the controls, requiring emergency action to regain control and save all their lives.[24] They alighted at 2:35 am. No explanation is offered, but the evidence simply points to fatigue.

Don and Lang never flew together again. It is not known if either advised Harris what had happened.

At first glance it may come as a surprise that Don mentions the incident but does not name Lang, given his usual absence of diplomacy when commenting on more senior officers he felt were lacking in some way or another. He had regard for Lang and saw this episode for what it was – something unrelated to Lang's competence.

Don returned to Dover with Mawdesley on 10 June and together they conducted the remaining ADEE experimental flights over the next month without major incident. The only real interruption was 'waiting to escort H.R.H. Prince of Wales across Channel'.

Back at Pembroke Dock, Don started taking up new pilots to introduce them to flying from the base. There was also some air-firing practise at sea markers. On the first day, they all had a go, but from then on it was all up to the air gunners.

Then, just weeks later, at the end of July 1933, Lang assessed Don's proficiency as 'Above the Average' and the axe fell.[25]

10

Back to the Future

Don was enjoying himself at No 210 Squadron, so what occurred came as a rude shock. Weeks out from his 23rd birthday, and just seven months after completing the Flying Boat Pilot's Course (FBPC) at Calshot, he was posted back to the School of Naval Co-operation and Aerial Navigation as a navigation lecturer. Just when he was beginning to flower as a flying boat captain, pilot and navigator, someone decided to stick him behind a desk, albeit on his feet with a piece of chalk.

His response was blunt and visceral: 'The idea of being a lecturer was repulsive to me.'[1] Maybe no one grasped navigational theory as well as he did, or he was the only suitable candidate available for the vacancy. Given Laddie Clift's view of Don's abilities, it is quite possible he had a hand in it. The reasons were immaterial because the RAF decided Don's postings and he had to wear it. Within two weeks of finishing at Pembroke Dock, he was back at Calshot. It was August 1933.

And so, the final two years of Don's short-service commission with the RAF began in a very unpromising and unhappy fashion.

Don delivered his first lectures to the pupils of No 19 FBPC. No doubt his students were more attentive than he had been back at Brisbane Grammar School. For his part, lecturing was his 'price of admission' to Calshot, an inconvenience to be endured while he got on with flying. He admitted teaching others helped his own learning and contributed to his seeing where the gaps were in the air-navigation syllabus. This would, in part, catalyse his desire to write a book on the subject.

By the end of September, he had flown regularly with 'Percy' Pickles, the instructor/examiner on Don's FBPC cruise to Scotland the previous November, and 'Crackers' Carey who, like Pickles, was now an instructor with the Seaplane Training Squadron. They had what Don wanted, so he persuaded Group Captain HR Nicholl, the station commander, to give him a dual role as lecturer at the school and instructor with the training squadron. For this, he obtained the Seaplane Instructor's Certificate.[2]

In this posting, Flight Lieutenant GIL 'Gill' Saye was his flight commander. Saye had vast experience as a flying boat pilot and just the style of leadership Don thrived under:

> Gill Saye was Flight Commander, one of the best I have ever had. He allowed a remarkable degree of freedom to both me and Crackers Carey ... The three of us remained together for a very long period ...[3]

As a superior officer, Saye, in giving Don freedom, was definitely more 'Duke' than 'Merer'.

In tandem with lecturing, Don now instructed the FBPC pupils in the air from the initial exercises in navigation, on to reconnaissance, search and shadowing patrols, interception exercises, night flying, etc., all the way through to the ten-day cruise at course end. Given a new six-month course began every three months or so, Don was instructing constantly. The highlight for him would always be the cruises. On 7 November 1933, less than three months after his return to Calshot, he wrote in his logbook: 'Commenced Training Cruise as Captain S1233 with three pupils of No 19 FBPC.' It was a year, almost to the day, since he had embarked on the last FBPC cruise, on that occasion as a pupil with 'Percy' Pickles as instructor/examiner. Pickles was now his equal.

This cruise did not disappoint. They headed for Scotland again. There were technical problems on the first day. The third day brought a forced landing at Douglas when the starboard engine failed. They waited for three days until a replacement arrived, which they fitted, allowing them to rejoin the cruise at Oban. There were oil pressure problems, then the W/T receiver was playing up, requiring them to drop into Stranraer for a replacement pair of headphones. On alighting in heavy swell back at

Douglas, three ribs were broken in the tailplane that they had to repair. A week after replacing the starboard engine, they had to do it again. Somewhere amid all that they did the prescribed exercises.

In all, Don instructed on seven FBPCs, including cruises, over the next two years, providing the 'backbone' of his time at Calshot. With the cruises, he was in his element because they tested everything – theoretical knowledge, mechanical ability, all-round expertise, temperament, planning, the aircraft and its equipment, a willingness to 'muck in' and, if you were captain, your leadership. When it came to navigational theory, it was the ultimate test of his theoretical and practical teaching, something he could observe in his pupils first-hand.

One month before that first cruise, on Sunday 8 October 1933, Don made his way up to the aero club at Whitchurch where he took a small monoplane on a ten-minute test flight. He bought it on the spot, paying £115, before flying it the 25 miles back to Southampton, with a 'scenic tour' flight time of 65 minutes.

This was quite some feat because the aircraft in question was a de Havilland DH.53, known as the Humming Bird. Designed and built to compete in the *Daily Mail* Trials at Lympne in October 1923, it 'was the first true light aeroplane produced by the de Havilland Aircraft Company'.[4] Only 15 were ever built. Its claim to fame was being underpowered, impractical and unreliable. Alan Cobham, then de Havilland's test pilot, demonstrated the problem with the 26-hp Tomtit engine on a flight from Stag Lane to the Brussels Aero Exhibition in December 1923:

> Although it was winter-time, Cobham elected to fly solo across the snow-covered fields of Flanders although he was forced by headwinds to abandon the return flight after finding himself in exceptionally low cloud. After a few minutes, however, he realised that in fact he was in a steam cloud, having been overtaken by a slow-moving freight train on the rail track he was following.[5]

An upgrade to a 32-hp Bristol Cherub 3 (as in Don's – G-EBXN) or a 35-hp ABC Scorpion engine did little to improve the power. The noise was considerable and the vibration strong enough to blur the instruments. Not that there was much instrumentation. The best that could be said for the cockpit was it offered a good all-round view as it rubbed the shoulders of the occupant. Even small mounds of earth could prove insurmountable for the tiny wheels.

Once in the air, however, the Humming Bird actually proved to be quite nimble '… with crisp, powerful ailerons and a responsive elevator and rudder … a most agile performer (provided the engine kept going) with loops, barrel rolls and steep, very small, diameter turns at low level'.[6] Don commented, '[It] was highly delightful and highly dangerous. It was as light as a feather, but it had a stall like the crack of a whip.'[7]

As early as 1924, the DH.53 had produced 'the very strongest feelings of no-enthusiasm' at de Havilland,[8] and yet Don loved his. For one thing, it was cheap to fly. It might seem an odd choice for a pilot wedded to technological advances, but therein lay the attraction and the challenge – it constantly tested his skills, giving him pleasure in mastering its flying characteristics, thereby getting the most out of it. His response to the Humming Bird was reminiscent of his commentary on the Siskin: both aircraft had significant and potentially dangerous shortcomings, but he had both a desire and a capacity to identify and test their limits. Whether this was due to intuitive feel or lots of practise is hard to say, probably both. Both were prefaced by strong theoretical knowledge and his ongoing love of the mechanical.

The only problem he ever had was the tail skid breaking twice. He flew often over the next two years and would only part with it when faced with a greater love, one that could not be accommodated in an already cramped cockpit.

Aside from the Humming Bird, the first few months back at Calshot produced a range of new experiences.

Days after Don's arrival, the Seaplane Training Squadron took delivery of a new type – the Saunders-Roe A.29 Cloud. Its dimensions were roughly the same as the Southampton II, but differed markedly in being a monoplane, faster, having a considerably higher ceiling of 14,000 feet and, crucially, an enclosed cockpit. Don took it for two test flights immediately. He flew it relatively regularly over the next two years, culminating in taking part in the RAF Display at Hendon in June 1935. The FBPC training, however, continued in the old Southamptons.

Supermarine itself was working on a successor to the Southampton II, the Scapa. Unlike the Cloud, it did not gain a foothold at Calshot as a training aircraft. Don flew it just once.

Beyond new aircraft, Don began participating in trials testing new methods and equipment. They went out a few times testing the 'four-point bearing' method of wind finding, among other activities. He escorted HRH Prince of Wales across the Channel again, did some general flying practice with Wing Commander TQ Studd, took visiting land-based pilots for some air experience in flying boats, and participated in a search for an aircraft lost from the liner SS *Bremen*. Elements of the flying boat pilot's repertoire were practised. Aircraft that had been serviced or fixed were tested. Don had an excellent working relationship with Crackers and Gill.

But it was becoming routine all too quickly and that was a problem; and it was not enough. New Year's Day 1934 ticked over and Don Bennett changed tack.

11

The Invisible Barrier Over the Horizon

The New Year of 1934 marked the 100th anniversary of the City of Melbourne. The Lord Mayor, Councillor H Gengoult Smith, and his Centenary executive had been contemplating various events to mark the occasion. An air race from England was mooted, leading to the commissioning of a confidential report that concluded the 'expense was insuperable'.

Undeterred, in early 1933, Smith approached Sir Macpherson Robertson, the wealthy Melbourne businessman and founder of the MacRobertson confectionery company, home to some of Australia's greatest creations like Freddo Frogs, Cherry Ripes and Columbines, who put up A£10,000. His prime motivation was stimulating the local economy, achieved by drawing the world's attention, and visitors, to the city. Smith could not have agreed more and the publicity machine swung into action.

The entire £10,000 prize would go to the aircrew who flew fastest from London to Melbourne in October 1934. Initially the Melbourne Centenary International Aero Cup for the MacRobertson Trophy, it was later officially known as the MacRobertson International Air Races or, more commonly, the Centenary Air Race:

> The Lord Mayor said that this would be the greatest air race in the history of aviation, and he was confident that the makers of aircraft throughout the world, and the world's foremost pilots, would compete.[1]

A critical early decision was taken to attract aircraft manufacturers and the international flying fraternity: the race was open to 'any make of machine, with any number of engines, and carrying a crew of any size'.[2] The Lord

Mayor saw this as stimulating innovation and invention that would advance aviation by five years. Essentially, manufacturers and aircrew were given *carte blanche*. It would not end up that way.

Sir Charles Kingsford Smith was asked immediately if he would be an entrant. He doubted it. He saw the prize doing a lot for civil aviation but 'such an ambitious project requires a lot of planning',[3] and for that reason was a likely non-starter.

Nine months later at RAF Calshot, New Year's Day 1934 arrived. As Don recalled, he was sitting at his desk in his bedroom 'solemnly' making resolutions. Chief among those was to participate in the race.

It was a fascinating development. He provided little explanation for the decision aside from a rather vague 'partly to get to know the world' in 1986.[4] There was likely no single reason. Aside from the Australian connection, there was the prestige; he could make a name for himself by following in the footsteps of his heroes (Bert Hinkler and Amy Johnson had performed the feat a mere six years earlier). There was also the challenge: such endurance races demanded the application of all your aviation expertise, making it the ultimate test of airmanship.

There were bigger issues at play. The end of his short-service commission was just over 18 months away, raising fundamental questions about his career. Would he try to stay in the Air Force or turn to civil flying? Would this decision be lived out in England or back in Australia?

The Air Force had given him a lot, but he was finding it too restrictive, even insular. In recent times, it seemed to have got it right, and then he was given the lecturing job. 'Gill' Saye gave him plenty of latitude and Calshot had offered new challenges, but all too soon things were becoming routine. The pace of change was not great, and civil aviation appeared to be innovating faster. The race definitely offered the opportunity to step into that world and, if he did well, opened up new opportunities and possibilities. Years later, in reflecting on those times, he admitted, '… I fully realised that a peace-time Air Force had its limitations, and if I were to continue to make real progress in aviation I must look elsewhere.'[5]

Don's decision to enter the air race created a problem: the unknown, 'impecunious' flying officer needed financial backing or the support of a manufacturer to get an aircraft good enough to go the distance. Despite having over 700 hours flying time in ten different types of aircraft, he did not see this as a sufficient distinction from other potential entrants, so he turned to his pet subject of navigation:

> I decided, therefore, that I must get my First Class [Air] Navigator's Licence, a qualification then extremely rare anywhere in the world, even in Great Britain.[6]

What he does not explain is this was a civil licence, and seeking it was not just a choice to distinguish himself in the context of the race, but the avenue for maximising his choices in the civil aviation world.

At the time, there were parallel military and civil air navigation qualifications. The RAF issued a 2nd Class Air Navigator's Certificate, the corresponding qualification to the civil Second Class Air Navigator's Licence issued by the Air Ministry. With virtually no specialist observers/navigators in the RAF, almost all obtaining the 2nd Class Air Navigator's Certificate were pilots completing a specialised navigation course, such as that offered at Calshot.[7]

To achieve the RAF's 1st Class Air Navigator's Certificate, an applicant had to be a 'holder of a 2nd Class Certificate for at least two years and not less than 200 hours of practical airborne experience as a navigator'.[8] This included submitting logbooks as proof of navigation hours completed, a near impossibility given logbooks did not specifically differentiate hours flown as navigator. Even if Don was eligible to apply, the 1st Class Air Navigator's Certificate was not an option because he needed a civil navigation qualification for civil aviation.

This is where Don's story has a twist. He records being awarded the civil Second Class Air Navigator's Licence (No 171), on 27 February 1933, having just completed the Flying Boat Pilot's Course at Calshot and been posted to No 210 Squadron.[9] He offered no indication he sat any additional exams or underwent any testing for the civil qualification, nor did he mention ever receiving the RAF's 2nd Class Air Navigator's Certificate at

85

Calshot. Further, he recorded on that very same day being awarded his civil Pilot's 'B' Licence (No 5230).[10]

Irrespective of the circumstances, he was aware the civil Second Class Air Navigator's Licence also had its limitations:

> In essence, any civil aircraft carrying passengers or goods for hire or reward on an international flight of more than 625 miles over the high seas or uninhabited terrain or at night was required to have on board the holder of a First Class Navigator's Licence.[11]

For most pilots and navigators in Britain, the 625-mile limit posed no barrier at all given flying was predominantly over inhabited land during daytime, meaning few needed a First Class Air Navigator's Licence. For Don, however, until he had that licence he was always going to be constrained within this invisible barrier over the horizon. Not for the race itself, because he would not be 'carrying passengers or goods for hire or reward', but he needed to consider the civil aviation realm beyond the race. Obtaining the licence guaranteed he could go anywhere; it was the key to opening up all future possibilities in civil aviation.[12]

Obtaining the licence was a gruelling exercise. The examinations were only held once a year, in March, leaving very little time.[13] Don started studying immediately. He was supposed to turn up 'dressed for dinner' five nights a week as required by the Station Commander – 'a lengthy procedure' – but found an ally in the President of the Mess Committee, PD Robinson:

> [He] was both human and intelligent, and he turned a blind eye to this requirement, whilst I cooked sausages on my open coal fire in my bedroom and kept at my study from the moment work ceased in the afternoon until 2 a.m. every morning.[14]

It wasn't just the limited amount of time Don had; it was the scale of the exercise. Over four days in March, he faced a combination of ten oral, practical and written examinations covering the following subjects: International Legislation; Form of the Earth, Maps and Charts, Tides; Meteorology; Dead Reckoning and Direction Finding W/T Navigation;

Earth's Magnetism and Compasses; Astronomical Navigation; and Visual Signalling.[15]

A simple pass was insufficient. Candidates were required to score higher than 90% for Visual Signalling, higher than 60% for all other subjects, and an average higher than 70% for all subjects aside from Visual Signalling.[16]

To help candidates grasp the scale of the mountain they were seeking to climb, the Air Ministry helpfully attached to its 'General Information and Conditions' a four-page appendix providing a detailed breakdown of the syllabus for the seven subjects. For instance, on the subject of Astronomical Navigation, *one* component specified the ability to provide definitions for the following (and it was laid out in this fashion):

> Celestial Sphere, Celestial Poles, Equinoctal, Ecliptic, Celestial Meridian, Parallel of Declination, Geographical Position, Declination, Polar Distance, First Point of Aries, Right Ascension, Zenith, Nadir, Zenith Distance, True Altitude, Observed Altitude, Apparent Altitude, Vertical Circle, Prime Vertical, Visible Horizon, Sensible Horizon, Rational Horizon, Dip, Azimuth, Amplitude, Semi-diameter, Refraction, Horizontal, Parallax, Parallax in Altitude, Hour Angle, Meridian Passage, Superior Meridian, Inferior Meridian, Superior Transit, Inferior Transit, Rising, Setting, Easting, Westing, Mean Sun, Apparent Sun, Mean time, Apparent time, Standard time, Sidereal time, Greenwich Mean time, Greenwich Apparent time, Greenwich time at Place, Apparent time at Place, Greenwich Date, Equation of time, the quantities R. and E., Mean Solar Day, Apparent Solar Day, Sidereal Day, Change of Date, Lunar Month, Calendar Month, Civil Year, Solar Year, Sidereal Year, Equinox, Solstice, Aphelion, Perihelion, Obliquity of the Ecliptic, Star, Planet.[17]

It all concluded with a General Note that read more like a warning:

> Candidates for the first class examination will be expected to display a more advanced knowledge of the theory of the subjects as set out above, but nevertheless practical work and experience will carry

most weight in the examination. Neatness, speed and accuracy will be expected, and particular attention should be paid to those methods where applicable which serve to eliminate all unnecessary work in the air. Errors in computation, of whatever nature, will be heavily penalised.[18]

Don had exactly 11 weeks before the first three exams on Monday 19 March. During the day, he was expected to fulfil his lecturing and instructing duties. On 4 January, the next Flying Boat Pilot's Course commenced. That would culminate in the cruise to Scotland from 20 February to 3 March, a precious ten days where he would have little time to himself to study.

On 19 March, he went up to London, spending the first three days doing written exams before heading to the Airport of London, Croydon, on the final day for the practical tests.

> … to my great surprise I passed. I was, I believe, then the seventh holder of the licence in the world.[19]

It was officially awarded on 20 July 1934. Like Harris's 'most efficient' comment, Don's claim to be the seventh holder in the world is ubiquitous. Unlike Harris's assessment, it is open to question. On arrival at Imperial Airways in early 1936, Don discovered that most, if not all, the airline's pilots had a First Class Air Navigator's Licence because they were flying international routes.

He did not offer a source for his claim, but he clearly believed it. Given most of his flying experience in Britain was restricted to the RAF, it is distinctly possible he was told he was the seventh in the RAF to gain the licence. It should also not be forgotten this was a British licence. There were many civil airlines operating international flights with their own complementary licensing arrangements, so however one defines 'seventh', it only pertained to Britain.

Nonetheless, the magnitude of this achievement for Don's career and his sense of self cannot be overstated. It undoubtedly carried prestige and made a profound contribution to consolidating his reputation for navigation. He said people in the RAF now began to seek his advice on navigation matters,[20] and he became known pre-eminently as a navigator, dovetailing neatly with his lecturing and instructing roles at Calshot. It fed his developmental mindset by helping him find 'gaps in training, equipment and methods'[21] when it came to air navigation. That, in turn, would lead the 23-year-old to decide in the coming year to plug those gaps by attempting a comprehensive book on the subject, a subtle blend of the precocious with his now officially-endorsed self-confidence.

That he could absorb the volume of material necessary to pass the exams in such a short period of time is the purest illustration of his prodigious memory, his capacity for understanding and for detail, something colleagues and subordinates would routinely find awesome or irksome.

Above all, this was his ticket to the world of civil flying. As a pilot with a civil First Class Air Navigator's Licence, the entire world was now within reach.

Concerning the Centenary Air Race specifically, all he needed was an aircraft and, most likely, a crew as well.

12

The Unshackling Begins

Don's life at RAF Calshot continued through 1934 largely unchanged from the closing months of 1933. He lectured and instructed. Successive Flying Boat Pilot's Courses (FBPC) came and went, each ending up in Scotland on the ten-day cruise; '… the cruises every three months were always an experience, if not a pleasure.'[1] He continued his very productive working relationship with 'Gill' Saye and 'Crackers' Carey at the Seaplane Training Squadron. His flying hours steadily increased. Amid all that, he openly executed his plan for escaping from the Air Force, with apparent encouragement from Saye.[2]

While still studying for his First Class Air Navigator's Licence, Don's superiors offered further recognition of his navigation prowess by appointing him a lecturer and instructor on the Long Navigation Course, known as the 'Long N'. Offered by the School of Naval Co-operation and Aerial Navigation, it was the only advanced post-graduate navigation course in the RAF.[3] This was a course for a select group who, at least in the early days, had been 'selected by competitive examination'.[4] Don's instructional role affirmed the course name – training pupils to undertake navigation over much longer distances, including at night. It gave him additional experience and confirmed his status as a navigator within the RAF. It did not, however, give him any pause for thought regarding his overall direction.

At the end of No 21 FBPC cruise, in late May, he took a month's leave. He went straight up to Eastleigh, on Southampton's outskirts, where he kept his Humming Bird. In 1932, Eastleigh had become the new home of the Hampshire Aeroplane Club after it moved from Hamble.[5] Reginald Mitchell, of future Spitfire fame, was a member and the president was Lord

The Unshackling Begins

Louis Mountbatten KCVO RN.[6] The Club played a prominent role training pilots in the local region. It had a number of aircraft, mostly Moths, but also a Spartan Three Seater I (G-ABKJ), a biplane registered to the Club's manager, WL Gordon.

On 3 June, Don commenced what was a two-step process to being licenced to carry civilian passengers. Under Air Ministry rules, every pilot holding a Pilot's 'B' Licence was required to undergo tests every time they wished to extend that licence to 'a further type of machine'.[7] Don chose to begin with the Spartan, under the instruction of K Winton, one of the Club's two instructors.[8] Five days later, he flew his first civilian passenger from Eastleigh on a much longer flight to Manston, near Margate, and back, again endorsed by Winton. The following week, over two days, he did 11 flights around Eastleigh carrying two passengers on each occasion. With those complete, Winton declared him good to go.

That was 16 June. On the 18th, he began moonlighting as a pilot with Jersey Airways. It was an extraordinary development. The airline had been in business less than six months. Formed in late 1933 to fly passengers and freight to the Channel Islands, its maiden flight between Jersey and Portsmouth had taken place on 18 December. It flew routes between Jersey, Portsmouth and Southampton, with periodic flights up to Heston, London, and Paris in the summer months.

Apparently, Don had responded to a newspaper advertisement by the founder of Jersey Airways, Bill Thurgood, who was looking for pilots.[9] Why Jersey Airways engaged Don for such a short period while he was on leave from the RAF is unknown. Perhaps it was a resumé showing 860 hours in 16 different aircraft types, including multi-crew flying boats as captain and instructor. He had done a lot of flying over the Channel, so was thoroughly familiar with local conditions. He lectured in navigation, including the 'Long N' course, and had passed the exams for the First Class Air Navigator's Licence. On all those measures, there were few more qualified for a 'try out'. What's more, if they gave him a go, he would do it for free.

On paper he looked good, but a short-term 'gig' was not without its risks. Principal among those was entrusting Don with passengers. To lose an aircraft so early in the company's life would be catastrophic for business.

And he was to carry those passengers in an aircraft entirely new to him, the de Havilland DH.84 Dragon.

The Dragon, a small twin-engine biplane with a fully enclosed cabin, was designed for short haul trips and much smaller and lighter than the Southampton II. It was relatively new, having first flown commercially in April 1933, but had quickly established itself as suitable for some international routes. Jersey had configured its small fleet of Dragons to carry eight passengers each.

It was a land-based aircraft, and therein lay another risk for Jersey Airways and a challenge for Don. Portsmouth, Southampton and Heston all had runways, not so at Jersey. The Dragons landed on a strip of beach at St Aubin, around the small bay from St Helier. Landing there was a feat in itself, made all the more stressful by having one of the highest tidal variations in the world – up to 40 feet – with the incoming tide rushing up the beach at a spritely vertical rise of three inches per minute. In summer months, this reduced to a more manageable tidal variation of up to 30 feet. Consequently, the Jersey Airways timetable was constructed to fit the tides.

If that were not enough, 'Each pilot did his own paperwork and everything else, including looking after the passengers and their baggage, and running the aircraft single-handed.'[10] It was just the sort of multi-faceted challenge Don found highly attractive.

It all happened very quickly. As with the Spartan, Don needed his Pilot's 'B' Licence endorsed for the Dragon. On 18 June, for Don's first flight in the Dragon (G-ACMO, *St Ouen's Bay*),[11] Jersey Airways arranged a full load of seven people and he flew around Heston practising heavy and light landings and flying on one engine. He passed, and three days later began flying passengers commercially. He did multiple trips for eight days straight, often with full, or near full, passenger loads.

The beach landings were tricky. On his fourth flight into Jersey, Don encountered 'strong, gusty crosswinds' offshore and on the beach. The following day (the 24th), he arrived at 6:45 pm to find the tide well in and the 'beach very narrow!'[12] 'Never was any aircraft lost to the tide, in spite of many close shaves, but the derelict bus which was driven down to the beach to act as our office was eventually lost.'[13]

The Unshackling Begins

All too soon, leave was over and he had to return to work at Calshot. He had clocked up 36 flights and nearly 30 hours of flying while learning plenty first-hand about how a civil airline operated. He reached an agreement to return to fly on weekends, which he did three times in July, more than doubling his time in the Dragon cockpit, but the adventure ended almost as quickly as it started, as he later recorded:

> … it came to an untimely end when some MP asked the Secretary of State for Air whether it was true that a regular Air Force officer was flying for a civil airline and thereby depriving a civil pilot of a livelihood. I had never taken any reward for my services, and I was, therefore, entirely in the clear. I felt, however, that discretion was the better part of valour, and discontinued my weekend flying.[14]

It was perhaps fortuitous because it was now the beginning of August 1934 and there was the small matter of the air race to Australia being just over two months away.

13

The Vega's Leg

With the First Class Air Navigator's Licence under his belt, Don had secured Air Force permission to participate in the Centenary Air Race, starting on 20 October 1934, and set about finding an aircraft and someone to fly with:

> Captain Baird [*sic*], of early Schneider fame, joined forces with me, and he managed to get an offer of a Rolls-Royce engine on loan if we were able to build an airframe around it. Unfortunately, this came to nothing, in spite of a number of interviews and journeys.[1]

Henry Biard, test pilot for Supermarine since 1919, had won the 1922 Schneider Trophy race in Naples in a Supermarine Sea Lion II, with a third placing the following year in a Sea Lion III. He had not competed further after crashing his Supermarine S.4 into Chesapeake Bay, off Baltimore in Maryland, during a trial in 1925.[2]

An early discussion took place on 5 May 1934 when Don flew Biard in a Moth up to Sherston in Wiltshire to meet with the aviation writer, Frank S Stuart, who captured what happened next:

> I shall not forget my first meeting with [Don Bennett]. He and Captain H.C. Biard, the Schneider Trophy winner and Chief Test Pilot for Supermarine, flew over to see me and landed in a small field behind my country house. They came to see if I could induce a British aircraft company to provide them with a machine for the MacRobertson Air Race to Australia.

Although we got as far as obtaining the promise of a then-secret British engine which, as it has since proved, would have had speed enough to win the race … unfortunately last minute adjustments could not be finished in time, and the project had to be dropped.[3]

Biard had, in fact, left his position at Supermarine in 1933 after Vickers-Armstrongs took over the company and assigned its own pilots to the test-flying roles. At the age of 42, he was looking for a new challenge.

Stuart offered an anecdote from Don and Henry's visit:

When Biard and [Don] went back to their little machine in that Wiltshire meadow after we had had lunch, it became obvious there would be take-off difficulties. The wind had dropped completely, and the meadow was very short for take-off even with wind; moreover it was crammed with village children.

We moved the children away, and the aircraft began its run. It cleared the stone wall at the end of the meadow by not more than six inches – and then I realised that it could not possibly clear a clump of high elms just ahead. My heart stood still as Bennett somehow banked around them. I do not think a dozen men living could have made that take-off that day.[4]

Even had they secured the engine, they still needed financing for a machine and the race more generally. As Don indicated, all their efforts came to nothing. This was hardly a surprise given the only real collateral on the table was Biard's good name. Potential backers would also have been aware of other race entries, including some teams with the brand-new de Havilland DH.88 Comet.

There are hints Don was casting around elsewhere by June. On successive days, just prior to his time at Jersey Airways, he tested two 'modern' aircraft, a Hawker Hart biplane (K2443) at Bicester and a Miles M.2 Hawk (G-ACOP) at Eastleigh. The latter, a two-seat monoplane, was just two weeks old when placed in his hands by K Winton, the flying instructor at the Hampshire Aeroplane Club. Don did not say specifically that either aircraft was tested for the purpose of the race, but it is worth noting another Miles Hawk did end up as an entry.

The race was just weeks away when Don was thrown a lifeline. '... I was happy but disappointed in having to accept a post as navigator with Jimmy Woods, an Australian, flying a Lockheed Vega with a single engine and a high wing.'[5]

As Jimmy Woods's story shows, this was a hastily undertaken marriage of convenience on both sides. Woods was not exactly Don's type, though perhaps, given the lateness of his entry into the race, Don could be excused for not having sufficient time to do his 'due diligence' and make an informed decision as to whether this really was a good idea.

Woods was actually born in Scotland in 1893. Like Don, he had a passion for motor mechanics. In early 1914, he emigrated to Auckland, New Zealand, to become a chauffeur.[6] Having developed a taste for aviation, he trained as a pilot before heading to Britain to serve in the Royal Flying Corps in 1917.

At a time when instrumentation was primitive, Woods showed a 'natural skill for what became known as "seat of the pants" flying'.[7] He returned to New Zealand after the war doing exhibitions and joy rides in a DH.6 before heading back to Britain to buy some second-hand aircraft to start his own business in New Zealand. Instead, he took a job doing 'pleasure flights' around Aberdeen. In 1923, he saw an advertisement in a newspaper placed by Major Norman Brearley who was in Britain in search of pilots.

Brearley was the owner of West Australian Airways (WAA). He had a government contract for the 'North-West route', flying passengers and mail to and from Perth up to ports in northern Western Australia. Woods performed a flying test for Brearley who offered him a job if he could make his way to Perth.

By late October 1924, Woods was flying DH.50s on the Perth to Geraldton and Carnarvon route. It was a peripatetic life that suited him quite well. While he got along with others, he was described as '... a loner, self-sufficient and independent'.[8]

'He was quite prepared to take the law into his own hands if necessary, unwilling to waste time waiting about for instructions, especially if he suspected they would require him to act in a way he didn't want.'[9] That said, he did not have a reputation as an undue risk-taker; rather, he was much like Don in that his participation in any venture was based on assessing it to be within his capabilities. Unlike Don, however, that self-assessment may not have always been undertaken with rigour or, once taken, followed by adequate planning.

People commented on an unruffled temperament and a laconic demeanour – the loner doing it his way and in his own time. Woods danced to his own tune, one far slower than Don's, at least when it came to getting on with things in the most efficient manner.

Suddenly, in late 1932, he developed an obsession with beating Jim Mollison's record-breaking flight from Australia to England – eight days, 22 hours and 25 minutes – set in March of that year. Until then, he had shown no interest in such record-breaking flights. He needed to talk to someone.

Enter Horace C 'Horrie' Miller. A First World War pilot and the same age as Woods, Miller had set up the Commercial Aviation Company in Adelaide in late 1920 with an Armstrong Whitworth F.K.8. In 1927, David Robertson, a long-time friend, introduced Miller to his older brother, none other than Macpherson Robertson, the Melbourne confectionary magnate, who financed the purchase of a de Havilland DH.61 Giant Moth, leading to the formation of the MacRobertson-Miller Aviation Company Ltd (MMA) in 1928, with Miller as managing director, chief pilot and chief engineer, based in Adelaide.[10]

In early 1933, Woods dropped in to see Miller to discuss his proposed record attempt. Miller did not like Woods's chances solo; he had a better idea. He suggested they team up for the Centenary Air Race. Woods eventually agreed, leaving them about 18 months to prepare. With several financial backers, including David Robertson, Miller commenced negotiations to buy a second-hand aircraft he had found for sale in England, with the intention of using it commercially in Australia afterwards.

Unfortunately, Woods remained obsessed with beating Mollison's record. In May 1933, he secured a Gipsy Moth and financial backing, but preparations were haphazard. He departed from Perth on 2 July 1933, leaving a lot to the last minute.[11] The trip itself was a nightmare lasting six weeks not eight days. He encountered everything: mechanical failures, monsoon thunderstorms, hunger and exhaustion, and, at one point in Iran, taking off with a cockpit full of red ants. He mostly shrugged it all off, showing occasional bouts of frustration.

In the meantime, Miller had bought the aircraft. It was the Lockheed Vega DL-1A, a one-off version of the DL-1, a single-engine, high-wing monoplane with a metal fuselage and a 450-hp Pratt & Whitney Wasp engine, built in 1930 and delivered to Lieutenant Commander Glen Kidston in London, who re-registered it in January 1931 as G-ABGK to incorporate his initials.

After Kidston was killed flying a Puss Moth in South Africa in May 1931,[12] the Vega was flown occasionally before being hangared at Hanworth and put up for sale in March 1933. Miller bought it not long after, specifically for the race. It remained at Hanworth, not flying again for over a year.

Developments in Australia in 1934 set the scene for Don's participation in the race. The Federal Government put Australia's airline routes under review in preparation for the forthcoming Empire Air Mail Scheme. Brearley's WAA had to re-tender for the North-West route despite having pioneered the work and performed it for over a decade. Miller's MMA decided to bid, submitting a tender it knew would undercut WAA. In a shock decision, it won, with commencement due in October 1934.[13]

Miller needed an experienced hand in Western Australia and turned to Woods, whom he signed on as MMA's route manager. However, with little time to get the service ready for launch in October 1934, Miller pulled out of the air race. Woods would go to England alone, find a copilot/navigator, get the Vega ready and make all the preparations to fly the race.

Mid-June, Woods departed for England via the US, visiting Lockheed in California to familiarise himself with the Vega type and make arrangements for spare parts. It had been a hasty departure from Australia.

There was never anything leisurely about Jimmy's departures. He rarely seemed to allow enough time to do all that had to be done, yet he never panicked; quietly – stoically perhaps – he dealt with each item or crisis as it arose.[14]

Arriving in England in August, he set about the mammoth task of getting everything in place for the race, just over two months away. After working on the Vega for a week he got it airborne on 15 August and the following day flew it to Rotterdam to have the engine overhauled and a new controllable-pitch propeller fitted by KLM at Waalhaven. Facing all the other race preparations on his return, Woods decided to head off to The Oval to watch the fifth test match before spending a week in Scotland with his family!

On 17 September, his diary entry was ominous: 'Been working hundred and one things.'[15]

The Vega had returned to Heston on 14 September but was impounded while HM Customs & Excise awaited the payment of duty on the newly-installed propeller and 'other alterations' made by KLM. Woods visited Customs on the 18th to clarify the issues, but a letter nine days later reminded him payment was outstanding and the Vega would be held at Heston until it was made.[16] By the time he paid the duty on 1 October and got the Vega to Hanworth, he was running out of time to do all the final modifications.

Just over a week earlier, and still without a copilot/navigator, he heard about Don. His first mention of Don is a diary entry on Friday 21 September: 'Wire Bennet [*sic*]'[17] Neither commented on the nature of the discussions leading to Don signing on. It is impossible to know whether Don formed any clear picture as to the state of the Vega or the preparations. His comment – 'I was happy but disappointed in having to accept a post as navigator with Jimmy Woods' – simply reflects his disappointment Jimmy had determined to do all the piloting.

After finishing flying at Calshot on 9 October, Don took leave and headed for Hanworth. He was not impressed with what he found. 'The preparations were extremely hurried, and Jimmy Woods was obviously harassed by lack of sufficient funds to do things properly.'[18]

It was not until 13 October, a week out from the race, that Woods obtained the Certificate of Airworthiness. This, in itself, created a headache. He wrote in his diary, 'Got my C of A today but only for 4,500 lbs [gross weight].'[19] It was the coda on a saga that had begun with a very bad assumption. When purchasing the Vega, Miller was told the Air Ministry had approved Kidston for 5,600 pounds for a record-breaking flight from England to Cape Town despite a 'Notice to Operators' issued by the US Department of Commerce that clearly stated: 'This aircraft has been approved as airworthy by the Department of Commerce at a gross weight not to exceed 4,500.'[20]

While in California, Woods had discussed the issue with Lockheed, who had submitted fuel loads and distance figures on the understanding 'but had no documentary proof'[21] the Air Ministry had approved Kidston for 5,600 pounds, and Woods expected similar treatment. He then organised the overhaul in Holland without checking. It was only on 10 September that he wrote to the Air Ministry to clarify, having to carry on preparing the Vega in the expectation it was a *fait accompli*. The Air Ministry took five weeks to advise Woods it would not approve a weight in excess of 4,500 pounds, irrespective of any modifications. This significantly impacted fuel loads and attendant speed.

Weeks earlier, Woods had received a letter from Miller advising receipt of some Race Committee papers. Miller made an interesting observation: 'I presume the one about petrol tanks is for handicap purposes but you want to be very careful of those Pommies. I heard that they intend to find some excuse to disqualify any American machine that wins.'[22]

The contest, comprising speed and handicap races, was being flown under *Fédération Aéronautique Internationale* (FAI) regulations[23] and according to the Competition Rules of the Royal Aero Club. As such, it was firmly in the hands of those 'Pommies', evidently fuelling rumours the Race Committee was taking every opportunity to ensure fellow countrymen had the best possible chance of taking the prizes.

On the same day as the Certificate of Airworthiness came through, Woods dropped into the Aviation Department of The Automobile Association to pick up the special air route maps and Admiralty charts Don needed to plan the navigation,[24] resulting in '… a few hectic days preparing maps …'[25]

They were due at Mildenhall on the 14th to complete final registration. Realising they were not going to make it, Woods called the Air Race Committee and got a 24-hour delay. Late that afternoon, they finally got the Vega out of the hangar for their first flight together, a 10-minute hop to Heston. Woods's demeanour worried Don: 'Jimmy was quite unperturbed.'

Why they had flown to Heston, not directly to Mildenhall, is not clear, but another warning sign was immediately apparent on landing when the hydraulics failed on the port oleo leg and it jammed:

> Even this did not interfere with Jimmy's tranquillity; he went off to London for some social reason and left me and one mechanic to try to rectify the trouble … We were well into the small hours by the time we got it back on the aircraft apparently serviceable.[26]

Don recalled leaving for Mildenhall the following morning, but in fact they didn't take off until 5:00 pm, cutting it very fine. The *Glendale News Press* recorded their arrival at dusk 30 minutes later:

> [Woods] had to arrive today to be eligible for the race. Officials had given up hope of his arriving tonight. They were about to close the hangars when the white shape suddenly zoomed from the skies …

> The slight figure in soft hat and short raincoat grinned cheerfully as he switched off the engine and climbed from the cockpit.

> 'That was certainly a moment.' He said, 'Just as I was coming down I saw what looked like a giant sausage in front of me. I only just managed to pull up over it. It was too dark for me to see what it was.'[27]

The following day both Don and Woods received official authorisation to fly in the race as crew.[28] That was the easy part. Woods noted, 'Rushing around the Air Ministry. Cabled Washington for extra 500 lbs all up load. Getting things together to take by plane too much rush.'[29]

The weigh-in occurred on the 17th: 'Aeroplane special no. Arrived back at Mildenhall. Bennett & machine weighed & only allowed 132 gals petrol & about 10 oil',[30] a significant reduction on the 170-gallon capacity. Wire seals were installed in front of the fuel gauges for the three tanks indicating the maximum fill for each, with this provision written into the Vega's official flight logbook. Checks would be made throughout the race to ensure compliance. They were also handed a time handicap of 2 hours, 31 minutes and 48 seconds.

With Friday 19 October being an open day for the public, all preparation needed to be finished the day before. Woods wrote in his diary: 'Having an awful job getting battery for machine, second one arrived and did not fit.'[31]

> The day before the start of the race, hordes of people converged on Mildenhall. They gazed at the aeroplanes, marvelling that some of these flimsy machines were capable of the long flight ... The unexpected arrival of King George V, Queen Mary and the Prince of Wales brought even more chaos.[32]

Woods noted:

> Last day before race. Was introduced to Prince of Wales & had him looking over my machine, great chap. King & Queen also came along. Bit of a message. Had to wait until after midnight then stayed at a farm.[33]

That afternoon they had performed a 20-minute air test and the Air Race Committee certified them ready to go. Adding together the air test and the two flights to get there, Woods and Don had flown a total of one hour together. Woods had only done a few flights in the Vega since arriving mid-August. Don had not piloted the Vega at all; he was just there as navigator. '... Jimmy Woods had implicit faith in my navigation, and was under the happy delusion that a navigator could work magic.'[34]

The massive crowds re-appeared before dawn on the Saturday morning, with a DH.88 Comet the first airborne at 6:00 am. Don sensed the excitement: '... it certainly was a magnificent sight to see the aircraft taking off in an English early morning on a race half-way around the world.'³⁵

They took off at 6:39 am for the first leg to Marseilles. Don noted the extensive cloud cover and that he navigated by dead reckoning because there was no radio (weighing in at 98 pounds, it had been left behind). They were on the ground at Marseilles for just 36 minutes before heading for Rome. They departed there at 2:05 pm for the longest leg of the day to Athens. Don recorded the last half of the leg as 'night flying'.

They contemplated continuing on to Aleppo in Syria, but were not sure they had enough fuel, so decided to refuel in Athens, get some sleep and head off in the early hours the following morning. The race rules only counted airtime towards the final race total. Arriving at Athens offered the first sign something was wrong. Don wrote in his logbook: 'Oleo leg jammed on landing.'³⁶

After a few hours' sleep at the Greek Air Force barracks, they took off for Aleppo at 3:25 am, passing over Cyprus. Flying the final hour into the rising sun, they touched down at Aleppo right on 7:00 am. Actually, 'touched down' does not quite capture what happened. That night, sitting in the Hotel Baron, Woods offered a full account in a letter to David Robertson:

Dear Mr Robertson,

... To begin with we made good times to here, and today we were managing 210 mph in places coming from Athens, gradually decreasing to 190 as the wind gradually worked round, and when landing here we made a good three pointer and after running along for a bit I was just preparing to draw up with the brakes when down she went on one side. The wing tip touched & over she went on her back with a crash. Luckily I switched the engine off & my main battery switch was also off which I think probably prevented a flare up.

I rolled round the cockpit and hot oil poured all over my head. I realised it was time to be outside but could not open the cockpit door, and the roof exit of course was on the ground. Bennett was already outside shouting was I alright so after a bit of pressure I forced open the door.

He was sitting at the rear of the cabin when she went over and got thrown forward. He had a nasty gash on his leg and I got one on the forehead however we went to the hospital & had them dressed & apart from a few bruises we are quite OK.

But when we realised things afterwards, it was plain the poor old Vega was badly wounded. The impact of the engine on the ground burst all the mounting and also pushed in the front of the metal fusilage [*sic*] which is badly sprung. The port wing tip is also gone including a piece of the spar. The fin is only slightly damaged. The cause of it all being the radius strut on the port side of the undercarriage breaking right in the middle. This undercarriage oleo leg has stuck in the out position several times, and we had it down at Heston a few days before the race and found the cylinder scored & worn, but could not do anything with it.

… I don't know if it was insured, but I cabled Horry [*sic*, Miller] about it and he replied that insurance was not available. I had the machine pulled on to its wheels after repairing the strut & have it over at the military hangars. I removed most of the instruments this afternoon, as they would certainly be stolen by these thieves.

… This is a hell of a place to be stranded in, they all talk French. Bennett leaves for London tomorrow morning via Port Said. I told him to send his passage expenses, as that was the least we could do. It is all very regrettable and I can assure you I am feeling pretty miserable tonight …

Yours sincerely

J Woods[37]

Don's own account differed somewhat:

On arrival at Aleppo, Syria, Jimmy brought the Vega in to land, whilst I took up my position as far aft as possible. He hit the ground

> with a fair wallop and the undercarriage collapsed; down she went, and the nose went in as we whipped over on our back. I was in the tail of the machine, and my velocity from one end of the cabin to the other was remarkable. Even more astounding was the degree of 'concertina-ing' of my body which took place at the far end.
>
> I rolled out into the dust of the aerodrome and then helped Jimmy Woods out with his forehead bleeding rather badly. He looked an awful mess, but he was not really as badly bent as I was. I had done a fair bit of damage to one knee and could not move my head or shoulders, due to what I subsequently discovered to be three crushed vertebrae. We were taken to a convent, and some Syrian nuns patched us up.[38]

Indeed, Don did depart the following day, as he said, '… [leaving] poor Jimmy Woods with practically no money, and a badly broken aircraft, waiting for funds to be cabled to him from Australia.'[39] He took a ship from Beirut to Naples via Alexandria, contracting a bad ear infection on the way, then flew back to England.

While Don gave considerable space in his memoirs to the race, he provided no commentary on Jimmy's performance, other than uncharacteristically oblique references to his lackadaisical approach to preparation and 'she'll be right' attitude. His 'disappointment' was restricted to being consigned to navigator. He gave no hint of having second thoughts after arriving at Hanworth and seeing the state of both the Vega and Woods. He offered no reflections on what the episode taught him, despite Woods seemingly epitomising everything one should *not* do in planning and preparation for any flight, let alone one halfway across the world. He seemed content to have been in the world-famous race and survived, or simply appreciated having the best use to date of his First Class Air Navigator's Licence. In a 1981 interview, he made a passing remark about his participation: 'It did keep me in the swim', meaning the race had played its part in helping him towards his goal of a career in civil aviation. Whatever the case, he was moving on rapidly:

> A week later, back at Calshot, I passed my annual medical with one stiff leg, three crushed vertebrae, an abscess in the ear and

immobility of my head and shoulders! It took me months of treatment with my specialist before my backbone was fully recovered … Within a fortnight of my crash at Aleppo, I was flying my favourite Southampton S1234 ex-Calshot.[40]

14

Ly and her Complete Air Navigator

Sometimes the best laid plans get waylaid; sometimes everything is under control until the moment it is not. The introverted, focused and ambitious pilot suddenly found himself in the same airspace as an extroverted, highly intelligent and capable young lady whose presence was unanticipated and immediately threatened to blow him off course.

Elsa Gubler, known to all as Ly, was from Zurich. It is no surprise, given her Swiss origins, that the initial catalyst for their meeting related to her becoming multi-lingual. Ly's version of events was she had spent two years in France to learn French, but 'had always wanted to visit England and learn the English language'.[1] She had arrived back in Switzerland from France to hear via family connections of a position as an *au pair* to the family of Group Captain William Callaway, the station commander at Calshot.

Ly's father, Charles, wrote the initial letter, resulting in an invitation to accompany his 19-year-old daughter on a visit to the Callaway home in Southampton. Ly was delighted with the Callaways and the feeling was mutual. So began the twin tasks of looking after the children and expanding her English language skills.

'Life in Southampton was beautiful, and I enjoyed making visits to Calshot when I got the opportunity. I'm not sure how it happened ...'[2] This led Ly to two recollections. One was that, due to her French, Callaway had asked her to fill the role of a young French girl in a play on the base. The other that she was in the Officers Mess when she laid eyes on a young officer across the crowded room who returned her gaze with a smile that 'made [her] heart melt'.

She later asked Callaway about this young officer which led to the two being introduced. Don picks up the story:

> It was at about this stage of my chequered career that a poor innocent girl from Switzerland, visiting England, found herself unable to avoid me. Inevitably and inexorably I ultimately fell for the job of teaching her English – and I have been doing it ever since.³

This hints that he found himself entangled in a relationship not of his own making, that he didn't fall in love so much as fall for the job of teaching her English, which somehow led to something more.

Ly neatly skewers Don's version of events by recounting that when she first enquired of Callaway about Don, the station commander revealed Don had already been making enquiries about her, placing in doubt the notion she was the only one demonstrating an inability to avoid.

If truth be known, she was simply better at this than him. This was quite foreign territory for Don to navigate – women, soft emotions, contemplating not being the loner who made decisions just for himself, and having someone who took him just as he was without having to prove himself in any way. She was smart, talented, independent-minded, and had (what must have been for him) enviable social skills. In cricket parlance, with Ly, Don was batting well above his average.⁴

On the basis of the events as described, it is not possible to know whose idea it was for Don to tutor Ly in English – his, hers, the station commander's or Mrs Callaway's – but it was a deft move. One can only speculate on how quickly Ly learned to correct Don's frequent spelling mistakes. What Ly did reveal is that from this time, around April 1935, Callaway, who 'was responsible for my care and safety ... would often write to my father giving him updates on my progress in English'.⁵

That surprising development was still several months away as 1934 came to a close. Following the air race, Don recommenced his Calshot duties on 3 November. Despite his injuries, he persuaded 'Gill' Saye to let him lead

No 23 Flying Boat Pilot's Course cruise to Scotland beginning nine days later. The cruise passed largely without incident until the final leg. They had left Pembroke Dock, rounded Land's End and were heading up the Channel for Calshot when the con rod went through the port engine sump, with immediate effect. They came down on the water within the vicinity of a destroyer who, Don noted, 'came alongside us with obvious glee'.[6] With a steel hawser attached, the destroyer commenced the tow. Don recalled:

> Soon a matelot on the stern of the destroyer signalled in semaphore: 'Can we go faster?' I was already disturbed at the strain, and thoughtlessly and abruptly said: 'Certainly not.' My wireless operator … signalled 'certainly', and before he could send the word 'not' to the destroyer the matelot had disappeared and in no time at all we were practically planing. The destroyer was really going as fast as she could, and never in my life have I seen a flying boat go so fast at the end of a piece of string as on that occasion.[7]

Don's concern the Southampton's bollard would break was unfounded and they made it to Portland post haste. The event is worth re-telling just for the visual image it creates, but it is his conclusion that reveals the most about how Don saw himself:

> They put us on a mooring and we then changed the engine, as was our custom, with one fitter, three pupils, one instructor and one wireless operator – and no outside help.[8]

Pilot, navigator, mechanic, instructor and works supervisor. Practical, hands-on, self-sufficient and self-reliant. Under his leadership, everyone contributed to the task at hand irrespective of their rank or specialist role, resulting in their being back at Calshot just over 24 hours later.

The year concluded with one final highlight, a couple of test flights in the Saunders-Roe A.27 London (K3560). The twin-engine biplane was a cousin of the A.29 Cloud that Don had flown several times a year earlier. The test flight was part of his ongoing desire to fly as many new aircraft types as he could lay his hands on. In the latter half of 1933, along with the Cloud, he had flown the Supermarine Scapa. In May 1934, he and 'Percy'

Pickles had a test run in the Short S.8/8 Rangoon, a one-off military version of the S.8 Calcutta Don would encounter at Imperial Airways the following year.

Then, on 11 February 1935, he was at Felixstowe as part of a crew on the Short S.14 Sarafand, piloted by Squadron Leader John Breakey of the Marine Aircraft Experimental Establishment (MAEE). A flying boat with an unprecedented six engines in tractor/pusher pairs, it was a one-off prototype and the largest aircraft in the UK at the time. His logbook simply states: 'Flying on strange type. Six Rolls-Royce Buzzards.' The Sarafand and the MAEE was the closest he had been to the 'cutting edge' of flying boat development, a brief foretaste of his coming days at Imperial Airways piloting Shorts flying boats.

The start of 1935 had caused a reappraisal, just as it had done a year earlier. Don now had experience as a civil pilot and had partaken in the air race. Flying hours were steadily building at Calshot. What was next?

At the end of January, just prior to flying the Sarafand, he did a stocktake and analysis of the aircraft he had flown in the previous four-and-a-half years. It would have taken some time to compile given the level of detail. In short, he had flown 18 types, everything from his Humming Bird to the fighters, to five different types of flying boat and two floatplanes, to the DH.84 Dragon with Jersey Airways and the one-off Lockheed Vega DL-1A.

This complete reassessment of his flying history had revealed an arithmetical error: instead of 1,157 hours 50 minutes total flying time, it was 1,157 hours 10 minutes. This was duly corrected! Nearly 50% of flying time had been on the Southampton. Night flying had occupied less than 10%, with that mostly on the Southampton, too.

He didn't indicate why he did it. It was not an exercise in nostalgia; these were cold, hard facts, constructed in his logbook and signed off by a superior officer. His efficient mind would have no truck with purposeless activity, especially one taking this much time. He had performed the exercise three times previously at approximately six-month intervals: 1 July 1932 when finishing at No 29 (Fighter) Squadron; 31 December 1932 after his first stint at Calshot; and 30 June 1933 when part way through his time at Pembroke Dock. Then nothing for 18 months.

Now, at the end of January 1935, he was just over six months from leaving the Air Force. Perhaps, therefore, this was an exercise in determining if he had any gaps in knowledge and experience in preparation for his entry into civil aviation.

In this regard, it came just over three weeks after he had been awarded the 'A', 'C' and 'X' Ground Engineer's Licences (No 2727) on 7 January.[9] He now had all the qualifications except for the W/T Air Operator's Licence. It was definitely a gap he needed to fill.

The first six months of 1935 raced along. He led two more Flying Boat Pilot's Courses and cruises to Scotland. There was the occasional night flying. He climbed out of the Southamptons semi-regularly to fly the Cloud. Aside from conducting a W/T exercise with a submarine on 21 February, it pretty much reeked of routine.

He flew his Humming Bird as often as he could get up to Eastleigh. He could take to the air alone, indulging in blessed solitude, piloting stripped back to the raw essentials. He kept his hand in at Jersey Airways by flying a passenger from Southampton to Weymouth in late February. It was something more than a refresher as he recorded negotiating cloud down to 40 feet with heavy rain.

His experience in the Vega DL-1A had given him a taste for aircraft of such ilk. Flying Officer Barry Littlejohn, his old mate from Point Cook, had purchased a Klemm Kl 32-V (G-ACYU),[10] a German touring plane designed for competing in the *Challenge International de Tourisme 1932*.[11] It was hangared at Hanworth, where the Vega had been, and Littlejohn gave Don 15 minutes at the controls on 17 March. A month later, he met up with Littlejohn at Ford, and flew it over the weekend, including a solo to Hamble and back. The next day, they got four people aboard the three-seater for a brief local flight.

It was June when everything happened.

With Ly and Don clearly 'unable to avoid' each other any longer, he proposed. Ly wrote to her parents, leading to a letter from her father to

Callaway, who demonstrated ongoing complicity by apparently replying 'he had never before seen such love between such a perfect couple, and that in his opinion, Ly was with an outstanding young officer and a true gentleman, who was one of his best staff officers'.[12]

The betrothal created an immediate problem. The Humming Bird was a single-seater, and a cramped one at that. Don turned to Barry Littlejohn. After more practice in the Klemm, Don drove Ly to Southampton in his Talbot on 23 June and took her aloft. She was duly impressed: in her opinion, 'his flying ability was outstanding'.[13]

The following day, he flew his beloved Humming Bird up to Chilworth near Wheatley to demonstrate it to a prospective buyer. With no room for Ly, his pride and joy had to go.

Don had begun practising in earnest for a two-stage climax to his first stint in the RAF – participation in its 15th Air Display at Hendon. This was the RAF's annual showpiece event that had started in 1920, not only as a demonstration of its growing strategic air-power capabilities independent of the older services, but a large-scale piece of aerial theatre for an inter-war public enamoured with, and awed by, flying machines and the technological future they embodied. Regular attendances exceeded 100,000:

> While the working and middle classes came in droves – the former by bus or Tube, the latter, increasingly in their motor cars – the event also came to take a place 'amongst the foremost of the functions of the London social season' with British and foreign royalty in regular attendance, along with political, social, industrial and military elites.[14]

The Hendon Air Display was visited annually by a young Leonard Cheshire who, in 1935, was still in school at Stowe.[15] Within a decade, Don would cross paths with the future Bomber Command luminary and VC winner.

The 1935 show was the only year the much-anticipated set piece 'mock battle' finale was replaced by a mere flypast.

This ... was, perhaps, intended as both a reassuring image and a threatening message after the recent revelation of Germany's illicit aerial rearmament and the start of the RAF's own, far from complete expansion ...[16]

Don, Flight Lieutenant Reg Chadwick and crew arrived in their A.29 Cloud (K2895) for two days of rehearsals at Northolt. The crowds gathered on Saturday 29 June for the Display, the day concluding with the flypast of nearly 80 aircraft. It was another case of Don's aviation career intersecting with history.

As it would again just over two weeks later, when he piloted Southampton S1235 as part of King George V's Silver Jubilee Review of the Fleet at Spithead on 16 July.[17]

The following day, 17 July, Don was awarded his W/T Air Operator's Licence (No 185). He had completed the oral and practical exams in radiotelephony and radiotelegraphy (four in total), covering a wide range of theoretical and practical knowledge, government regulations, and the use of the Marconi A.D.6M transmitter and receiver, the apparatus used in civil aircraft.[18] The practical exam in radiotelegraphy placed an emphasis on showing competency in the internal workings of the Marconi, including tracing and repairing faults. There was little that gave him more pleasure and satisfaction than understanding how things worked, then pulling them apart and putting them together again. The exam also required demonstrating a high degree of capability in Morse code. It became common knowledge Don could send and receive Morse as fast as trained operators.[19]

It would be a mistake to believe Don thought his W/T and engineer licences were secondary or inferior to his pilot and navigator licences. For one thing, obtaining them increased his options and employability post-RAF. They were, as he said, 'in preparation for the day when I would enter civil aviation'.[20] At a deeper level, studying for these immersed him in his core interests – mechanics and electronics. While the adult Don demonstrated great competence as a pilot and superlative skills in navigation, it was mechanics and electronics where the 'boy inside' could play.

One cannot discount the possibility obtaining these licences held an even deeper significance for him. In early 1928, he had arrived back in Brisbane from 'Kanimbla' to tell his parents he was casting aside the traditional career paths followed by his brothers for one in the newfangled world of aviation. He had not finished school and there was no university degree in the offing. Now, just over seven years later and having recently turned 24, he had the full suite of licences, placing him officially among the world's most highly qualified aviators. With over 1,300 flying hours, he was a highly experienced one as well. It was proof to himself, as much as anyone that, in his chosen field, he was the equal of his brothers in their professions. And his mother, with whom he was very close, could once again refer to him with some pride as 'the last hope of the Bennetts'. She would demonstrate this in a very pointed fashion within a few short years with a small gift to his old school.

All the 'loose ends' were getting tied up.

Back on 12 July, Don headed up to Blackpool where, at Squires Gate, he handed over his DH.53 Humming Bird to its new owner, Mr J Gillett of Preston,[21] for £130, fifteen more than he had paid for it. It would contribute to his financial buffer for the forthcoming period of unemployment.

Then, at 11:20 pm on 8 August, he took off from Calshot on a searchlight exercise involving 'attacks on Hythe, Soton [Southampton], Newport and Ryde', alighting in the early hours for the last time.[22] Squadron Leader J McFarlane, Officer Commanding the Seaplane Training Squadron, assessed Flying Officer Bennett's performance as 'Exceptional' and his first period of time in the RAF concluded on 11 August 1935.

He was now ready to embark on the next stage of his career. He hinted his departure was not all smooth flying. 'When I left the Air Force there were many regulars who regarded me as a thoroughly bad type for not wanting to stay in the Service.'[23] He admitted to being 'both happy and well rewarded … tremendously grateful for all the R.A.F. had done for me'[24] but that the Air Force had constrained him. Nothing had transpired since New Year's Day in 1934 to make him change his mind.

At the same time, he described himself as having 'very mixed feelings', 'determined', but also 'scared and happy, for ... I was to be married'.[25]

Back in civvies, there was no time to waste. The plan was to get married, visit Ly's relatives in Switzerland and then sail for Australia to repeat the exercise with Don's. He departed Calshot and headed for Southampton where Ly would begin packing up at the Callaways'.

With Don's priorities absolutely clear, he headed to the aero club on 14 August with Ly and took her on a jaunt in the Spartan Three Seater I to deliver newspapers down to Cowes.

One week later, on 21 August, Don and Ly were married at the Registry Office at Winchester. With his strong Methodist background, it was perhaps surprising there was no church wedding, but there was probably no point given neither family was able to attend.

They departed immediately for Switzerland. Facing having 'to be inspected thoroughly by all [Ly's] relatives',[26] the time in Switzerland was, for Don, mercifully short. They were back in Southampton by early September. On the 10th, the couple squeezed in another jaunt in the Spartan down to Cowes before boarding the passenger ship SS *Hobson's Bay* at Southampton the following day for the trip to Australia.

Don genuinely appeared not to know if they would return to England. He had been pursuing employment options for some months and had secured an interview with New England Airways, a Lismore-based company run by GA Robinson and Keith Virtue, which flew the Avro 618 Ten between Brisbane and Sydney. It is intriguing because he must have known, or suspected, his prospects for a civil-aviation career in Australia matching his abilities were limited. He had left the Service for that very reason. He also had a Swiss wife to think about. There is no hint of homesickness.

The only logical explanation is that he was expecting the offer of an 'executive position', thus attracting a degree of status. Coming just over seven years after telling his parents he was off to have a career in aviation, it would have a made an emphatic statement to the family.

What complicates matters is his admission he departed England knowing Imperial Airways was keen to engage his services as a flying boat pilot.[27] New England Airways operated land-based aircraft only. Nonetheless, he did take the Australian option seriously. Prior to departing England, he applied to the Controller of Civil Aviation for clearance to fly in Australia. On arrival in Brisbane, he received notification he could do so until 31 January 1936.[28]

All that was yet to come. Despite him being a self-described 'gentleman of leisure' and on his honeymoon, the six-week passage to Australia tested the limits of his patience. Being productive 'was a struggle with a ship full of passengers violently and aggressively engaged in the waste of time'.[29]

He was trying to be productive about writing a textbook on air navigation. This was not on a whim or the product of boredom. Prior to their departure, he had secured the commitment of Pitman's to publish it. His reason for doing so extended all the way back to sitting for the First Class Air Navigator's Licence exams in early 1934. He felt there were gaps in the syllabus. This was reinforced by his experience at Pembroke Dock and then back at Calshot. The release of the revised edition of the Air Ministry's 'bible' on the subject, the *Manual of Air Navigation (AP1234)* issued in April 1935, had clearly not filled those gaps adequately.

This he would now do by producing a volume whose title proclaimed both his goal and, with considerable youthful *chutzpah*, his sense of self – *The Complete Air Navigator*. With all the time wasting aboard *Hobson's Bay*, he only managed to get the first chapter written. It would have to wait.

Arriving at Brisbane docks in on 25 October, they were greeted by a large family contingent. Newspaper reports recorded Don as the 'young Brisbane man returned from England to engage in commercial flying here after serving as a flying officer in Royal Air Force'.[30] Just four years after his departure, it must have been a triumphant return. His father George's real estate business was doing well. He had a Ford Model A Rumble Seat Coupe, which Don borrowed to take Ly up to the Sunshine Coast, away from all the attention she was getting in Brisbane.

They were back in time for Don's job interview. On 19 November, Tom Young,[31] flew them on a scheduled trip down to Sydney in one of New England Airways' Avro trimotors (VH-UPI), stopping off at Lismore. The timing was propitious. While Don and Ly were en route from England, a new company had been formed – Airlines of Australia Ltd – amalgamating New England Airways and the Larkin Aircraft Supply Company in Melbourne. The intent was to expand the airline beyond the Brisbane–Sydney route to encompass the eastern seaboard from Melbourne to Cairns, commencing in early 1936. GA Robinson was staying on as managing director with Keith Virtue appointed flying superintendent.

Don recalled Robinson offering him the position of navigation superintendent. It was superficially attractive because this was an 'executive position' in an expanding airline on home turf. Robinson was keen, but inevitably he failed. Don had the Imperial Airways offer in his pocket for comparison. They were flying international routes, in flying boats, and were willing to pay twice as much for him to commence as a first officer. Airlines of Australia was not in the same league. Don simply could not take on the world from Australia; he had outgrown the place. They would return to England, which must have come as a great relief to Ly.

There was just one thing left to do. On 25 November, the Bennetts took off from Archerfield Aerodrome in a borrowed DH.60G III Moth Major (VH-URL) for a trip to Townsville to see Don's favourite brother, Aubrey.[32] Ly saw the splendours of the Queensland coastline from on high as they made landings at Maryborough, Rockhampton and Mackay.

Leaving Ly in Townsville the following morning, Don flew Aubrey to Ingham and back, providing conclusive proof to Aubrey that his 1928 letter, written at the behest of their parents to dissuade Don from embarking on his aviation career, had failed miserably. There was just enough time for lunch and farewells before Don and Ly headed back. By mid-afternoon on the 27th, they were back in Brisbane and preparing to sail for England.

On the return sea journey, time was of the essence because it would be busy at Imperial Airways, so Ly and her complete air navigator got to work and finished the book. In the preface, he laid out the book's scope (and thereby its target audience) along with a brief summary:

> ... to cover the whole syllabus of the Air Ministry First Class Navigator's examination.³³
>
> Briefly, Air Navigation includes any or all of the three different methods of observation (astronomical navigation, air pilotage, and D/F wireless), together with, in all cases, a certain amount of calculation (dead reckoning). Then, with this, it is necessary to ally knowledge of the weather, the tides, signals, law and regulations, etc.³⁴

He was then at pains to point out: *'But common sense is a vital factor'* [his italics]. Common sense was derived from constantly accumulating practical experience, not merely from mastering the content of textbooks or manuals, even one as good as his. In this vein, he laid out the priorities for those studying for the licence: they 'must be thoroughly exercised in working problems' and 'develop a thorough background in theory *in addition* [author's emphasis] to their practical knowledge' because 'the First Class Navigator should ... be a person having both practical and theoretical navigational knowledge of the very highest order'.³⁵

'Common sense' was a recurrent theme for Don. Aside from seeing it as a key *modus operandi* for accumulating practical flying experience, in Don's thinking it applied more broadly. He observed countless instances in bureaucratic hierarchies where common sense was not 'a vital factor'. In general terms, he would encounter decision making where the theoretical overrode the practical, but the wider impression he leaves is that 'common sense' was not replaced by mere 'lack of common sense' but rather the sharper-edged 'non-sense', better rendered as 'nonsense'.

Thus, with the respective families out of the way, the final weeks of the honeymoon were utilised in the most efficient manner. On disembarking in England, the book was submitted and published six months later, becoming, as Don had hoped, the standard work on the subject, passing through many editions. With the book off his chest, he was now free to get out from behind the proverbial desk and go flying.

15
Imperial, Empire and 'Empires'

The distance between RAF Calshot and Imperial Airways' land base at Croydon was just over 66 miles as the crow flew, but it may have been a world away as far as Don was concerned when he arrived in January 1936. Imperial was on the cusp of generational change in its fleet of flying boats. Don was anticipating using the latest technology and exploring new horizons. It was where records were being set.

Don was issued with his Imperial Airways uniform and his *Handbook of Instructions and General Information* (Copy No 184)[1] which, at over 250 pages, covered all aspects of the airline's operations.

On the first few pages of his new small three-ring binder, he noted down key positions in the hierarchy in ink, then filled in the names of the occupants with his trusty pencil as he was briefed, introduced, or asked around. Aside from the General Manager, 'Colonel Burchall', it included other key players in his upcoming life, including the General Manager Technical, 'Major Robert H. Mayo', and the Air Superintendent, 'Major Brackley'.[2] He would fill the rest of the pages with technical matters, including some wonderful drawings of flying boats, or parts thereof.

He received no special treatment, beginning, like all new pilots, as first officer on land-based aircraft. There were several commercial flights to Continental destinations such as Paris and Rome for a month from mid-January in a Handley Page H.P.42, de Havilland DH.86 Express or Short L.17 Scylla. He did a couple of flights around Croydon to secure his 'B' Licence endorsement for the little-known Boulton & Paul P.71A (G-ACOX), a mail plane developed for Imperial.

Orientation complete, he was dispatched to Imperial's hub at Alexandria to commence duties on the eastern Mediterranean leg of the

119

Britain to Egypt route, flying the Short S.8 Calcutta and S.17 Kent flying boats. He departed Croydon with a promise from Brackley he would shortly assume command of a flying boat. Brackley would rapidly endear himself to Don:

> So far as 'Brakles' [*sic*] was concerned, captains had the first and last word concerning all decisions with regard to cancellation; their word was law. ... Brackley ... protected and preserved the integrity and the authority of his captains of the line.[3]

Don came to see this as conferring on him a freedom to make decisions in the widest possible sense.

Around the time the Bennetts moved to Alexandria, Ly discovered she was pregnant.

Don's arrival at Imperial Airways at the start of 1936 came at a time of remarkable flourishing of international and intercontinental air travel. Fierce competition for well over a decade had spurred continuous development to address a series of challenges – routes and navigation, speed and distance, aircraft types and their limitations, refuelling issues, a vast array of international meteorological and climatic conditions, and national politics – all with a touch of nationalistic fervour that saw records of all types being set, challenged, and then bettered as nations, fledgling airlines and individual pilots took to the skies.

In early 1936, Imperial had been operating for nearly 12 years. It flew land-based planes from Croydon to various Continental destinations, but its principal focus was on extending and consolidating the empire routes and, as aircraft technology evolved, tackling the holy grail of the Atlantic.

Imperial's core empire route was from England south-east across the Mediterranean to its hub at Alexandria, known to all as Alex. Two routes emerged from there. The Eastern route ran through the Middle East, across into South Asia and onto the Far East, connecting up with Qantas into Australasia.[4] The African route went south of Egypt through to the terminus at Kisumu in Kenya on the shores of Lake Victoria.

Imperial, Empire and 'Empires'

Imperial's strategy for reaching to the edges of empire[5] lay with the development of flying boats. At the time Don joined, they plied the Eastern Mediterranean between Alex and Brindisi in Italy, connecting with land-based aircraft at both ends. Establishing land-based routes, with their multitudinous trans-national stops, had produced a multitude of problems. Flying boats decreased the need for infrastructure because there were many more 'landing strips' on water. They could also carry greater payloads than their land-based counterparts, which were hampered by primitive airfields.

What Imperial needed to develop, and keep developing, was flying boats that could go further, faster, more efficiently and with greater payloads. This would mean fewer stops and greater profitability, ultimately eliminating the need for land-based machines altogether on the empire routes.

Critically important, the payload priority in the early days was freight not passengers. Flying was prohibitively expensive for most. Throughout the 1930s, most passengers were on commercial, government or military business. The remainder was largely made up of wealthy individuals undertaking family business or leisure activities.[6]

For Britons generally, the Continent was a short boat trip away, with trains on the other side of the Channel. To reach the empire or the Americas, they went by ship. Even if they could afford to fly, concerns persisted over risk, particularly when air disasters featured prominently in the press. By the late 1930s, flying had become much safer, which, along with decreasing airfares, greatly increased passenger demand and, as will be seen, a headache for Imperial.

Freight, more particularly mail, was a different matter. Dependable air communication greatly assisted governing the empire and binding its people together; the faster and more reliable the better. The limitation of mail by sea was obvious: the voyage to Australia, for instance, would take six weeks on a limited number of ships whereas a reliable flying boat service could deliver mail in a fraction of that time, and more regularly.

A major issue was that mail was the province of the Post Office. Imperial Airways, government-backed like the Post Office, carried mail, but with a surcharge. As Imperial developed its routes, the government increasingly saw the benefits of air mail. For its part, Imperial need extra revenue for expansion. This laid the groundwork for the Empire Air Mail Scheme (EAMS).

Originated by Imperial, the EAMS concept had gone to the British Cabinet in March 1933 in a memo from Sir Eric Geddes, Imperial's chairman.[7] It led to a government announcement by Sir Philip Sassoon on 20 December 1934 that Imperial would deliver all first-class mail within the empire without surcharge. Ultimately, customers would be charged at the standard rate of 1½ pence per half ounce letter and a penny for postcards.[8] The scheme had to be heavily government subsidised. For its part, Imperial would be paid an 'economic rate' as carrier, providing the much-needed revenue stream.

Originally, EAMS was scheduled to begin on the African route out of Alex in early summer 1937, then the Eastern route by the end of that year, and finally to Australasia by early 1938. Negotiations with countries in the empire caused these dates to slip, but not by much. The government had laid out its expectations of Imperial:

> Sassoon suggested a schedule of just over two days to India and East Africa, four days to the Cape [South Africa] and Singapore and seven days to Australia. Moreover, new aircraft would be required to sustain a flying rate of four or five flights per week to India, three per week to Singapore and East Africa, and two per week to South Africa and Australia.[9]

It could only be achieved with a new class of flying boat, one Imperial needed quickly to meet EAMS arrangements. Many months before, Imperial's technical manager, Robert Mayo, had submitted a specification for a new flying boat to Short Bros, Imperial's supplier of choice for flying boats. Arthur Gouge, head of Shorts' design team, would produce the revolutionary and legendary S.23 'C-Class', known as the 'Empire' (a story told shortly).

This combination of new flying boat (including variants) and delivering large payloads of mail quickly and efficiently throughout the empire and across the Atlantic was the twin foundation on which Don would build his reputation in civil aviation and become a household name.

Imperial, Empire and 'Empires'

Short Bros' first flying boat for Imperial was the S.8 Calcutta. It had first flown in early 1928 and been delivered to Imperial in the August for use out of Alex. Imperial initially had a fleet of five, but fatal accidents in 1929 and December 1935 had left three.

The critical aspect of the Calcutta in Don's story is that, in terms of aircraft, it provided the one point of continuity between his military and civil service. Its design was the closest Imperial had to the Southampton. They were generationally alike, but the Calcutta was more advanced. It was a biplane with an open cockpit, like the Southampton, but had three engines not two. It was larger and could carry more while flying faster, higher and further. Don's only experience had been a single flight in the military version – the S.8/8 Rangoon – back in May 1934.

Despite its superior characteristics, the Calcutta was already courting obsolescence in early 1936. Its speed was relatively slow, its range was short, it had limited space for passengers and freight and struggled with the Mediterranean weather, all issues quite unsatisfactory for an airline trying to offer dependable schedules.

Its shortcomings had been partially overcome by Shorts delivering the S.17 Kent, an upgraded version of the Calcutta. Larger, and with an additional engine, it was not designed for extra passengers but additional freight, especially mail. Its characteristics meant it was only ever going to be a marginally better performer in the Mediterranean climate. At least Shorts had given the pilots an enclosed cockpit.

Imperial took delivery of three Kents, flying them commercially from May 1931. One was lost in a fire at Brindisi, just two months before Don arrived in Alex. It was not replaced because the S.23 Empires were expected in the latter half of 1936.

The typical route between Alex and Brindisi included two stops – Mirabella (Crete) and Athens. Depending on conditions, the two remaining Kents, *Scipio* (G-ABFA) and *Satyrus* (G-ABFC),[10] and the more elderly Calcuttas, might alight in Corfu. Headwinds disrupted westbound schedules, even causing aircraft to turn around. Bad weather at the destination might result in lengthy flight postponements.

Weather appears the likely catalyst for a short entry in Don's memoirs. He had commenced as first officer from Alex on 27 February, mostly flying the Kents. On 18 March, Captain Frank Bailey, with Don, crew and a full passenger list, had arrived in Brindisi but could not return to Alex for several days, with poor weather at their destination the probable cause. With time on his hands, Don got to work:

> On one occasion I had five days in Brindisi, and in that time wrote a complete book – on the handling of flying boats (*The Air Mariner*).[11]

The remarkable nature of this lies not just in his assertion to have written the book (100-plus pages) in such a short period, but *when* he did it. *The Complete Air Navigator* was still awaiting publication. He had been flying Imperial's flying boats for less than a month and only as a first officer.

One can't help speculate what the senior Imperial pilots thought of the precocious young officer producing a basic textbook on flying their aircraft. He was no neophyte, having arrived at Imperial with 817 hours on flying boats, including many as an instructor, but the vast majority of these were on the Southampton. Depending on your view of Don, either he was setting out to fill a gap in the textbook market and not concerned with what they thought, or this was an early, intentional demonstration to all of his talent and experience, one designed to build his profile in the world of civil aviation. It would have been unsurprising if his new colleagues exchanged a few raised eyebrows.

The Air Mariner was not published for nearly two years due to significant developments in flying boats. Although it passed through five editions, it never achieved the prominence of its predecessor.

On 3 April 1936, after probationary flights under several senior captains, Don captained his first Kent, *Scipio*, from Alex to Brindisi.

The road to taking command had been short but Don had to fight for it, despite Brackley's promise at Croydon. The base was in high anticipation of the arrival of the S.23 Empires in coming months and 'the competition to be first in command of these boats was already intense'.[12] Some of his contemporaries, such as John Burgess and John 'Jack' Kelly-Rogers already had acting commands on the Kents and competition was fierce:

> All were paid on the basis of the amount of flying they did, and the keenness to get flying was almost 'cut-throat'. I pushed into the roster with considerable difficulty and consequent unpopularity.[13]

In June, Don found it worthy to note he flew Sir Miles Lampson, British High Commissioner to Egypt, and his wife, along with Air Chief Marshal Sir Robert Brooke-Popham, Air Officer Commander-in-Chief, Middle East, and their entourage, on a scheduled westbound flight.

He did some night-flying instruction for first officers on the Calcutta, but mostly he flew back and forth between Alex and Brindisi in the Kents. Then disaster struck. Arthur Wilcockson crashed *Scipio* into Mirabella harbour at Crete in difficult conditions. Two passengers perished. Aside from the tragedy itself, it created repercussions for Imperial because it left just one Kent – *Satyrus* – on the eastern Mediterranean route. With the Empires still undergoing testing and accreditation in England, timetables were reworked for the Calcuttas to fill the gap. Without seniority, Don was back on them four days after the accident.

Just after his 26th birthday, he did his 'B' Licence endorsement test on the land-based Handley Page H.P.42 and commenced flying the first legs of the African route under the supervision of Captain Harrington in *Horsa* (G-AAUC). As Don noted, it was 'part of the company's policy of giving as wide experience as possible to all its captains, wherever practicable'.[14]

Flying the African route was the very definition of exotic for the Queenslander, as it was for all on board. Cruising at a leisurely 96 mph (83 knots) and at just a few thousand feet, passengers were afforded the best possible views from the H.P.42's wide windows. The route took them south past Cairo, tracking the Nile to Luxor. Then it was on to a series of stops in the Sudan – Wadi Halfa, Karima, Kosti, Malakal and Juba – before crossing the border into Uganda and landing at Entebbe. From there they tracked across the top of Lake Victoria to Kisumu in Kenya. Other aircraft would then carry passengers, mail and freight through southern Africa and on to Cape Town.[15]

The H.P.42 was Imperial's land-based flagship. It offered a first-class service for 18 passengers, with luxurious interiors featuring inlaid wood panelling, plenty of leg room, complimentary liquor, and meals served on bone china with silver cutlery. Overnight stays were in the best possible hotels, and passengers, in keeping with the premium service, were well dressed.[16]

On his return, Imperial promptly sent Don on the first legs of the Eastern route with Captain Woodhouse in *Hadrian* (G-AAUE). The three days took them from Alex to Gaza, then into Iraq and on to Baghdad for the night. From there they flew down the Tigris valley to Basra before the short hop to Kuwait City. With the Persian Gulf on the port bow, it was off to Bahrain, before turning east, out over the water, to reach Sharjah (near modern day Dubai) for their second night. Departing around dawn, they flew across the Gulf of Oman and on to Gwadar and finally Karachi in what was then India.[17]

Don must have been ecstatic: Karachi was halfway to Australia. Until then, the furthest east he had been was Aleppo in the Vega. These two trips out of Alex further confirmed the decision not to return to a civil airline job in Australia.

Don was back in Alex on 26 October to witness the highly anticipated arrival of the first S.23 'C-Class' Empire, *Canopus* (G-ADHL), under the command of Frank Bailey, with Air Superintendent Brackley on board. Within days, *Canopus* performed Imperial's first scheduled S.23 flight to Brindisi, carrying mail from, of all places, Brisbane. The letters and parcels had left Don's hometown just nine days earlier.

The Empire flying boat is one of the most famous and celebrated aircraft. Short Bros' designer, Arthur Gouge, had delivered the design to Imperial in June 1934, a full six months before Sir Philip Sassoon laid out government expectations of Imperial under the Empire Air Mail Scheme.

Imperial, Empire and 'Empires'

In January 1935, Imperial gave the go ahead, wanting deliveries to begin from April 1936 (hence the tie-in with Don's recruitment in the January).

The airline was in a hurry. Shorts wanted to develop a prototype of the S.23 but Imperial ordered 14 in April 1935, doubling that to 28 five months later.

Shorts was under considerable pressure, working with a brand-new design that produced a plethora of construction challenges and innovations. The biplane era departed with a high-wing cantilever monoplane of all-metal construction powered by four 920-hp Bristol Pegasus Xc engines. The fuselage was enlarged, allowing for a double deck, and the planing hull meticulously redesigned to reduce drag.

Canopus only had its first official flight on 4 July 1936. *Caledonia*, the second, was used for Certificate of Airworthiness tests, with *Canopus* delivered to Imperial on 7 September for fitting out, final testing, pilot certification and some familiarisation flights for luminaries such as the Minister for Air, Lord Swinton, and the Chief of the Air Staff, Sir Edward Ellington. Within weeks it would be in Alex to commence commercial flights, the vanguard of the Empire revolution. A new Empire would roll off the production line approximately every two weeks.

Of greatest importance to Imperial was its game-changing performance. Designed to carry 3,000 lbs of mail with 24 daytime passengers, or 16 by night, Imperial could load more than 4,000 lbs of mail by reducing passenger numbers to 17.

Passengers and freight were delivered to their destinations at a cruising speed of 150 mph. That, combined with a range of 760 miles, went directly to Imperial's desire to reach the edges of empire in the shortest possible time. Sassoon's expression of the government's desire to reach Australia in seven days was not achieved, but just over nine to Sydney was deemed extraordinary when compared with 40 days by sea.

Imperial and Qantas ultimately operated a common fleet between England and Australia. Hudson Howse, who flew Empires for Qantas, offered a glimpse of the flying experience for passengers who occupied the one-class service – first:

> There was an 11 foot ceiling on the lower deck and a 9 foot ceiling on the upper deck … Getting out of his chair, a passenger could

walk about and, if his seat was in the main cabin, stroll along to the smoking cabin for a smoke, stopping on the way at the promenade deck with its high handrail and windows at eye level, to gaze at the world of cloud and sky outside, and a countryside or sea slipping away at a steady 150 mph ...

The passengers were invariably upper class, people with good or private incomes. Most of the passengers were bound for extended tours of the Continent. We often carried parents taking their daughters to London for their debutante season. Mind you, a lot of them treated the flying boat pilots like gods because, in their minds, they probably were.[18]

The in-flight service and overnight stays matched the expectations, and wallets, of the well-heeled customers.

The successful introduction of the Empires in late 1936 allowed Imperial to pursue the next phases of its international strategy. They would now fly the entire Mediterranean route between England and Egypt, with Imperial establishing a new base at Hythe, Southampton, from March 1937. The African route would be extended south from Kisumu in Kenya to a new terminus at Durban. The entire Eastern route through India to South-East Asia could be progressively revamped.

As the Empires arrived, Imperial's senior captains were tasked with testing their capabilities. Despite EAMS not officially being underway, *Caledonia* left England on 13 December to deliver five-and-a-half tons of mail to India to meet the Christmas demand. Five days later, a similar load was delivered to Egypt by *Centaurus*. Work was well underway to modify two of the Empires for long-haul flights across the Atlantic.

For weeks, Don was left to wait his turn. He went on leave, or was invited to do so, because Ly was shortly to give birth. On 14 November, Noreen Daphne Bennett was born. With wife and daughter doing well, he was back on duty from the 17th.

At dawn on Christmas Day, Don flew an Empire for the first time, with Lawrence Egglesfield in command. They reached Brindisi in just over eight hours, including a single 50-minute stop in Athens. Don's most recent efforts in the Kent, excluding flights with mechanical issues, had taken anywhere between 11 and 13 hours.

The game-changing performance was highlighted on the final leg of the return journey from Athens to Alex when he recorded 'Av. G/S [ground speed] = 162 knots' (186 mph). Given the Kent had chugged along at an average 91 knots (104 mph), he could hardly believe what he was experiencing. This was stupendous, revolutionary.

He was sent back to the old aircraft, including another trip to Kisumu.

As 1937 dawned, Don's impatience surfaced over the issue of seniority.

Brackley had introduced a system of assigning a captain and crew semi-permanently to each new Empire as it was delivered.[19] While Don found the arrangement attractive, it intensified the competitive pressures among the pilots. For the Empires sent to Alex, Bailey retained command of *Canopus*, while *Centaurus* (G-ADUT) went to Egglesfield, and then *Castor* (G-ADUW) went to Howard Alger. Don was peeved: '… to my annoyance as he [Alger] was a newcomer to flying boats.'[20]

Many of the pilots, including Frank Bailey, Griffith 'Taffy' Powell, and Arthur Wilcockson were among the original group of 19 pilots employed by Imperial back in 1924.[21] Between them, these pilots had done developmental work, flown a variety of aircraft and introduced new ones, and established the routes Don was now flying. Many had instructed Don. All had First Class Air Navigation Licences and the 'A' and 'C' Engineer's Licences, and most had the Wireless Operator's Licence.[22] Most, if not all, had accumulated flying hours well in excess of Don. In fact, none of Imperial's 'originals' had less than 8,000 hours flying experience.[23] After a year at Imperial, Don had just over 2,500. Alger had been flying with Imperial since 1928, two years before Don joined the Royal Australian Air Force.

In this context, Don's reaction to Alger beating him to the command of an Empire shows, at best, his impatient self-belief in the superiority of his abilities vis-à-vis others. His argument was that flying hours on the relevant aircraft (that is, flying boats) should be the principal determinant of who was prioritised for command on the Empires and therefore, by implication, not 'seniority' in terms of years served, nor that Alger had more flying hours overall.

The episode is given a puzzling twist by his admission that he gained command of the very next Empire delivered to Alex – *Cassiopeia* (G-ADUX) – just weeks later. Why Brackley offered him command of the fourth Empire is not known, but Don would have felt it vindicated his argument. It is unlikely he saw the irony of leapfrogging more senior pilots with considerably more hours on flying boats.

'Taffy' Powell conducted Don's 'B' Licence endorsement tests on *Cassiopeia* and on 1 February he flew through to Marseilles, with Powell in oversight. Then, after a second trip under Powell, flying the entire new route from Alex to Southampton, Don took command of *Cassiopeia* on 14 February 1937 as captain.

Ultimately, despite Don's wholehearted support, Brackley's system proved unworkable, ironically because Brackley discovered the Empires could fly more often than pilots and their crews and the scheduling was thus inefficient!

By May 1937, Don had been at Imperial for over a year and had done no pioneering work. It was the more senior pilots who forged the revised routes and tested the Empire's capabilities. From mid-February, for three months, he had just flown back and forth between Alex and Southampton, a veritable 'aerial bus driver', the pejorative term his parents had used to describe Qantas pilots years earlier.

In late May, he found an excuse to put his name up in lights: the late arrival into Alex of a service with passengers, freight and mail from Australia. To get the timetable back on track, and his passengers to England on schedule, he decided to fly them the entire Mediterranean route in a day. It was unprecedented, but justifiably within Imperial's guidelines.

Under 'General Instructions – Captain and Crew' in Don's handbook, his top three 'Responsibilities' read as follows:

1. *Aircraft* – The safety of the aircraft and its passengers is to be the Pilot's first consideration.
2. *Passengers* – The Captain will invariably act as host to his passengers and see that they are attended to and their comfort duly considered until he is able to hand them over to the Manager or Agent at an aerodrome.
3. *Punctuality* – The punctual performance of all services is essential, consistent with paragraphs 1 and 2 above.[24]

The third one would do nicely as the excuse. His second responsibility was immediately in question with his decision to depart at the unholy hour of 2:40 am on 23 May. It caused some grumbling among his already weary passengers, but with Southampton having no night-flying facilities, they needed an early start and a smooth passage to make it.

They encountered headwinds all the way to Brindisi. He was starting to worry. On arrival at Marignane (Marseilles, France), and with nearly four hours left to fly, he thought the record had eluded him. His first responsibility now came into question but, buoyed by the weather forecast and the now-enthusiastic passengers, he took off.

> Half-way across France the passengers were in such a state that I had to report ground speed every quarter of an hour, and even then they sent the steward up to wait for more news long before we were ready to give them any.[25]

He recorded reaching Southampton with just 20 minutes of daylight to spare. Given to rarely writing in the 'Remarks' column of his logbook, except for technical and mechanical matters, he decided that setting the record should be accorded the full treatment:

> IW.548. Night take-off. This flight was the first carrying passengers between Egypt and England in one day.[26]

More importantly, it led to a syndicated press report that found its way to Australia.

The Argus in Melbourne was one outlet to print it:

EGYPT-ENGLAND IN 15 HOURS

Flying-boat's Record

LONDON, May 23

'Egypt at sunrise and England by sunset' is the record of the Imperial Airways flying-boat *Cassiopeia*, which arrived at Southampton at 8.20 p.m. today with 14 passengers, a crew of five, and 30cwt. of mail and freight.

Australian passengers and mails were taken on board at Alexandria on Saturday. Halts were made for refuelling at Athens, Brindisi, Rome and Marseilles. The actual flying time was 15 hours at an average speed of 150 miles an hour for 2,300 miles.[27]

It was not front page, and it failed to mention the captain's name, but it was in the press because it caught the public's imagination. Don now officially qualified as a pioneering aviator, albeit on a small scale.

In June, he commenced flying the entire route from Southampton to Durban. They navigated the Mediterranean leg with a single stopover in Rome. Flying the new route through Africa reads like something from a *Boy's Own Annual*. From Alex they flew to Cairo and on to Luxor. At Wadi Halfa, where they refuelled, Don records handling being 'extremely tricky' if dealing 'with a five-knot stream running and a twenty-knot wind'. They lost a local to crocodiles on one occasion when he fell off a refuelling barge.[28] Tracking further up the Nile took them to Khartoum for a night's stop. From there to Malakal, then Port Bell (Kampala), before flying across to Kisumu in Kenya for night two.

Previously, he had turned around there, but now they headed on to Mombasa on the coast in southern Kenya, then south to Dar es Salaam for night three. The final day and a half took them down the east coast of Africa – Lindi in Tanzania, then Quelimane, Beira, Inhambane and Laurenço Marques (now Maputo) in Mozambique, before the final hop to Imperial's new base and terminus at Durban.

When flying this route over coming months, Don records them taking the opportunity to give passengers their money's worth by flying low over herds of elephants and giraffes, hippos below the Murchison Falls, and rhinoceros. On the Kisumu–Mombasa leg, a slight deviation in course took them past Mount Kilimanjaro.

Turning passengers into 'incidental air tourists' was a common occurrence at Imperial, because of its marketing value. Over the Mediterranean, passengers would see the remains of the Greek, Roman and Egyptian empires. Down through Africa, they were enthralled by pre- and post-colonial sights, vast and spectacular natural landscapes and, as Don highlighted, an exotic array of animals. Making herds stampede for passenger enjoyment occurred. 'The pages of one air-travel narrative (Crile, 1937) teem with giraffe, elephant, rhino, lion, eland, hartebeest, gazelle, hyena and warthog ... Spotting wildlife had never been as easy, or as fruitful.'[29]

The in-air sightseeing was matched by the many stopovers punctuating the journey. In Africa, these ranged from Shepheard's Hotel and the Continental Savoy in Cairo, frequented earlier by the likes of General Gordon and Lawrence of Arabia, through to less salubrious colonial outposts. There might be time to visit local sights and experience something of local culture in places that would never have been visited otherwise.

This, of course, applied as much to Imperial's aircrew as it did to passengers.

Come September, Don decided it was time to establish another record, one he, again, felt worth mentioning in his memoirs.

Imperial's route schedule for the last few legs of the African route to Durban saw arrival in Dar es Salaam early afternoon, giving passengers the rest of the day to mosey around or rest up before the scheduled departure just after 5:00 am the following morning. They would then fly to the next night stopover at Beira, arriving mid-afternoon for the evening at leisure. The final leg out of Beira to Durban had a scheduled departure close to dawn, with passengers alighting at Durban just after lunch.

Don appears to have found this pace was far too leisurely and inefficient. It was within his authority as captain to change things should he deem it necessary. On 19 September, in Dar es Salaam, he advised passengers they would depart at 2:00 am, with the plan to reach Durban that day, avoiding the night stopover in Beira. The passengers had no choice.

The problem he created for himself was at Lindi in Tanzania. This was a small administrative post of the British Empire, located at the mouth of the Lukuledi River. On the two previous occasions he had flown in there, he alighted and took off around dawn. Now, he would arrive and depart either side of 4:00 am. They had no flare path, meaning alighting on the estuary in the darkness. Don had a solution:

> … to do this I used the passenger launch as a single datum, and 'felt' the boat down with a little engine on to a calm sea.[30]

Clearly there were few, if any, lights on in Lindi at that hour to provide additional points for getting his bearings, making it quite tricky to get his approach right. He was unfazed. Fortunately, the weather was kind and the stop was negotiated without incident.[31]

At every stop that day he shaved precious minutes off the turnaround time, arriving into Durban just before 5:00 pm.[32] The passengers might have been totally exhausted, but he had got them there 20 hours ahead of schedule. No doubt some were grateful; others might have thought a refund for the lost night's accommodation at Beira was warranted.

Following just months after the Alex to Southampton record flight, this achievement had significance. Brackley, by giving his captains full authority to make decisions, had created an environment for Don to push the boundaries. And this is where Don's decision in this instance might have attracted Brackley's attention. In addition to outlining the responsibilities of captain and crew in the General Instructions in the Handbook, Brackley had issued the more specific 'Appendix 1 – Instructions to Captains by the Air Superintendent'. Two were pertinent:

> 1.(a) *First Care* – The Captain's primary responsibility is to ensure the safety of his passengers, freight and aircraft, and nothing in these Instructions or Notices issued from time to time by the Company relieves him therefrom.

8. *Expeditious Passage* – The Captain must on no account whatever imperil the safety of his aircraft for the purpose of making expeditious passage.[33]

These did not limit Don's authority. All he had to do was weigh the risk to 'safety' in both instances against his self-assessed ability to alight at 4:00 am in Lindi and then fly the whole day to Durban with minimal turnaround times. Having concluded the risks were low, he was left to consider the comfort of his passengers from whom he was requiring 16 hours of continuous travel time. That was subjective. And as for punctuality, the 'most efficient airman' was undoubtedly convinced delivering his passengers much earlier than scheduled was a win for all concerned.

The official Imperial response was to leave the schedule unchanged.

For Don's part, the 'Dar es Salaam–Durban in a day' logbook entry received an extravagant flourish of an underline, signalling the conclusion to another episode of what carried the appearance of self-gratification. He had now seized two opportunities to set records and developed a taste for it.[34]

16

Mercury and the Atlantic

Don Bennett had only one love affair while married to Ly and that was with *Mercury*.

Aside from the two episodes setting records, 1937 and early 1938 passed without incident or highlight, apart from the news Ly was expecting their second child. Don kept flying the Mediterranean and African routes. Twice he flew the Eastern route through to Karachi. He accumulated another 1,200 hours flying time, which only served to reinforce the routine nature of the work. While he was making up the roster on the empire routes, the real developmental action was centred on the Atlantic.[1]

The Bennetts had moved back to England and, as 1938 dawned, Don once again started looking for new opportunities. His specific interest fell on the Mayo Composite 'pick-a-back' aircraft. It was being developed as part of Imperial's strategy for tackling the Atlantic, and was undergoing testing by Short Bros at Rochester.

The Atlantic was proving a significant challenge. Two routes were under consideration. The northern route, between Ireland and Newfoundland, was the most direct but also the most problematic. It involved a large stretch of water, prevailing westerlies, and typically bad weather, especially in winter. Botwood, the preferred seaplane base in Newfoundland, also froze over in winter. The southern route, the one favoured by the Americans, was longer but utilised bases in Bermuda and the Azores as refuelling stops. Imperial believed the harbour at Horta in the Azores posed issues for its flying boats, especially if fully loaded and heading west.

The airline was running two technological paths in parallel. One was developing a system with Sir Alan Cobham's company, Flight Refuelling Limited, for in-flight refuelling of the forthcoming S.30 Empires, an upgraded version of the S.23. Cobham had been evolving in-flight refuelling techniques since 1934. The S.30s had the potential for freight and passengers. The other was the Mayo Composite, a freight/mail only option.

In evolutionary aviation terms, the Mayo Composite was something of an oddity and reflected airline priorities at the time it was conceived. Major Robert Mayo, Imperial's technical general manager, first mooted the idea of a larger aircraft carrying aloft a smaller one to maximise the range of the latter in a 1932 memo. 'Crossing the Atlantic was the goal.'[2] He produced Specification 13/33 for Shorts, which commenced building the aircraft – S.20 *Mercury* (G-ADHJ) and S.21 *Maia* (G-ADHK) – in 1935. They were not officially launched at Shorts until September 1937.

Throughout development, Imperial had lobbied the Air Ministry to pay half the costs, only succeeding in 1937. Government support was only ever half-hearted. The Sassoon Committee, formed in mid-1935 to look into how the British were competing against the Americans for establishing a transatlantic service, had formed the view flying boats were a better option than the Mayo Composite. The committee also considered a land-based aircraft for the transatlantic service but had great reservations because a forced landing at sea needed to be considered, and that meant restricting its uses to mail only. This showed a combination of passengers and freight for the service was always preferred, placing the Mayo Composite at a distinct disadvantage.

From July 1937, while the Mayo Composite was still in Shorts' workshops, Imperial commenced experimental flights on the northern route using two S.23 Empires – *Caledonia* and *Cambria* – which had been modified by installing long-range fuel tanks. While not carrying passengers or freight, they demonstrated the viability of a commercial transatlantic crossing, while also testing a host of issues such as communications, direction-finding equipment and meteorology data.

The ultimate subordination of the Mayo Composite to developments with the S.30s was not just about 'no passengers', but the limitations of its design. *Maia* was a bespoke Empire requiring extensive and expensive modifications, including a wider beam to accommodate having *Mercury* perched on top with the resultant higher centre of gravity. The engines were also moved further outboard and the wing area increased.

Mercury was a new all-metal seaplane design. A monoplane with four engines and fixed-pitch propellers, it was all about maximising economical cruising once it achieved in-flight separation from *Maia*. It had a crew of two, a pilot and a navigator/radio operator. Mayo envisaged a fuel load of 1,200 gallons would carry a 1,000-lb payload in the fuselage and modified floats over 3,800 miles in still air. In effect, if you could launch from *Maia* on the west coast of Ireland with a full load of fuel and freight, you could theoretically reach New York non-stop.

However, for *Mercury* to realise its full potential, *Maia* was required at every base it flew from because its floats were not designed to operate independently of *Maia* with a full fuel load. Imperial could have multiple *Maias* at various ports or *Maia* could make its way independently to whichever foreign base *Mercury* had arrived at, creating a logistical nightmare. Either way, all ports utilised required suitable onshore hoisting equipment. Without *Maia*, *Mercury* had to return to the UK under its own steam, necessitating lesser fuel loads and alighting at various ports on the way. Logically, if Imperial could develop a reliable in-flight refuelling system for the S.30s, the Mayo Composite would struggle for strategic relevance, financial viability and government support.

Even if these limitations were apparent to Don at the beginning of 1938, they were trumped by what he found attractive with the project. It posed unique technological challenges, an excellent antidote to boredom. Beyond that, he wanted one role in particular and applied 'formally in writing to be put in command of *Mercury*'.[3] *Mercury* would be doing the real work of flying across the Atlantic, not *Maia*.

The most complex part of the Mayo Composite was the coupling mechanism. Mayo's patented design saw *Mercury* hoisted by crane onto

Maia and secured via six 'ball spigots' (Don's description) – two on the fuselage and two for each float. There were three latches for decoupling. Each pilot controlled one, coordinating when to release via a telephone link. The third latch would then release automatically once the two aircraft were at 140–150 mph and having a lift differential of 5,000 lbs, allowing *Mercury* to commence its solo flight.

The aerodynamics were tricky. Although all eight engines assisted on take-off, *Maia* bore most of the load. Once they reached cruising speed, however, *Mercury* carried a proportionately greater load than *Maia*. This was crucial because at the final point of release, *Mercury* needed to ascend and *Maia* descend at sufficient rates to allow both pilots to take control of their aircraft and avoid colliding. Until that moment, *Mercury*'s controls were locked. Effective communication between the pilots was paramount. A patented barometric control ensured separation occurred above a calculated minimum altitude.[4]

Mayo's grand experiment took to the air at Rochester in early September 1937 with separate flying tests for the two aircraft. The Composite flew for the first time on 20 January 1938 and achieved the first separation on 6 February. It occurred without incident, but also without much of an audience. Not so the second separation, a staged event witnessed by the press and a selection of dignitaries. It was captured by *Movietone News*[5] and ignited public interest.

A significant development from Don's perspective was that Arthur Wilcockson had taken the role of *Maia*'s pilot. He had conducted Imperial's first experimental transatlantic flight in *Caledonia* in July 1937[6] and was now manager of Imperial's Atlantic Division, with the Composite under his operational control.

That day, Don was in *Clio* somewhere between Port Bell and Khartoum, returning from his most recent trip to Durban. It would be his last on regular routes for nearly a year. There is no mention as to whether he knew he had the job as pilot of *Mercury* during this trip or found out on arriving back at Southampton on Sunday, 27 February. He headed immediately for Rochester where, five days later, he participated in a series of flights testing forward and aft centre of gravity limits without separating from *Maia*.

Don got busy with his personal notebook. Aside from a single page covering *Mercury*'s key specifications, he sketched the all-important fuel system. He also did a detailed drawing of the coupling mechanism, accompanied by written details of how it worked. Later, he would add a chart on which to plot various 'limiting speeds' when coupled with *Maia*, based on variations on *Mercury*'s weight and engine revolutions.[7]

Two weeks later, the Mayo Composite was flown across to the Marine Aircraft Experimental Establishment (MAEE) at Felixstowe for official testing and certification. Despite now being *Mercury*'s designated pilot, Don got just 55 minutes in the cockpit doing take-offs and landings.

His principal role at MAEE, at least in his opinion, was making up for Robert Mayo's lack of pushiness. He admired Mayo for his intellect and his contribution to Imperial but, according to Don, he displayed a reticence when confronted with MAEE's 'laborious processes' and 'governmental inefficiency' due to his shyness and kindness.[8] Don believed the certification process would have taken at least two years had he not come to Mayo's aid. Instead, it ran for just two months.

At one point, Don recalled, the Air Ministry and Aeronautical Inspection Directorate convened a meeting to discuss *Mercury*'s range. According to the 'lashings of civil servants' present, the MAEE fuel-consumption figures rendered the westward Atlantic flight barely achievable and the return journey impossible. He was able to point out, no doubt with disdainful precision, that Felixstowe had performed their tests 'ignorant of the peculiarities of the Exactor hydraulic controls',[9] meaning the wrong fuel mixture was used.

Too many people in the room, not across the technical detail, wasting his time making collective pronouncements based on ignorance and incorrect data, was the perfect recipe for eliciting a barely controlled Don broadside. 'To me it was a new experience, the pattern of which repeated itself in my life many times in the years that followed.'[10] His brutally direct responses, delivered with a barely concealed sarcasm, would not soften with age.

Extraordinarily, only two flying tests were performed at MAEE. On 6 May, the Shorts pilots demonstrated a full-load take-off and landing without separation. Then, three days later, the MAEE pilots performed a separation with the Shorts pilots flying alongside. That done, the Mayo Composite returned to Shorts at Rochester on 19 May. One reason was to replace the Napier Rapier V engines on *Mercury* with the newer VIs. These weighed less, offering increased power at a slightly improved, but nonetheless vital, fuel consumption.[11]

Over at Imperial, it was decided *Mercury*'s first attempt on the Atlantic would also be, officially, Imperial's first commercial flight, carrying a full load of mail. Wilcockson, Don and team were making plans.

On 28 May, Ly gave birth to Torix Peter. Whether Torix's arrival generated any discussion between his parents about the timing of *Mercury*'s attempt as the plans took shape in June is a moot point.

One key planning decision was Don's choice of AJ 'Jimmy' Coster as his wireless operator and fellow navigator. They had flown together as recently as Don's last trip down to Durban. Don referred to him as '"Faithful" Coster', a rare term of endorsement, even endearment, for someone whose competence had earned Don's complete trust.

Mercury was going to follow the same route Wilcockson had taken on Imperial's first experimental flight in *Caledonia* a year earlier. They would depart from Foynes, where Imperial had built a seaplane base specifically for the transatlantic flights, and fly the Great Circle route, to Botwood in Newfoundland. After refuelling, they would track across eastern Canada to Boucherville, Montreal, then down to Port Washington, New York.[12] The fundamental difference was that Don and Jimmy were now attempting the crossing in a much smaller, untried, fully laden floatplane (not flying boat), with recently installed engines and no first officer.

Perhaps it was this, more than anything else, that made the final preparations seem underdone. On 2 July, Don and Wilcockson arrived in Rochester to fly *Mercury* and *Maia* to Imperial's base at Hythe. Don took the opportunity to run some fuel-consumption tests on the new engines.

The following day, the starboard inner engine was vibrating and the fuel-mixture control went unserviceable. It wasn't a good start. After two days working on the problem, all he could conclude on resuming the consumption tests was the mixture control was 'doubtful'. A much longer flight on the 6th finally satisfied him he had control of the fuel mix. Two more consumption test flights followed, just to be sure.

He and Wilcockson took off mid-afternoon on 12 July to perform their very first separation as a team, but also as individuals. It went as planned. Don captured the magnitude of his relief with a large, succinct 'OK!' in his logbook.

He then embarked on a long-haul flight of over 12 hours, starting with their second separation at Hythe before tracking across to Foynes and out over the Atlantic. Everything was tested and he returned satisfied with the fuel-consumption tests.[13]

Despite the minimal preparation, they decided to attempt the Atlantic crossing on the 20th, weather permitting. That was less than a week away. It was enough time to show off the Mayo Composite at a display marking the opening of Luton Airport. They performed their third separation to the crowd's delight. The low altitude allowed an amateur photographer to capture a few precious seconds on film.[14]

The scene was set on 19 July when Don flew across to Foynes from Southampton with Robert Mayo on board so he could witness the start of the record-breaking journey.

The following day, *Mercury* was craned onto *Maia*, fuelled up and loaded with freight. This included UK newspapers, mail and newsreels weighing 900 lbs. At 6:50 pm, the Mayo Composite took off, separating at just under 2,000 feet, and *Mercury* was on its way.

The weather was expected to be challenging. They were flying at a steady 500 feet when the rain began about 90 minutes into the flight, continuing for the next six-and-a-half hours and forcing Don to ascend above the cloud.[15] The flight was further hampered when they encountered much stronger headwinds than forecast about 300 miles east of Newfoundland.

The weather wasn't the only problem. Shortly after uncoupling from *Maia*, Don discovered *Mercury*'s 'fuel gauge – a vital item – stuck at the upper limit of its travel'. This, he noted, 'caused some concern' and only began functioning normally after six hours. The long wing fuel tank caused some lateral instability as fuel was used, a problem inadequately handled by the autopilot. And, to top it off, 'A minor but irritating trouble occurred due to the fact that some of the windows of the control cabin were not water-tight. In rain, everything in the cockpit therefore became wet – including charts, log sheets, etc.'[16]

Despite these challenges, they were untroubled navigating. They took bearings, as planned, from ships along the route as well as astro readings when above the cloud during the night. Belle Isle and Botwood broadcast on medium wave D/F (direction finding) to help guide them in:

> With a compass excellently sited, and very accurately swung, I found *Mercury* a wonderful aircraft to navigate. I had an old Bigsworth chart-board on my knees, my sextant was in a 'bed' behind me, my Bygrave slide-rule in a container by my side, with a place for almanac, pencils, etc., conveniently at hand. Never before nor since have I experienced such an excellent navigational arrangement, nor have I ever achieved more accurate results.[17]

Approaching Botwood, Don decided there was sufficient fuel to continue directly to Montreal, a significant decision because his calculations showed it added 800 nautical miles, about 46%, to the trip.[18] They set down at Boucherville at 3:24 pm, after 20 hours and 24 minutes in the air, and were greeted by the Canadian Minister for Transport, Clarence Howe, who had a passionate interest in expanding aviation in Canada.

They had covered 2,477 nautical miles at a ground speed of 141 mph (122.52 knots) and, in so doing, set three world records: the first commercial payload across the Atlantic; the first non-stop east to west flight from the UK to Montreal; and the fastest east-to-west Atlantic crossing, as measured from Foynes to Cape Bauld, recorded as 13 hours and 29 minutes.

Don was in a hurry to get to New York, his ultimate destination. They were on the ground for two-and-a-half busy hours. His official report was succinct:

Customs and Port clearances were obtained, freight for Montreal was off-loaded and 250 gallons of petrol taken on. An engineer was also picked up for passage to New York.[19]

This did not quite capture the scene. There had been much anticipation of the flight in the Americas and *Mercury* arrived to 'a tumultuous press reception'.[20] Unused to being besieged by the media, tired, and wanting to press on to New York, Don was irritated. 'I made myself very unpopular by ordering all the press out of the customs shelter until I had completed the required formalities.'[21]

At 6:00 pm, they were away and Don stepped up the pace despite low cloud and rain, averaging 161 mph (143.5 knots), touching down at Port Washington just two hours and 19 minutes later. The press reception was even greater, but this time he embraced it. Surrounded by reporters and onlookers in the dozens, and fronting mounted microphones with camera flashbulbs popping, Don gave an account of the trip.

The following day, *Mercury* and its crew were front-page news. *The New York Times* captured the essence of why the flight attracted so much interest: 'Pickaback plane arrives', 'Young pilot in unwrinkled suit and "not yet tired", sets *Mercury* down at Port Washington with London papers of the day before', 'total hours to Port Washington 25 hours and 9 minutes'.

News of *Mercury*'s flight across the Atlantic took little time to reach Australia's shores. The Australian media had been following developments for months and the syndicated articles now appeared in newspapers across the country.

On 22 July, Brisbane's *Courier Mail* carried a lengthy article on its front page from the Australian Associated Press (AAP) correspondent in London. It began:

Brisbane Pilot Flies Atlantic

PICK-A-BACK RECORD

London, July 21 – Captain Donald Bennett, a native of Brisbane, in command of the Mercury, the upper component of the Mayo Composite aircraft, flew across the Atlantic today, from Foynes

(Ireland) to Cape Bauld (Northern Newfoundland) in the record time of 13 hours and 29 Minutes.

Something of the daring of the venture was captured by the Montreal correspondent:

Rain, Fog and Rough Sea

Montreal, July 21 – A wireless message from the Mercury at 9 o'clock last night stated she was 1,000 miles from Foynes, and was passing through continuous rain and turbulent seas.

At midnight, the airship reported she was 800 miles from Botwood, and was flying at an altitude of 5,000ft, through heavy rain and thick fog.

Service crews were waiting in the St Lawrence, 10 miles from Montreal, for her arrival.[22]

The *Courier Mail* added a further front-page article offering a potted biography of Don under the heading: 'Pilot Born in Queensland – Colourful Career.' On page three were adjacent pictures, one of Don in his captain's uniform and the other of his father George on the phone. It was his Real Estate Institute photo repurposed with the caption: 'Mr G.T. Bennett, of Auchenflower, receiving the news last night of the record Atlantic flight by his son, Captain Donald Bennett (see page 1).'

The following day, Don's mother, Celia, decided it was time to act. Despite the fact Don's premature departure from Brisbane Grammar School was due to his tardiness, and his father's frustration with him, Celia wanted to let the school know her youngest boy was doing very nicely indeed, thank you. Rather than Brisbane's *Courier Mail*, she chose the more august broadsheet, the *Sydney Morning Herald*. She carefully cut out the article headlined, '"Pick-a-Back" Plane. Two records. Eire to Canada Non-stop.'[23]

She then went to Barkers Book Store in Edward Street and ordered a copy of *The Complete Air Navigator*.[24] When it arrived several weeks later, she pasted the *Herald* article inside the front cover and then wrote on the opposite page:

Presented to
The War Memorial Library
Brisbane Grammar School

by Mrs G.T. Bennett
the mother of the Author

2nd September 1938

Her none-so-subtle gift was an emphatic repudiation of the school's official final assessment that Don, now firmly on the international stage as a record-setting pilot and leading author on air navigation, was only just 'of fair ability and steady industry'.

Without *Maia*, a return journey non-stop from Newfoundland to Foynes was impossible because *Mercury*'s floats were not designed to withstand a full fuel load, so Don planned to refuel at Horta in the Azores, then in Lisbon. He calculated two routes.[25] One was to retrace his steps through Canada. The other was to make use of Imperial's base in Bermuda, which had been established for flights by the Empires back and forth to New York. He decided on the former. After two-and-a-half days being feted in New York, they returned to Montreal before alighting at Botwood in Newfoundland. After refuelling, they flew to Horta in the Azores, then on to Lisbon and back to Southampton, landing just after lunch on the 27th.

The AAP London correspondent had been following every step, producing almost daily syndicated articles. On 28 July, *The West Australian* carried his article under the heading 'Atlantic Flown Twice. The Mercury Returns to England', which included:

> ... Captain Bennett, the Brisbane-born pilot of Mercury, was present at the departure from Southampton today of the first flying boat to carry air mails to Australia without surcharge. He was warmly congratulated on his successful flight by the High Commissioner for Australia (S.M. Bruce). The chairman of Imperial

Airways (Sir John Reith) said: 'You have done a splendid job. I think you ought to take the day off.'[26]

This suggestion would have been welcomed by Ly. Torix was just two months old, and during the past few weeks she had endured Don wrestling with *Mercury*'s technical issues, the first Composite separations, and then the transatlantic flight itself.

It was not just the public and the politicians who recognised the magnitude of the achievement, his peers did too. In England, the Honourable Company of Air Pilots incorporating Air Navigators, awarded Don its Johnston Memorial Trophy. Since its inception in 1931, previous winners of the annual award included Francis Chichester, Bert Hinkler, Jim Mollison, Jean Batten and, most recently, Arthur Wilcockson.

Back in Don's home country, the Oswald Watt Gold Medal had been awarded annually since 1921 by the Royal Federation of Aero Clubs of Australia for 'a most brilliant performance in the air or the most notable contribution by an Australian or in Australia'. It now went to Don for the *Mercury* flight. Bert Hinkler and Charles Kingsford Smith, two of Don's boyhood heroes, had shared it four times each between 1927 and 1934. Just four years later, Don had officially joined them in the pantheon of Australian pioneer aviators.

117 West Street, Toowoomba, Queensland. Bennett family home and birthplace of Don on 14 September 1910. (Queensland Air Museum)

Bert Hinkler and his Avro Avian (G-EBOV) in Queen Street, Brisbane, March 1928. G.T. Bennett and Co. had its real estate offices in Queen Street. (Wikimedia)

De Havilland DH.60X Cirrus Moth (A7–9, later VH-UAO). Don performed his Moth test with Squadron Leader George Jones in this aircraft at RAAF Point Cook on 30 November 1930. (Civil Aviation Historical Society, Barry Maclean Collection, via Phil Vabre)

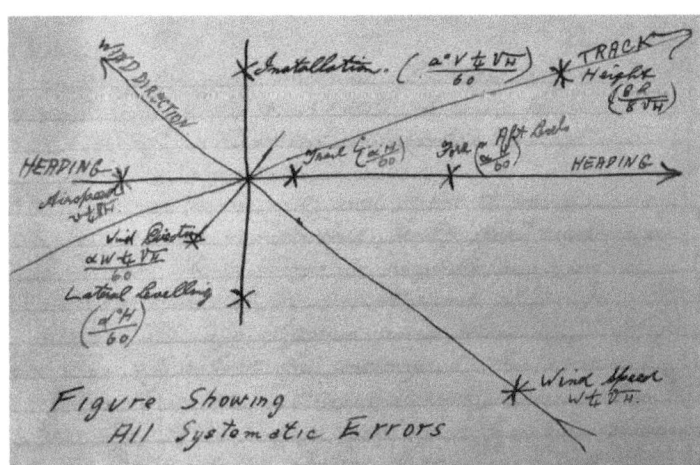

Don's drawing of bombing parameters for use with a Course Setting Bomb Sight Mk.IIB. Taken from RAAF Point Cook lecture notes, 1931. (Queensland Air Museum)

Don after the awarding of his wings, aged 20, June 1931.
(Queensland Air Museum)

Armstrong Whitworth Siskin IIIA (J8959). One of several Siskins Don flew at RAF North Weald, September 1931 – August 1932. (Bill Pippin Collection, via 1000AircraftPhotos.com)

Supermarine Southamptons Mk.IIs (S1233 and S1234), School of Naval Co-operation and Aerial Navigation, RAF Calshot. Don says S1234 was his favourite Southampton. (Wikimedia)

Acoustic mirror at Denge, one of many built along the English coast. A predecessor of radar, constructed by the Air Defence Experimental Establishment (ADEE) to pick up the sound of approaching aircraft. (Ian Castle)

Don, 'Gill' Saye and 'Crackers' Carey at RAF Calshot. A rare photo, taken from Don's logbook. (Queensland Air Museum)

Supermarine Southampton Mk.II (S1124), the aircraft flown by Squadron Leader Frank Lang and Don on the night of 1 June 1933. (P.H.T. Green Collection, via John Evans, Pembroke Dock Heritage Centre)

De Havilland DH.53 Humming Bird (G-EBXN), Don's first private aircraft. Photo taken on 26 April 1936 at Squires Gate, Blackpool, when owned by Mr John Gillett of Preston, who had purchased it from Don the previous July. (via Stuart McKay, de Havilland Moth Club UK, Richard Riding Collection)

De Havilland DH.53 Humming Bird (G-EBXH) in flight. (via Stuart McKay, de Havilland Moth Club UK)

Jersey Airways' de Havilland DH.84 Dragons on the beach at St Aubin, Jersey, 1934. Note the bus used to deliver passengers to the aircraft.
(Dick Flute, UK Airfield Guide)

Lockheed Vega DL-1A (G-ABGK) flown by Jimmy Woods and Don in the Centenary Air Race. (Peter de Jong Collection, via AirHistory.net)

The Vega was in no position to continue the race.
(James Woods Collection, via Ron Cuskelly)

Ly Bennett. (Queensland Air Museum)

Jimmy Woods standing outside the Vega following the crash at Aleppo, Syria, on 21 October 1934. Photo taken by Don. (James Woods Collection, via Ron Cuskelly)

Imperial Airways' Handley Page H.P.42 *Hadrian* (G-AAUE) at Basra, Iraq, May 1936. Don flew this aircraft on Imperial's eastern route out of Alexandria. (Frederick G. Clapp Collection, University of Wisconsin – Milwaukee Libraries, via AirHistory.net)

Short Bros S.8 Calcutta (G-EBVH) *City of Alexandria* (Wikipedia)

Short Bros S.17 Kent *Scipio* (G-ABFA) at Imperial Airways' Alexandria base. Arthur Wilcockson crashed this aircraft into Mirabella harbour, Crete, on 22 August 1936, killing two. (US Library of Congress)

Short Bros S.23 Empire *Cassiopeia* (G-ADUX), Don's first command in Empire flying boats, commencing 14 February 1937. (Spirit Aerosystems Belfast, via Paul Sheehan)

The crews of the Mayo Composite with its inventor Robert H. Mayo. Arthur Wilcockson is standing to the right of Mayo with Don to the left along with his radio operator, Jimmy Coster. (Foynes Flying Boat & Maritime Museum)

The Mayo Composite uncoupled: S.20 *Mercury* (G-ADHJ) and S.21 *Maia* (G-ADHK), July 1938. (Foynes Flying Boat & Maritime Museum)

Oblique view of Mayo Composite coupled at Rochester
(Short Bros Archives via Paul Sheehan)

An excellent view of the Mayo Composite from the front. (Irish Independent Archives (Media Huis Ireland))

Don and Jimmy Coster, *Mercury*'s first crew. (DCT Bennett, *Pathfinder*)

Mercury being hoisted on to *Maia* at Foynes on 19 July 1938 prior to the world record transatlantic flight. (Foynes Flying Boat & Maritime Museum)

The Mayo Composite crews prior to the world-record transatlantic flight, Foynes, 20 July 1938. Don and Jimmy Coster are on either side of Arthur Wilcockson, *Maia's* pilot. (Irish Independent Archives (Media Huis Ireland))

Mercury and *Maia* separate in flight, Rochester, 1938. (Alamy)

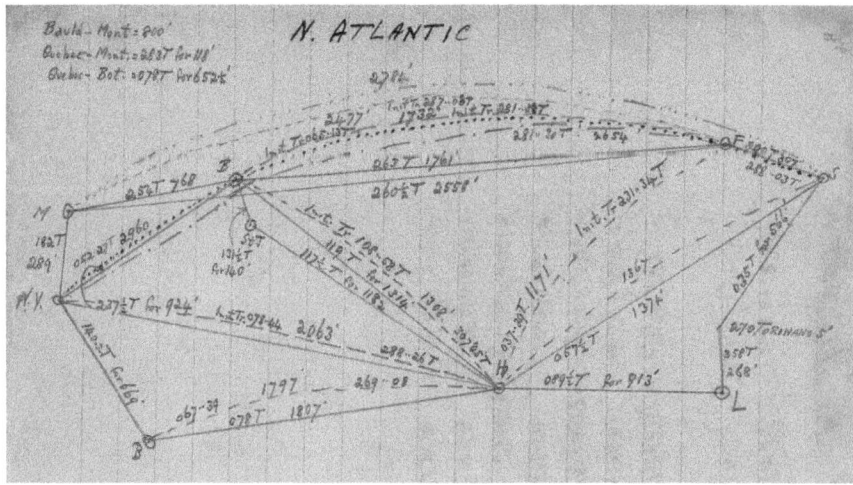

Don's hand-drawn navigation chart in his Imperial Airways private notebook for the world-record transatlantic flight on 20–21 July 1938. (Queensland Air Museum)

A rare photograph of the Mayo Composite crews at Dundee in October 1938 just prior to the non-stop flight to Orange River, South Africa, to establish the world distance seaplane record. Don and Ian Harvey in the middle, Arthur Wilcockson to Don's right. (K. Munson, via Paul Sheehan)

Public demonstration of in-flight refuelling on 30 July 1939. Don and crew in *Cabot* (G-AFCU), with a Handley Page Harrow of Alan Cobham's Flight Refuelling Limited. (via Paul Sheehan)

The Willingdon Bridge at the northern end of the Akra Reach, Hooghly River, Calcutta. Don allegedly flew under it in an Empire flying boat. The Imperial Airways base is in the foreground adjacent to the first span. (Alamy)

Lockheed Hudson Mk.III (BW769), repainted as T9422, the Hudson flown by Don on the first transatlantic flight with the Atlantic Ferry organisation on 10–11 November 1940. On display at the North Atlantic Aviation Museum, Gander, Newfoundland. (Sandra Seaward, NAAM)

General Władysław Sikorski, Polish Prime Minister and Commander-in-Chief of the Polish forces, who Don flew back into France to rescue his General Staff.
(US Library of Congress)

The crews of the seven Hudsons making the inaugural transatlantic flight of the Atlantic Ferry organisation, 10–11 November 1940. Don is on the far right.
(DCT Bennett, *Pathfinder*)

Consolidated LB-30A Liberator B.Mk.I (AM258), the first production Liberator. Flown by Don at San Diego on 13 February 1941, just four weeks after its initial flight. He flew it across the Atlantic on 4 May, inaugurating the Return Ferry Service. (Ron Dubar, Aeroplane Photo Supply, via 1000aircraftphotos.com)

Section of No 77 Squadron photograph, January 1942. Don centre, second front row. To his right is Wing Commander John Embling, then Squadron Leader DFEC 'Dixie' Dean. To Don's left is Squadron Leader George Seymour-Price. (Paul Markham, 77 Squadron Association, held at Yorkshire Air Museum)

Tirpitz with camouflage netting in Fættenfjord. (Public domain)

The tail section of Don's crashed Halifax Mk.II 'B for Baker' (W1041) in Norway.
Photograph is by his rear gunner, HG 'Mick' How.
(via Linzee Duncan, archieraf.co.uk)

The Lady Constance Malleson.
(McMaster University Library, Hamilton, Ontario, Canada)

Don in Sweden, May 1942. (Queensland Air Museum)

17

Mercury and an Obsession with Records

Don's flight in *Mercury*, and Imperial's focus on the Atlantic generally, were part of broader developments in aviation. Due to Hitler's aggressive militarisation and an increasing sense of an impending war, the Germans were drawing attention.

Like the English, the Germans were very interested in flying to the Americas. In 1935, Dornier had introduced the Do 18, a flying boat that could be catapulted from seaplane tenders.[1] In September 1936, a Luft Hansa Do 18 had flown from Horta in the Azores to New York, a distance of 2,270 miles, in 22 hours and 15 minutes. This had been repeated.

Luft Hansa followed this on 27–29 March 1938 with a non-stop flight by a Do 18 (D-ANHR), launched from a catapult on SS *Westfalen*, from Start Point in Devon to Caravelas in Brazil. Despite failing to reach its destination of Rio de Janeiro, it claims to have established the straight-line distance seaplane record of at least 5,214 nautical miles in 43 hours.[2]

Four months later, Don was in New York at a Pan American Airways dinner celebrating *Mercury*'s transatlantic flight when he found in the room, to his great surprise, Captain Joachim Blankenberg and his crew of the Dornier Do 18 *Nordmeer*, who had flown in from Azores after a 17-hour, 42-minute flight. Mooring opposite *Mercury*, Blankenberg had alighted to announce the Germans had in mind a regular air-mail service across the Atlantic:

> This is the third time we have come to New York and we now feel ready to fly here regularly. We hope the day is not far off when we shall be able to carry mail.[3]

Now, at the dinner, Don was reminded about the Dornier's world seaplane record in March. He hinted at being annoyed, perhaps even offended, by Pan Am's judgement in inviting the Germans. He had, like many in Britain, been thoroughly disturbed by Hitler's increasing militarisation. He determined to have a go at the record, 'the "powers-that-be" permitting'.

The opportunity presented itself immediately on his return to Southampton when he attended Imperial's ceremony inaugurating the first non-surcharged air-mail service to Australia. Launching the service was the Secretary of State for Air, Sir Kingsley Wood. Having been lauded at the ceremony, Don seized the opportunity to approach Wood about attempting to beat the German seaplane record.

With Wood favourable, Don discussed it with Mayo to determine if it was feasible. Mayo took it up with Imperial's management; the government added its approval.

The timing coincided with yet another German aviation record and propaganda victory. On 10–11 August, the Germans flew a prototype four-engine airliner, the Focke-Wulf Fw 200 Condor *Brandenburg* (D-ACON), from Berlin to New York non-stop. Despite bad weather, it completed the flight in 24 hours and 56 minutes, the first four-engine land plane to achieve the crossing. The crew reduced the time by five hours on the return flight two days later. They were congratulated by Hitler on their return.[4]

The Fw 200 Condor was designed originally for long-distance flights from Europe to South America, with specifications that equalled or bettered Imperial's flying boats and their Pan Am counterparts. The strategic shift was in it being a land-based aircraft; the Germans betting the proliferation of runways would render flying boats obsolete in international passenger and freight transport.

At Imperial, rather than Don's idea of attempting the world long-distance seaplane record, flown from Southampton to Cape Town, they would attempt the absolute distance record owing to *Mercury*'s better-than-expected fuel-consumption figures across the Atlantic. This record – 6,305 miles (10,148 km) – had been set by a Russian crew in a Tupolev ANT-25 just over a year earlier. Mayo decided the attempt should be overseen by *Fédération Aéronautique Internationale* (FAI), the world governing body that set the rules for verifying record-breaking flights.

Mercury and an Obsession with Records

They postponed *Mercury*'s second flight across the Atlantic on 17 August, and Don flew to Shorts at Rochester for modifications. The fuel tank capacity was doubled with additional tanks fitted in the floats. New electric pumps were installed to pump the fuel into the wing tank, along with a small wobble hand pump as backup.

> Before they departed Hythe [*sic*], Bennett, and his engineer, Geoff Wells carried out the pre-flight inspection. All appeared well, until Bennett saw fluid dripping from a wing; he sampled the fluid but only Wells could see one of the handling party relieving himself from the cockpit.[5]

Maia and *Mercury* flew separately to the River Tay, Dundee, on 21 September. Given the length of flight, a second pilot was needed, so 'Faithful' Coster was replaced by First Officer Ian Harvey, who would also be radio operator.

Initially, the launch was set for 23 September. The first hint that the weather was to become a major problem occurred on that day as *Mercury* was being craned onto *Maia*. Hit by a sudden gust of wind while hovering on a rope over *Maia*, *Mercury* rotated and a wing nearly hit the crane while the mechanics scrambled to control it. The launch was postponed.

It was just the start. The weather blew from the south for two weeks, making a launch impossible. Then the Air Ministry put the attempt on hold due to uncertainties leading up to Prime Minister Chamberlain's meeting with Hitler in Munich. With the Munich Agreement signed on 30 September, the attempt was finally given the all clear.

By 4 October, the winds had reached gale force, gusting to 80 mph, making even mooring on the Tay difficult, but as they abated the following day, launch was set for the 6th. Despite *Mercury*'s greatly increased weight (about a third heavier than the transatlantic flight),[6] no thought was given to a trial separation; *Maia* taxied out and they got underway having calculated the higher speeds required for take-off and separation. As Don said, '… there was therefore something of an element of doubt about the proceedings.'[7]

All went to plan right up to a successful separation when the possibility of breaking world records was immediately put in jeopardy. A cowling came loose and tore off one of the engine nacelles, increasing drag. Not yet knowing what had occurred, Don found his pre-flight calculations for power required, fuel mix and fuel consumed rapidly becoming obsolete. He was wrestling with maintaining an optimum airspeed. It took him an hour before he reached an altitude where he felt able to switch to the weaker fuel mix, though he would have to continue at full throttle.

At this point, he discovered the cowling problem. He considered aborting the flight, but the two weeks of delays due to the weather, a political situation that might prevent a second attempt, and having to fly for at least 12 hours if he jettisoned enough fuel to reach landing weight, all convinced him to press on.

He had plotted a route based on the so-called Constant Course, 'a theoretical optimum for long-distance flights … [which he derived from] calculating the mean forecast winds for the whole of the journey'.[8] This took him across Britain, France and the Western Mediterranean, before crossing the Algerian coast at Bougie (now Béjaïa).[9] Having navigated the Tell Atlas at the eastern end of the Atlas Mountains, he would continue south-south-east across the vast Sahara Desert in first Algeria and then Niger. Over Nigeria, he would pass to the east of Kano, heading for Bafia in Cameroon, the closest he would get to the Gulf of Guinea. From there he would maintain a direct course for Cape Town, through Gabon, touching on the western edges of the Belgian Congo, before entering Angola, then South West Africa (now Namibia). At times he would be brushing the coast but without ever flying over the ocean except on approach to Cape Town. Aside from passing over the English Channel and the Western Mediterranean, *Mercury* would be flying entirely over land.

Every hour, Don gave Ian Harvey the position for transmitting to Imperial headquarters, so the flight could be tracked. It was originally planned that every four hours Don would add to the signal a 'code group indicating our estimate of progress and chances of success'.[10] This confidential code was a masterpiece of efficiency. It consisted of just five letters. They were as follows:

> 1st letter: *State of Fuel Left in the Tank*: 'A' was 'as estimated'; 'B' to 'M' indicated 'less than estimated' in 10 gallon increments; 'N' to 'Y' indicated 'more than estimated' in 10 gallon increments.

2nd letter: *Time Taken*: 'A' was 'as estimated'; 'B' to 'N' indicated 'more than estimated' in 10 minute increments; 'N' to 'Y' indicated 'less than estimated' in 10 minute increments.

3rd letter: *State of Engines*: 'A' was 'All O.K.'; 'B' to 'Z' addressed a plethora of potential problems on an engine-specific basis such as vibrating, missing, losing power, oil pressure issues, and failing completely.

4th letter: *State of Aircraft*: 'A' was 'All O.K.'; 'B' to 'M' addressed potential issues with the Sperry autopilot, the fuel and oil pumps, the wireless equipment, fuel gauges, etc.

5th letter: *General*: 'A' was 'All O.K.'; 'B' to 'Y' addressed issues such as altitude variations, weather conditions being encountered, his estimate of the chances of reaching Cape Town, along with a list of places he might aim for should Cape Town not be possible.[11]

If everything was going according to plan, Don would send 'AAAAA' every four hours from 4:00 pm (GMT) on the first day. His memoirs indicate he sent them every hour, a decision he probably took after separation. Either way, 'AAAAA' was never sent. Don noted five particular codes he did send. The first of these, just over ten hours into the trip, was 'HAAAX'. Decoded back at base, Mayo, Wilcockson and the team read: 'We have 70 gallons less than estimated in the fuel tanks; the time taken is as estimated; engines OK; state of the aircraft OK; unlikely to reach Cape Town.'

The code offered a good summary, but the lack of detail was problematic for those on the ground; they were left to make assumptions. Don was a master of the fuel system, as demonstrated in the Atlantic crossing, so if the engines were fine and *Mercury* was still on time, why were the fuel tanks down this much so early in the flight?

At 3:00 am, he sent the code 'HBAAX', indicating they were now ten minutes behind estimated time. Don said these were among a number of pessimistic codes that:

… apparently caused no end of panic in London, and Bob Mayo and Wilkie rushed back to Headquarters, Imperial Airways, and had in mind sending me a signal ordering me to return. By morning, however, my reports were fortunately much more favourable.[12]

This may have been true, but eight-and-a-half hours later, at 11:32 am, the code 'HCADY' indicated *Mercury* was now 20 minutes behind schedule ('C'), the 'electric fuel pump was u/s [unserviceable]' ('D'), and Don had shifted the possibility of Cape Town from 'unlikely' to 'highly improbable' ('Y').

What this did not reveal was both electric fuel pumps had failed during the first night as they flew across the Sahara. Ian Harvey had tried everything to get them working again but without success, leaving the only option of the wobble hand pump. It was an exhausting exercise to pump the estimated 1,400 gallons from the tanks in the floats up into the wing tank while at 12,000 to 13,000 feet, so they took turns. When Don was pumping, he relied on the Sperry autopilot, leaving an increasingly tired Harvey in the cockpit to monitor for anything untoward.

The pump failure was not happening in isolation. Although they had experienced good weather over the Sahara, the weather turned on the afternoon of the second day as they flew into the equatorial areas of sub-Saharan Africa. They encountered 'very heavy turbulence and torrential rain'.[13] While this posed no immediate threat to *Mercury*, it made pumping more strenuous and tiring. Don now confronted the possibility of having insufficient fuel in the wing tank to make it through the second night.

At 6:54 pm, he sent a disturbing code: 'ZXAAY'. The state of fuel was 'unknown', reflecting his candid assessment he had no idea how much fuel they could pump up into the wing. Despite the time taken now being 20 minutes less than expected, he still thought making Cape Town was 'highly improbable'.

As *Mercury* had to land on water, he was forced to consider turning west to make a night landing somewhere along the Gulf of Guinea, but 'there were no seaplane stations down the whole of that coast'.[14] Still, he had to plan for dire circumstances, with his maps showing he considered Douala in Cameroon, and Libreville and Port-Gentil in Gabon.

It meant re-doubling their efforts to pump at least sufficient fuel into the wing tank to last until dawn when he could at least see his chosen landing spot. The situation became so serious that Harvey started hallucinating, including seeing another person on board. Don put it down to fatigue, the result of altitude, fumes and relentless exertion in rough weather.

Despite this, they succeeded in pumping enough fuel not only to make it through the night but for Don to be confident of continuing the journey. That said, he was certain the overall fuel situation put Cape Town out of reach. Much earlier, he had decided they could not break the absolute distance record, but the long-distance seaplane record, his original goal, was well within reach. In fact, he wanted to beat the old record by a considerable margin. It was just a matter of calculating where he could reach with the remaining fuel.

In pre-flight planning, they had considered three safe-harbour possibilities. The best-case scenario was Saldanha Bay, just over 50 miles from Cape Town. Failing that, they would aim for the mouth of the Orange River, which marked the border between South Africa and South West Africa, preferably alighting on the South African side. Finally, there was Luderitz Bay, much further up the coast in South West Africa, some 520 miles short of Cape Town.

All three offered sheltered waters and were incorporated into the fifth letter of the code group. The team in England knew *Mercury* wasn't going to make Cape Town and were awaiting a code with a fifth letter announcing Don's goal: Saldanha Bay ('O'), Orange River ('P') or Luderitz Bay ('Q').

As dawn broke on the third day, *Mercury* was 80 miles inland of Walvis Bay,[15] itself some 250 miles north of Luderitz Bay. The wobble pump was sucking air, but they still pumped for every last drop of fuel. Based on *Mercury*'s position, clearing weather and the remaining fuel, Don ruled out Saldanha Bay, setting course for Alexander Bay at the mouth of the Orange River.

Right at the last, Don considered passing over Orange River and carrying on to Port Nolloth, nearly 50 miles further down the coast. It had not been considered as an option during pre-flight planning because it was open to the sea, producing a difficult swell for *Mercury*. He had a map of Port Nolloth among his many maps on board. It showed an additional hazard of a multitude of reefs and shoals immediately offshore.

A significant problem was the map itself. Originally drawn up in 1871 by the Admiralty, it had undergone only 'small corrections' over the years, looking little changed from the original.[16] This was part of a broader challenge Don had during this flight. Many of his more detailed maps of specific African localities, like Port Nolloth, were relatively primitive.

It was best not to take his chances and he finally called it quits. 'We had broken the world's long-distance seaplane record, we were at the South African border, and there seemed little justification for risking danger to the aircraft …'[17]

They set down on Alexander Bay, Orange River, after 42 hours and 26 minutes non-stop flight. Alighting on the South African side avoided the headline, 'Scotland to South West Africa.'

In due course, the FAI certified the flight as the world record, with an entry in its official records:[18]

Class C-2: Seaplanes

Capt. D.C.T. Bennett and First-Officer I. Harvey, pilots, in seaplane Short-Mayo "Mercury", four Napier motors of 370 hp, from Dundee (Scotland) to the mouth of the Orange River, near Port Nolloth (South Africa) on 6–8 October 1938. 9,652 kms.

Don recorded it as 5,208 nautical miles, that is, 9,645 kms. It is intriguing that it was awarded the international long-distance seaplane record given the Dornier Do 18 flight seven months earlier claimed to have flown somewhere between 5,214 and 5,278 nautical miles. It appears the German flight was not sanctioned by the FAI, explaining the discrepancy in the recorded distances it flew. It proved Mayo's wisdom in ensuring *Mercury's* flight had such FAI oversight. Rather than compare himself to the Germans, Don was able to compare his achievement with the standing FAI record: '… we had flown just on six thousand miles against the previous existing record of just over four thousand miles.'[19] The record still stands.

After refuelling with the assistance of workers at the Orange River Diamond Mine, they took off for Cape Town where they received a rapturous welcome, including an official mayoral reception. The celebratory nature of the next two days was only sullied by a radio interview on the first night where his performance reflected no sleep for 60 hours!

Without *Maia*, *Mercury* had to return to England under its own steam. They flew first to Imperial's terminus at Durban for the engineers to perform some maintenance. Once again, they were given a big welcome. They then proceeded up Imperial's African route to Alex and then across the Mediterranean back to England, landing at Southampton on 20 October, a leisurely six days after departing Durban.

After the reception they had received in New York on the previous flight, then similar at Cape Town and Durban on this one, Don was nonplussed and ultimately angry by what happened on their arrival home:

> After this flight, which was one of the major world air records, I received no official reception, no trace of recognition of any sort from the Government and no celebrations from Imperial Airways, from Short Bros., the makers of the aircraft, or from Napiers, the makers of the engine. I wonder if such a flight could have occurred in any other country without at least some form of celebration on the return?[20]

It says something of how Don's sense of self had been bolstered in recent times by his achievements in *Mercury*. Nonetheless, the failure to fete him for the achievement was surprising given both Imperial and the Air Ministry had been involved in the preparations. Don hinted the response was a cultural one, and that may have played its part, but the real reasons probably lay elsewhere.

Mayo had taken *Mercury*'s Atlantic crossing in July as the Composite's proof of concept and begun talking in August about a land-based version. Come autumn, Imperial was hoping the government would order a fleet for a mail-only transatlantic service. Part of the plan was to increase mail capacity by using the new technology of airgraphs (microfilmed letters weighed much less), but the Post Office objected on the basis it breached privacy. No order was forthcoming, with the Air Ministry continuing to harbour reservations about the Composite concept.

Seven weeks after the record flight, the Mayo Composite was thrown a lifeline. With the launch of the non-surcharged mail service to Australasia in late July 1938, Imperial was to receive a government subsidy of £750,000 per annum for delivering the mail, lasting for three years, after which it would progressively decrease. The volume of mail now soared. Unfortunately (or fortunately depending on how you looked at it), so did the number of passengers. Due to volume irregularities in the mail, Imperial now encountered a multitude of scheduled flights unable to carry the demand for both mail and passengers. With the subsidy being for mail, passengers were being bumped off flights. This was causing bad press and a threatened exodus of disgruntled passengers.

With Christmas approaching – the period of the year with the highest volumes of mail – the problem was only going to get worse. It was decided to deploy *Mercury* on the Southampton to Alex run, and potentially through to Karachi, to reduce the demands on the Empires. Here was Don's opportunity to once again step into the limelight.

Mid-afternoon on 29 November, *Mercury* was launched from *Maia* over Southampton loaded with 2,180 lbs (nearly one ton) of mail, arriving at Alex just 14 hours and 40 minutes later after a non-stop night flight. It was the first commercial non-stop flight from the UK to Egypt; another feather Don could put in his cap. He repeated the trip on 12 December.

This seemed to generate hope within Imperial. Mayo proposed to the Air Ministry a land-based version launched from modified Armstrong Ensigns with a small fleet of *Mercurys* costing £200,000.[21]

While the future of the Mayo Composite lay in the hands of the Air Ministry, Don was not giving up. In fact, he had grand plans. On 2 January 1939, he submitted a memo, 'Long Range Routes', canvassing six route options for breaking the absolute distance record. Top of the list was South Africa to Australia.

> This route, on account of the Westerlies of the Roaring Forties, gives the best opportunity of breaking the Long Distance record. Topographically it is good, being completely free from high mountains. The most suitable time for the flight would be the immediate future.[22]

Mercury would be launched from *Maia* at Cape Town, not Durban, and he had in mind five destination options once he reached the Australian coast

– Fremantle, Adelaide, Melbourne, Sydney and, at the furthest distance, his home town of Brisbane, a distance of 6,302 nautical miles (7,257 miles). His map shows a flight path crossing the Kerguelen Islands deep in the southern Indian Ocean and closer to Antarctica than either Africa or Australia.[23]

He canvassed other options: Southampton to the Philippines; Southampton to Borneo in the East Indies (now Indonesia); Southampton to South America, with a range of destinations from Rio de Janeiro to San Antonio in Argentina; Southampton to Vancouver; and Alex to the East, including non-stop to Darwin.

When Don took *Mercury* through its Certificate of Accreditation on 7 February 1939, no decision had been made. He flew it again two days later to run some tests, after which *Mercury* stayed on the ground.

In support of his plans, he commenced bench tests on 100-octane fuel, discovering it 'greatly improved fuel consumption [when combined with] a much more advanced ignition setting'. Results showed a percentage improvement for different settings of anywhere between 5 and 10.4%.[24]

On 7 March, he attached these fuel test results to a follow-up memo, entitled 'Proposed Route for Long Distance Test Flight', which revealed he had reduced the options to two. The memo carried a hint of desperation as he spent more time arguing the attractiveness of each option to Imperial commercially rather than just the demands of the flight itself.

His preferred option now was Alex to the East, with two possible destinations in mind: Koepang in Indonesia (5,904 nautical miles) and Darwin in Australia (6,342 nautical miles). Winds, he argued, were favourable, with the return trip over territory thoroughly familiar to Imperial, making organisation easier. On the downside, it would have to start from Alex, and the last part of the outward journey was over territory with no suitable base should problems eventuate.

The 'favourable winds' argument came with a bit of a hurry along: '… this statement is only true if the trip was to be undertaken in the next three weeks. I can see no reason why this would not be possible if an early decision is obtained.' The timeframe hints at frustration.

The other option was Southampton to South America. He offered no specific destination, instead choosing to focus on selling its benefits to both Imperial and the Air Ministry:

> The greatest advantage in this route lies in the fact that it starts from our own base at Southampton and should cause absolutely no delay in organisation. A second advantage is that both from an Air Ministry and, probably, I assume from the company's point of view it might provide an economy to combine with a long range trip a survey of a projected route. All the likely bases on the east coast of South America could be visited at the end of the long range trip. A large variety of bases along this coast makes it possible to terminate the trip due to lack of fuel at almost any distance.[25]

This option, too, had the hint of desperation. Even if Mayo was in favour, the proposed routes made no sense commercially. Don was not proposing further developments to the mail-only flights of the Christmas period; this was solely about claiming the absolute distance record. Imperial's focus was now totally on in-flight refuelling.

Even if Don had generated some interest at headquarters, the Air Ministry was having none of it. In March, it delivered an unqualified 'no' to Mayo's proposal for a fleet of land-based composites. In doing so, it effectively signed *Mercury*'s death warrant from a commercial perspective:

> They argued that [the Composite] would soon be outclassed by the competition, it was not sufficiently advanced, its usefulness was impaired as it did not carry passengers, and production would affect other contracts being fulfilled at Shorts Bros.[26]

Truth be known, ever since the Sassoon Committee of 1935 had expressed its doubts, the Air Ministry had never fully committed to the Mayo Composite.

Even at the death, Don was not going to give up; he decided on one final push. In a conversation with Mayo on the evening of 23 March, they discussed his draft proposal for a long-distance test flight from Southampton to South America. All previous options had been eliminated. Don's preferred final destination was Buenos Aires.

His draft proposal gave the primary purpose of the flight as a test for *Mercury*'s use of 100-octane fuel.[27] Secondarily, it would be a flight of 'considerable political significance', while also being of value should

Imperial consider establishing routes to South America. To this end, he now submitted a detailed return route that would survey seaplane bases up the east coast of South America to the Caribbean.

Tucked in at the end of his list of reasons for justifying the flight was his suggestion that 'it would provide an opportunity of extending [*Mercury*'s] seaplane distance record'. A handwritten amendment to the draft, not in Don's handwriting and most likely Mayo's, made the point the Germans had already attempted to beat *Mercury*'s record, without success.

The final memo Don sent to Mayo on 24 March contained no mention of seaplane records or the Germans, aside from a passing reference to Luft Hansa, along with Air France and Pan Am, those airlines already engaged with the South American routes. Instead, Don focused on testing *Mercury*'s range on 100-octane fuel and, more particularly, seeing it as a survey exercise that would also 'provide a valuable opportunity for "showing the flag" in South America and the West Indies'.

His conclusion was abrupt and, in being so, revealed his palpable annoyance at whoever or whatever was getting in his way. 'From a meteorological point of view it is recommended that the flight should be undertaken immediately. There seems to be no reason from the point of view of organisation why this could not be done.'[28]

While Mayo may have been receptive, Imperial Airways and the Air Ministry had moved on.

Don chose not to address these late developments in his memoirs, perhaps reflecting his frustration and deep disappointment that his plans for further daring deeds in *Mercury* had come to nought. He neatly segued, commenting that the Atlantic in winter prevented further flights by *Mercury* and, apart from flying a few scheduled trips in the Empires:

> ... I carried out development work, for I was still officially the development pilot for the company, and as such I had a good deal of testing of various sorts to do.[29]

For the most part, his 'development work' was introducing into service the very aircraft that had rendered *Mercury* redundant – the S.30 Empires. Don was given responsibility for conducting delivery flights from Shorts, performing tests, doing night-flying trials, and instructing and testing

Imperial's S.23 captains (including, no doubt to Don's delight, Howard Alger, who had beaten him to the command of an S.23 back in the Alex days).

On 5 May, he showed off the S.30 with three flypasts at the opening of the new airport at Guernsey by the Minister for Air, Sir Kingsley Wood.

The real action, however, was working with Alan Cobham's Flight Refuelling Limited on in-flight refuelling for the transatlantic service. Four of the S.30s were fitted out for this purpose, including installing extra fuel tanks.[30] The plan was for Cobham to deploy his modified Handley Page Harrows to both sides of the Atlantic – Ireland and Newfoundland – from where they would refuel the outgoing and incoming S.30s, greatly increasing their range.

Through the early months of 1939, Imperial's dates for launching a transatlantic service kept slipping due to problems with the in-flight refuelling but also, among other things, questions about the suitability of the S.30's Bristol Perseus XIIC engines for flying the Atlantic.

This latter issue saw pilots Gordon Store, 'Jack' Kelly-Rogers and Don, and crews, being given the job of flying *Connemara* (G-AFCW) around and around the UK to accumulate 400 hours on the engines so they could then be stripped to assess suitability. In-flight refuelling was not part of the exercise. They got underway in the second last week of May, Don doing 11 flights of differing lengths before the whole exercise came to a screeching halt on 19 June when *Connemara* was destroyed, ironically, in an on-water refuelling mishap, killing one and badly injuring a number of others.

By late June, Shorts announced the hull strengthening and in-flight refuelling system was ready for the first of the Atlantic-designated S.30s – *Caribou* (G-AFCV) – and on behalf of Imperial, Don headed for Rochester where, on 3 July, he flew on the very first S.30 in-flight refuelling exercise with Shorts' Chief Test Pilot, John Lankester Parker.[31] He sketched a detailed schematic of the in-flight refuelling system in his notebook to gain a clear image in his mind of how it all worked, along with another of the S.30 fuselage and where it had been strengthened to carry the additional fuel load.[32] A week later, he and his crew had *Caribou* to themselves at Hythe, and in-flight refuelled to the maximum 53,000 lbs.

With in-flight refuelling now performing to Imperial's satisfaction, Don did a press demonstration in *Cabot* (G-AFCU) on 30 July, which was captured on film for posterity.[33]

The scene was set for the launch of Imperial's first scheduled transatlantic service. Don recalled:

> In August 1939, I set out from Southampton and Jack Kelly-Rogers set out from New York on the inaugural two-way regular service of the Atlantic, which Imperial Airways (now BOAC) had laid on for that summer.[34]

This, unfortunately, conflates two separate episodes. Don had, in fact, been passed over for the inaugural flight from Foynes to New York, via Botwood and Montreal, in favour of Kelly-Rogers. Kelly-Rogers and crew departed Foynes on 5 August in *Caribou*, carrying 1,055 lbs of mail. The in-flight refuelling after take-off proceeded with one minor issue, watched on by a coterie of dignitaries. The crossing was achieved without incident and Kelly-Rogers and crew were met by the now obligatory gaggle of press on arrival in New York. On 9 August, *Caribou* made the return journey.

Don captained the second scheduled flight on 12 August in *Cabot*. The flight would have garnered little attention except it carried two birds no longer able to fly of their own accord. It was the 'Glorious Twelfth', the opening day of the English and Scottish seasons for shooting red grouse, and early that morning on Helmsley Moor in Yorkshire, a keen shot had dispatched a brace of grouse that were hurriedly flown to London before arriving in Southampton by train in time for loading aboard *Cabot*.

They were a present from Imperial Airways for President Franklin Roosevelt and his wife, Eleanor. If the flight went according to plan, the presidential couple would partake of English grouse a month ahead of the start of the American season. There was one small logistical complication: the president was vacationing aboard the cruiser USS *Tuscaloosa*.

By the time Don arrived in New York after 5:00 pm on 13 August, *Cabot* had been aloft for 25 hours, plus stops at Botwood and Montreal. The grouse were now loaded aboard an internal flight to Boston, then transferred to the president's plane along with his daily mail for conveying to their final destination.³⁵ The condition of the grouse and their consequent reception by the presidential couple is not known.

It was a novel story that generated public interest but, in truth, Don's flight, along with *Caribou*'s a week earlier, had been consigned to sideshow alley by none other than the Americans themselves. Back on 20 May, while Imperial was still wrestling with technical issues, the Americans stole a march by launching their own scheduled transatlantic service using Pan Am's Boeing 314 Clipper flying boats. It was a mail-only service, but on 28 June, Pan Am introduced a passenger service, conveying 22 passengers via the Azores.

Not only that, by mid-August 1939, the world's attention was well and truly elsewhere.

18
War

At 5:34 pm on 3 September 1939, Don took off from Foynes in *Caribou* on a scheduled flight to North America. All aboard was normal, including the in-flight refuelling off the coast. They would be chasing the setting sun for the first part of the trip, with most of the near 16-hour flight in darkness.

The wider world was anything but normal. Earlier that day, Don and Ly were driving to the base at Southampton for the short hop across to Foynes when they heard the broadcast by Prime Minister Chamberlain declaring war on Germany. They both experienced a sense of foreboding, but he saw no reason to cancel, especially as he would be flying west. The crew was similarly troubled by developments. Once again, the decision was made to proceed.

At 7:40 pm the passenger liner SS *Athenia* was torpedoed by *U-30* south of the Rockall outcrop in the North Atlantic and shortly after issued a distress call that she was sinking. The call was picked up by *Caribou*'s radio operator and passed urgently to Don. Checking their position, some 150 miles from *Athenia*, his initial thought was to head for the ship to assist in the rescue efforts but realised quickly they could do nothing in the darkness. 'We relayed her messages and tried to be useful, but in fact were relatively impotent in this calamity.'[1] Of the 1,418 on board, 117 died.

After returning from the Americas on 8 September, Don was in limbo for nearly two weeks. He had not flown any of Imperial's scheduled routes since January, having spent most of the year working on delivering the new S.30s or engaging in the transatlantic venture. What to do with him?

It was decided to deploy him in *Mercury* on 'special duties', his first direct contribution to the war effort. He hadn't flown *Mercury* since early February but now returned to Hythe and, after a week of testing, flew to Foynes to commence anti-submarine patrols.

He began looking for U-boat activity off the west coast of Ireland on 29 September 1939. Being a clandestine venture, each of these flights was recorded in his logbook as 'Test flight'. He ranged up and down the coast on flights typically lasting between three and four hours, though on consecutive days – 13 and 14 October – he flew for over nine. By coincidence, on 14 October, Günther Prien penetrated Scapa Flow in *U-47* and sank the veteran battleship HMS *Royal Oak*.

Don was only on the job for a week when he was recalled to Poole to report in. 'I used to be required to meet in a pub in the West End of London or some country inn … to meet someone from naval intelligence.'[2] It was all 'cloak and dagger' and 'story-book stuff' which bemused him because he didn't think it fooled anybody.

After a further week of patrols, he was once again recalled, but this time was switched off *Mercury* and onto *Cameronian*, an S.23 Empire. Back at Foynes on 30 October, he commenced a series of mostly shorter flights, only to be withdrawn again less than two weeks later, signalling the end of his sub-spotting career. He was not unhappy with the outcome because he summed it up as 'not very worthwhile'. Aside from running some performance tests on *Cameronian*, including flying on the two starboard engines, nothing happened and it bored him.

As was the case for the next four months, putting his patience threshold under considerable pressure. He did little but test components of Empires in for maintenance, ferry aircraft, perform Certificate of Accreditation tests and do a bit of instructing. He was neither at the front line of the war nor at the front line of Imperial's operations.

In these final few months of 1939, with his career in stasis, Don turned his restlessness elsewhere.

The Aeronautical Society of Great Britain had formed in January 1866, becoming the Royal Aeronautical Society (RAeS) in 1918. Its stated objective was 'for the advancement of Aerial Navigation and for Observations in Aerology connected therewith'.[3] From its inception, its principal interest was heavier-than-air flight. Over the years, pioneers such as Sir Hiram Maxim, Lawrence Hargrave, Percy Pilcher and Graham Bell were members. Its membership included inventors, engineers, scientists and meteorologists.

After 1910, the membership grades of Fellow and Associate Fellow were created. With an eye on ensuring the maintenance of high professional standards, examinations for Associate Fellow were introduced in 1922. Applicants were also required to nominate referees (typically more than two) who could attest to their suitability for membership. 'Especially during that period, Fellowship was pretty hard to get, though it did not come with a citation.'[4]

Don sat the exams and was elected an Associate Fellow on 30 November 1939. It is not known who his referees were.

This was a different form of recognition from the Johnston Memorial Trophy and the Oswald Watt Gold Medal awarded by his peers for the transatlantic flight in *Mercury*. The RAeS comprised people whose ethos was one of fostering exploration and innovation in aeronautics through advances in different scientific disciplines. Membership carried gravitas and allowed him to mix with a range of likeminded people.

It is worth noting that, from 1927, the RAeS incorporated the Institution of Aeronautical Engineers, no doubt just one of the attractants for Don.

The biggest development during the 'Phoney War' period was that Imperial Airways ceased flying. The British Overseas Airways Corporation (BOAC), an amalgamation of Imperial Airways Ltd and British Airways Ltd, was the brainchild of Sir John Reith. After 16 years as chair of the BBC, he had commenced as the chair of Imperial on 4 July 1938, the end result of government discussion following the Cadman report of four months earlier, which had been critical of Imperial's leadership, especially that of managing director, Woods Humphery.⁵ Imperial's view was that Reith was imposed on them. He was regarded as one of the country's top bureaucrats:

> It was his belief (and his alone, it would appear) that Imperial Airways and British Airways should be amalgamated into one world-wide company. He also believed that such a company should be free of any private share-holding and should be run, like the BBC, as a public Corporation. Service to the public rather than profits and dividends should be the driving force, thus laying down the ethos which ... would hinder BOAC for the next 20 years and beyond.⁶

The BOAC Bill was tabled in parliament in June 1939. BOAC was established on 24 November, with operations set to commence on 1 April 1940. 'And on this date BOAC became the sole chosen instrument for international overseas routes and entirely at the disposal of the Secretary of State for Air.'⁷ A critical development was the formulation of a 'War Book' outlining policies, plans and priorities for the airline during wartime. This would directly impact Don's last flights for BOAC.

His response to the amalgamation was thoroughly predictable, particularly strident and somewhat one-eyed. Don believed Imperial was a going concern (albeit with a 'small subsidy'), whereas British Airways was not. In his view, the British Airways leadership had 'personal animosity' towards Imperial, preferred foreign aircraft, and exercised undue influence over the government of the day. They secured the key management positions and took 'complete control' in an exercise of 'unhappy political viciousness'. The consequence of these machinations, according to Don, was that the leading lights of Imperial – Woods Humphery, Burchall, Mayo and Brackley – received no recognition for their efforts:

> ... the heads of Imperial Airways had done great things, not only for all British people but also for world civil aviation. It is as deplorable as it is typical that the country of their birth gave them no honour.⁸

The merits or otherwise of his case (expressed in these terms in 1958) are not the issue here; it is how this outburst encapsulates his thinking. He regarded 'nationalisation' and 'government control' as stalking horses for bureaucracy, with all its labyrinthine processes and political interference. It put an end to a leadership which had largely allowed him to flourish.

His comments about lack of recognition for his overseers at Imperial reflect another quite personal element of his make-up. Public recognition had become an important reward benchmark for him, especially for those engaged in 'grand pioneering work' like Imperial's leaders. In his eyes, Britain had a cultural bias against publicly recognising and rewarding its pioneers, as he himself had experienced with the failure to acknowledge his world-beating achievements in *Mercury*. (He sets aside his dissatisfaction at Imperial's leadership itself for not feting him after his record flight to South Africa.)

He chose not to see the overall picture of Imperial's performance vis-à-vis the competition, which led to his curious and lasting addiction to British-made aircraft. He criticised British Airways for using 'almost entirely foreign aircraft – Dutch, German, American'.[9] By implication, they lacked the requisite patriotic spirit, one even more necessary in the context of war. For someone so dedicated to technological advances in aviation, it is odd that his stance didn't create in him any intellectual dissonance. Did he really believe British aviation was at the forefront of aircraft development internationally? Did he not see that foreign aircraft may be the best strategic and financial choice for an airline?

Don's final commercial flight with BOAC proved eventful. He departed Poole on 12 May 1940 in the Qantas-registered *Cooee* (VH-ABF) through to Karachi. He was, at last, on his first (and ultimately only) trip to Singapore. After a four-day layover at Karachi, he was switched to *Canopus*, arriving in Singapore on 26 May.

Like all Australians who have flown the 'Kangaroo Route' from England, Singapore must have felt like 'almost home', a sense enhanced for Don given it had taken two weeks to get there and having flown *Cooee* part of the way. He had a six-day layover to savour the experience.

There were other highlights. Flying both east and west, the Empires stopped over at Calcutta, alighting on the Hooghly River. Adjacent to BOAC's base at Bally and at the northern end of the 2.5-mile landing strip – Akra Reach – was the Willingdon Bridge. One of Don's biographers, Alan Bramson, relates the following:

> It was a hot, windless day and the water was like glass, clinging to the hull of the flying boat and refusing to yield the great aluminium bird to the sky. After a long, protracted run across the water they finally unstuck to be confronted by the Willingdon Bridge, a large and remarkably solid looking edifice. There was not enough room to climb over it – so Don flew underneath.[10]

Phillip Sims, in his history of the Empire flying boats, offers a complementary piece:

> Calcutta was a difficult alighting as the Akra Reach was narrow and the spans of the Willingdon Bridge dominated the approach. It is said that on one occasion Captain D.C.T. Bennett (to become AVM 'Pathfinder' Bennett) flew under the bridge.[11]

Neither author provides a source for the story. The accounts are problematic. Don only went through Calcutta on this one trip to Singapore so, if it did occur, it was on either 25 May or 4 June 1940.

BOAC's pilots knew the Akra Reach, immediately south of the Willingdon Bridge, was challenging.

> The Hooghly was treacherous at the best of times. The fairway, always full of debris, was surrounded by small craft, fringed with large dockside cranes and bounded by the Willingdon Bridge.[12]

Immediately north of the Willingdon was worse: the river narrowed and turned, creating a multitude of additional challenges for any pilot attempting to take off under the bridge.

This, of course, assumed you could get an Empire under the Willingdon on take-off in the first place. Constructed in seven 350-foot spans of equal height resting on eight piers, the bridge offered an approximate 29-foot clearance at the high-tide mark, increasing to 45 feet at low tide.[13] With the Empire's tail approximately 25 feet above the waterline, clearance was somewhere between four and 20 feet provided, of course, the Empire was still on the water as it passed underneath. Bramson's account indicates Don was negotiating that space just after becoming airborne.

Things were further complicated by the Indian Ports Act specifying all water traffic heading north was required to pass underneath the single span between piers Nos 3 and 4. This included seaplanes which, according to definitions in the Act, included flying boats (presumably when taxiing, if required).[14]

The Hooghly River itself posed additional problems. During Calcutta's monsoon season (May to September), water flows increased markedly, creating issues with currents, changes to the riverbed, and increased debris. Choosing to take off to the north, against the current, increased the risk of hitting debris at speed. Several Empires had been significantly damaged at various ports by hitting items in the water. If, as Bramson suggests, the weather was calm, prudence suggested taking off to the south, away from the Willingdon, with a long clear run and moving with the current, thus reducing the risk of debris damage.

It is difficult to see why Don would attempt a take-off to the north, unless he was unconcerned about the debris and felt he had given himself sufficient runway to clear the Willingdon. It was only when it was too late to abort the take-off that he exercised the one remaining option!

Don, himself, made no mention of the event in his memoirs. Why subject it to such intense scrutiny? Does its veracity matter? A mythology has grown around Air Vice-Marshal Don Bennett, created by those who have come to see him as performing feats no one else could do, or would try, based on his having exceptional capabilities well beyond mere mortals. Stories about Don have been told and retold without question, accepted as true, and even embellished in the re-telling. It is the way of myth making, but it does Don no service if fact crosses into fiction because it impedes a true understanding of the man and his capabilities.

In this case, despite its utter improbability, it is impossible to prove Don did *not* fly under the Willingdon Bridge, either by design or necessity. If he did, it is reminiscent of Frank Stuart observing Don take off from the Wiltshire paddock in the Moth on 5 May 1934 and facing a clump of high elms: 'My heart stood still as Bennett somehow banked around them. I do not think a dozen men living could have made that take-off that day.'¹⁵

It was on the return journey to Southampton that Don got his first taste of the war close up. From Alex, he was captaining *Cooee* again for the Mediterranean leg back to England. He recorded flying into Rome on the day Mussolini declared war. It was, in fact, the day before – Sunday, 9 June 1940 – but no less dramatic for being so.

Germany had commenced its attack on the Netherlands, Belgium, Luxembourg and France 30 days earlier, making stunning progress. The evacuation from Dunkirk took place between 26 May and 4 June, with the Germans commencing Operation *Fall Rot* ('Case Red') on the 5th to take the remainder of France. Within days, Paris was in danger of falling.

Despite having signed a pact with Hitler a year earlier, Mussolini was yet to declare war. His armed forces were still grossly unprepared. Yet the pace of the German advance caused him considerable alarm. Having engaged in 'warlike propaganda' for many years, Mussolini saw himself being sidelined and humiliated. Count Ciano, his foreign minister, recorded, 'He wants to create enough claims to be entitled to his share of the spoils. He wants his part of the booty.'¹⁶ Mussolini's decision to declare war on France was imminent, but as late as Saturday, 8 June, the Italian public was still unsure it would happen. The whole country was on edge.

It was into this maelstrom Don arrived in Brindisi from Athens around 8:20 am on Sunday the 9th to the news that, given conditions in Rome, it may not be safe to alight there. He spent much of the 47 minutes on the ground talking to BOAC's Rome manager, during which time they agreed he would not only go to Rome but evacuate as many British staff as possible.

Don had to calculate very quickly how many passengers and their freight he could uplift, while also considering the risk to the passengers and freight he already had on board. The Rome manager's task was much more complex. BOAC's base was at Lake Bracciano, an hour outside the city. Don's scheduled flight time from Brindisi meant he was due into Lake Bracciano around 11:30 am. The turnaround would be quick. For staff, aside from any personal possessions, it was a matter of packing up BOAC's office as best they could.[17]

Don set *Cooee* down on Lake Bracciano at 11:29 am. While refuelling occurred, designated personnel from the Rome office clambered aboard with their luggage, and he took off for Marseilles at 12:10 pm. The Germans had been bombing Marseilles for days, so the turnaround there was achieved in 50 minutes and they headed for Saint-Nazaire on the Bay of Biscay for the night. They arrived back at Hythe the following morning. Mussolini declared war. Paris fell four days later.

It was BOAC's last scheduled commercial flight on the Mediterranean route. Kelly-Rogers in *Clyde*, tasked with returning Air Chief Marshal Sir Robert Brooke-Popham and his staff to England urgently, was in Malta on the 11th, bypassing Italy altogether. They bypassed Marseilles too, flying directly to the French flying boat base, Les Hourtiquets, at Lake Biscarosse.

Biscarosse would be the scene of Don's next, and even closer, encounter with the war.

In the meantime, he had a rather melancholic duty to perform: handing over his beloved *Mercury*. He had not flown her since performing the anti-submarine patrols off the coast of Ireland in late September and October. He had returned her to Hythe on 24 October and she hadn't flown since.

Now, on 17 June, he took her up at Hythe for 25 minutes to test everything and the following morning flew her across to Felixstowe.

> On arrival at Felixstowe I handed *Mercury* over to a Dutch Float seaplane unit, who were going to use her for reconnaissance work. It was a sad goodbye, for *Mercury* was very much a part of me in many ways. I had been her only operating pilot.[18]

He was there the following day when he got the phone call to get to Poole post haste.

19

Rescuing the Polish General Staff

Don's stealth operation to rescue the Polish General Staff in France from the fast-approaching Germans on 20–21 June 1940 was highly classified. Even BOAC's flight records offer no indication of his destination or the purpose of the mission.

The complete story of what happened during those 24 hours will probably remain forever a mystery. There are multiple accounts, some containing embellishments in unnecessary attempts to increase the sense of drama. Don himself was somewhat muted about the event in his memoirs.[1] Somewhere between the lavish and the sparse is the truth.

Most accounts pay no heed to context, yet it is the German invasion of France, the catalyst for the mission, that makes the story most compelling. The rescue marked the culmination of a week of extraordinary events, boasting a fascinating cast of characters, with everything moving at breakneck speed. It has been described as audacious. Of that there is no doubt.

On 14 June, the German Army entered Paris. The French government had retreated to Tours and then Bordeaux. The Germans maintained their blistering pace across northern and eastern France. They took Orleans on the 16th, then Dijon and Le Mans on the 18th, and were on the outskirts of Brest. The situation was so fluid it was nigh impossible to locate the front line. Panic was everywhere. The roads were chaotically clogged with fleeing civilians and retreating Allied forces.

Rescuing the Polish General Staff

Following the evacuation at Dunkirk, Operation *Ariel* began on 15 June: a large-scale evacuation by sea from France's western ports of British, Canadian, Czech and Polish units, along with various dignitaries and civilians. Within days, Cherbourg, Saint-Malo and Brest had fallen, focusing the exodus on the south-western ports along the Bay of Biscay, including those in and around Bordeaux.

The following day, the French Prime Minister, Paul Reynaud, presiding over a Cabinet split on whether France should continue to fight, resigned. Taking his place was Marshal Pétain, leader of those in favour of an armistice. Pétain approached the Germans, setting in train the drawing up of terms. The German armed forces maintained pressure on the French by keeping up the pace.

The Polish government-in-exile had fled Paris and was in the vicinity of Bordeaux. It was headed by Prime Minister and Commander-in-Chief of the Polish forces, General Władysław Sikorski. On the 17th, while Pétain was making overtures to the Germans, Sikorski issued a declaration that the Poles intended to continue fighting.

At this juncture, the seeds for Don's mission were planted by Dr Józef Retinger. He had known Sikorski for many years. On becoming prime minister in October 1939, Sikorski had offered Retinger a job:

> Retinger's status and functions were left vague. For lack of a better definition he was called advisor to the Council of Ministers, though he continued to live in London and not in Paris where the Government was. He mainly concerned himself with contacts, propaganda, external relations of all sorts.[2]

In May 1940, just after the Germans commenced their western offensive, Retinger visited Sikorski in Paris. He persuaded him it was in his interests to develop a personal relationship with Churchill. Now, in mid-June, as the catastrophe unfolded in France, Retinger, back in London, was starved of information about Sikorski, his entourage, and the Polish forces generally. He approached the Air Ministry to fly him to Bordeaux to find Sikorski and determine what was going on. It is most likely permission was granted due to the Polish ambassador, Count Raczyński, who was actively seeking Churchill's intervention in the Polish forces issue.

Late on 17 June, Retinger was told to be at Blackbushe Airport for a 4:00 am flight. The Lockheed Hudson (N7318) of No 206 Squadron was under the command of Flight Lieutenant William Biddle who had been awarded a Distinguished Flying Cross just four days earlier and was coming straight off a night operation to fly Retinger to Bordeaux.[3] Retinger recalled that Biddle was asleep for most of the time on the ground in Bordeaux!

On arriving in Bordeaux early on the 18th, Retinger found the place in total chaos. Everyone was rushing around yet paralysed, hoping for evacuation by sea. Nobody knew anything and official communication channels had broken down. By sheer luck, he got a tip Sikorski was at Libourne, some 20 miles away. This, apparently, was where the Polish government had set up its headquarters, though no one knew for certain. He and Biddle bribed a French driver and set off, navigating along roads congested with thousands of people with their cars, barrows, bicycles, and horse-drawn carts, or just shuffling along on foot, weighed down with their worldly possessions.

At Libourne, they found Sikorski at the Federal Government Office at 8 Avenue de Verdun. Sikorski was as surprised to see Retinger as Retinger was to find him:

> General Sikorski asked me what I had come for. I told him I had a plane and that I was going to take him to London that evening to arrange the evacuation of our forces to England. He replied without hesitation, that he would come back with me on two conditions: that I could assure him he would be back within two days and that he would see the British Prime Minister the next day.[4]

Retinger records that, having struck a deal, they 'went to lunch at a small hotel, where about twenty-five of General Sikorski's nearest collaborators were present'.[5] Sikorski, Count Tyszkiewicz (his adjutant), Retinger and Biddle then drove back to Bordeaux, where Retinger observed that panic had increased still further during the day. They took off just after 6:00 pm, with Biddle flying over occupied France. It was well after 10:00 pm when Churchill was advised Sikorski was in London.

Sikorski met Churchill at midday on the 19th,[6] after which he announced by radio the transfer of the Polish government to London while

Rescuing the Polish General Staff

also directing the widely dispersed Polish forces to move towards south-west France for possible evacuation.[7]

Due to Ambassador Raczyński's efforts, Churchill had agreed to a flying boat returning Sikorski to France to evacuate key members of his General Staff.[8] BOAC had commenced operations on 1 April 1940 and was 'entirely at the disposal of the Secretary of State for Air'.[9] The 'War Book' had established seven priorities for BOAC, including: '2. Would carry important loads such as passengers and freight at the instructions of the Secretary of State.'[10]

BOAC was now presented with the problem of finding, at ultra-short notice, an aircraft that could reach the flying boat base at Lake Biscarrosse to the south-west of Bordeaux, without passing over the French mainland, and then return without refuelling. The only suitable Empires were the S.30s with their larger fuel tanks and, of those, only *Cathay* was available.

The 'War Book' had provided for the possibility of BOAC moving from Southampton, with one potential base being Pembroke Dock.[11] BOAC had begun utilising it on 14 June. *Cathay* had only just arrived there from Poole on the afternoon of the 18th. It now did an immediate about turn, returning to Poole the following morning in preparation for the operation.

Meanwhile, BOAC had to find a suitable crew. Some versions of events indicate Don volunteered. Don himself simply recorded, 'I was given the job.'[12] Coverdale offers the most detailed account, allegedly quoting Don:

> On the 19th June 1940, Wilkie [Wilcockson] phoned me at Felixstowe, telling me to meet him at B.O.A.C. sea flying boat base, Poole, Dorset. There he told me I had been selected for a very special job if I wanted it, and I said, 'Oh yes, what is it?' Wilkie replied, 'There isn't anyone else for the job. The powers that be believe you are the best qualified.'[13]

Don's experience on S.30s and his navigational prowess would have played a part in his selection but, in reality, many senior BOAC captains met these requirements. The most likely 'qualification' was Don's ability to assess situations quickly and decisively. This was essential because he had just hours to be briefed, gather a crew and map out the flight. Beyond that was his appetite for risk or, put another way, his willingness to push the occasional boundary should he deem it necessary.[14] Such was the level of risk in this instance, some at BOAC harboured the view *Cathay* would not return. It would not stretch the imagination to think Don relished the opportunity to prove them wrong.

Whether he received sufficient information to make an adequate assessment of the risk is not known. The fluidity of the situation on the ground suggests no briefing, no matter how detailed, was ever going to be adequate. It is highly unlikely he would have said no, short of confirmation the Germans had occupied Bordeaux and the surrounding countryside. His reputation is reflected in Sims's recording the formation of the crew:

> The most audacious evacuation was that flown by, who else but, D.C.T. Bennett. An impossible task was set before BOAC, the rescue of the Polish general staff from under the noses of the Germans … With his apologetic smile, Bennett bumped into Morgan in the corridor and asked him to join the crew of First Officer Tommy Farnsworth, Radio Officer Jimmy Armitage and Steward W. 'Dinty' Moore.[15]

Bill Morgan was the purser, though whether he was recruited by Don with an 'apologetic smile' must be the subject of considerable doubt. With no disrespect to Morgan, Tommy Farnsworth was the more important choice. He was Don's regular first officer for the transatlantic flights.

On the morning of Thursday 20 June, Sikorski and Tyszkiewicz didn't commence the trip down to Poole until 6:45 am due to a last-minute meeting with Admiral Świrski in London. They were further delayed because accompanying them was one Major Peter Wilkinson, who had been given incorrect instructions about where to find *Cathay*.

Wilkinson, who would end up heading the Special Operations Executive in 1943, had been based in Paris since early April with Number 4 Military Mission, a clandestine group liaising with the Polish General Staff on mobilising and equipping the Polish underground in Poland and Czechoslovakia.[16] He knew Sikorski well. After Paris fell, he had returned to England essentially to warn the War Office of potentially 30,000 Polish troops in France who would need evacuation if the country fell. He found the War Office preoccupied with the evacuation of British troops. 'Nobody knew what was happening, or what to do, or really, frankly, *cared* very much about the Poles.'[17] He headed for the Foreign Office where confusion also reigned.

Wilkinson regretted not going to the Polish Embassy first because he soon discovered Count Raczyński was already operating at prime ministerial level. Sikorski was in England and Raczyński had secured the flying boat to return him to France. Sikorski now personally invited Wilkinson to join him and Tyszkiewicz on the rescue mission due to his on-the-ground knowledge of the Polish forces, and he could act as an interpreter.

Their late arrival at Poole meant being greeted by 'a pretty irate captain' because they had 'nearly missed the tide'.[18] *Cathay* didn't depart from Poole until 9:29 am. Don made no mention of Wilkinson, perhaps due to his intelligence role, though the Air Ministry account records there being a 'British liaison officer' aboard.[19]

As they were taxiing out for take-off, Sikorski summonsed Wilkinson to the back of the aircraft and asked him to pass on a message: 'Tell the pilot I have complete confidence in him.' He thought it a very uncharacteristic remark by Sikorski but made his way to the cockpit shortly after take-off and delivered it. 'And, I must say, what the pilot said about the General had better not be recorded! I took back a suitable expression of mutual esteem.'

Also on board was a French colonel, whom Don was told 'was anxious to search for his family'.[20] This was most likely a cover. Evidence suggests he was Colonel Rozoy, one of three French military representatives to the Supreme War Council. He was attached to RAF Headquarters representing the French Air Ministry.[21] He had been in a meeting between Sikorski and de Gaulle at 6:00 pm the previous evening, suggesting he was now under instructions from de Gaulle to take a message back to the French government at Bordeaux.[22]

It was a four-hour flight, taking them out over the Atlantic to avoid the German fighters at Brest, before turning south-east into the Bay of Biscay. Wilkinson had been napping when *Cathay* suddenly banked steeply:

> I went and looked out the window and there was … it must have been a tanker … I mean the smoke-stack was well aft … entirely broken-backed in the middle of the Bay of Biscay with a huge oil slick all around it, and one empty lifeboat which, I must say, had a very ugly red stain, which looked as though it had been bled over to say the least, and shot up.
>
> We circled around this for, I suppose, a quarter of an hour, around the crippled ship. There was absolutely no sign of life on it whatever. … It was a dreadful sight to see a broken-backed ship like that, with a bloody lifeboat – we went down quite low – with bullet holes. I was feeling quite emotional.[23]

They continued on to Les Hourtiquets at Lake Biscarrosse, arriving just after 2:30 pm. An 'extremely grubby' French officer rowed out, offering a curt 'No' to Wilkinson's questions about where the Germans were and, in fact, whether he knew anything at all, so Wilkinson got the officer to row Tyszkiewicz ashore to make some calls. An hour later, a car arrived to pick up Sikorski.

Don had been given 'two conflicting requirements': 'I … had strict orders that I was to do everything possible to assist General Sikorski and his colleagues, but that I was not to lose the boat under any circumstances.'[24]

This played out in blunt fashion during the hour-long wait as Wilkinson acted as intermediary between Sikorski and Don. Don asked Wilkinson to convey to Sikorski that his orders were to take-off should he face the 'slightest risk to the aircraft':

> So I did explain this to the General, and it was not at all well received. The General said, 'But I order him to stay.' This, being translated to the captain of the aircraft, was not all well received either. He said that he had received his orders from the Air Ministry and therefore was in no position to receive orders from General Sikorski. So, hoping to calm the waters a bit, I said rather self-importantly that, in any case, I would be there. Whereupon General Sikorski said rather tartly that he did not care very much whether I was there or not, the important thing was to have the *aircraft* there.[25]

Rescuing the Polish General Staff

It was agreed Wilkinson would stay behind as some sort of assurance to Sikorski, who promptly departed with Tyszkiewicz when the car arrived, each carrying a pistol, the parties having reached a provisional agreement Sikorski would be back by first light. (Rozoy went ashore alone, returning later. His movements remain a mystery.)

Don's first instinct was to check the weather, contacting the Met office at Les Hourtiquets.[26] They told him there were forward columns of Germans in the area. 'Quite frankly, I did not expect to see General Sikorski again.'[27]

The more immediate concern was being spotted from the air. The previous night the Germans had bombed Bordeaux for the first time 'in a move designed to put pressure on the French government to sign the ceasefire'.[28] Now, at 3:40 pm, an air raid warning reached Les Hourtiquets. It apparently emanated from Landes to the north of Bordeaux and sounded for 35 minutes.[29]

Wilkinson recalled an immediate fear Don would take off. They had a brief discussion about that possibility, which would see Wilkinson being sent ashore, before agreeing the best option was to stay put.

Don's logbook offers a different, perhaps complementary, account – 'taxying (air raid warning)' – noted as lasting 15 minutes. He also noted they found a spot where they could get 'close in to the shore and actually aground, so that she was partly hidden by the trees'.[30] It was a deft move because they subsequently heard a German aircraft but weren't spotted.[31] Darkness fell, providing air cover, but they spent a restless night with someone always on watch. Wilkinson recalled: 'At about two or three o'clock in the morning I woke up and heard a sound, which, if you've ever heard it, is unmistakeable, and that is of armoured fighting vehicles on a metalled road.'[32]

Sikorski's trip had taken nearly six hours, arriving at Libourne around 8:15 pm. He immediately went into a meeting with his generals and staff officers. By 11:00 pm, all those designated to return with him to London had commenced the return journey to Biscarrosse, driving through the night.[33]

Back at Biscarrosse, as dawn approached, Sikorski was nowhere to be seen and Don was restless. They rowed ashore, arriving just as the general and his entourage emerged from the gloom in four cars that, as Don recalled, were 'loaded to the brim'. This was no surprise because they had managed to fit in Sikorski, Tyszkiewicz, General Sosnkowski, five Polish colonels, three lieutenant colonels, two majors, and a woman.[34] Some reports suggest she was Sikorski's daughter, Zofia Leśniowska, his personal secretary at the time, but the passenger manifest gives her name as 'Mlle Tomaszkewska'.[35] The manifest also indicates Rozoy was on board, but whether he returned with the Poles cannot be established.

They were transferred two at a time to *Cathay* in 'the greatest silence' and Don took off at 5:47 am, just 30 minutes before sunrise. John Maynard, in his book on Don and the Pathfinders, recorded: 'As Bennett began to climb he saw the first German tanks approaching …'[36] The Air Ministry's account is that *Cathay* '[passed] over a group of German tanks almost immediately she was airborne'.[37] Archie Jackson, one of Don's biographers, claimed: 'Moments after take-off, the crew looked down on the startled faces of a German tank crew refuelling near the Biscarrosse base.'[38] All these appear to carry more than a touch of hyperbole given the early hour, and that *Cathay* took off to the south over sparsely populated land with a single road before immediately turning west and heading off the coast. More credible is French historian, Marie-Paule Vié-Klaze, who claimed Farnsworth 'reports having seen a column of tanks not far from Biscarrosse'.[39] Don corroborates none of them. On the contrary, he commented 'everything seemed peaceful and quiet', undoubtedly to his surprise and relief.

The only hostile action occurred as they passed to the west of the Gironde estuary. They were shot at by the skittish crew of a British warship, prompting Don to comment, 'Fortunately their shooting was more or less of the usual standard, and we were not hit.'[40]

They headed out over the Bay of Biscay, tracking north-west for Ushant, off the coast near Brest. The latter was now in German hands. Don found the journey depressing. They saw two damaged ships and also some lifeboats, one filled with bodies. Near Ushant they observed a column of black smoke rising to about 20,000 feet which Don put down to the French sabotaging their oil refineries.

Rescuing the Polish General Staff

Cathay set down on Poole harbour at 9:40 am on 21 June. 'Nobody there, of course, knew exactly what our job had been, but they seemed reasonably surprised that we had returned!'[41]

That very afternoon, Hitler travelled to Compiegne where armistice discussions began at 3:30 pm. The following day the French signed the armistice in the same railway carriage where the Germans had capitulated at the end of the First World War.

For a week, Don had little to do before an opportunity arose for another trip to the Continent.

The Republic of Portugal had planned an 800th anniversary celebration of the founding of the Portuguese state in 1140 and the 300th anniversary of its restoration of independence from Spain. Held on the banks of the Tagus River at Belem, it was promoted as the Portuguese World Exhibition, a trade fair with exhibits from Portuguese colonies described as 'a shameless piece of propaganda by the relatively new and weak Portuguese "New State" government'.[42] It opened on 23 June.

At the outbreak of war, the Portuguese government had decided to remain neutral. Notwithstanding this, the government confirmed the long-standing Anglo–Portuguese Alliance would remain intact. The British affirmed the decision, it being in their strategic interests for Portugal to remain non-aligned.

The British decided it was a good diplomatic move to send a high-level delegation to the Exhibition. The Duke of Kent, King George's brother, would represent His Majesty, accompanied by two elderly barons of significant military stature – Field Marshal Birdwood and Admiral Chatfield. The latter had been Minister of Coordination of Defence until two months earlier.

Don flew them to Lisbon on 28 June in *Champion*, remaining until the delegation completed its work. He had only one encounter with the 'enemy', a Luft Hansa crew was on the same floor of his hotel.

Late on the night of 1 July, he flew the delegation out of Lisbon back to England. He wasn't to know it was his final flight as a captain at Imperial Airways.

20

Fast Tracking the Ferry

Max Aitken, First Baron Beaverbrook, or the Beaver, as everyone called him for his Canadian origins and unlimited energy, was the sort of person Don admired. Churchill had appointed him as Minister of Aircraft Production in May 1940, removing aircraft supply for the war effort from Air Ministry control. 'There was no time for red tape and circumlocution … [Beaverbrook's] personal force and genius, combined with so much persuasion and contrivance, swept aside many obstacles.'[1]

This applied to establishing the Atlantic Ferry: whatever it took to get results, and quickly. Central was Beaverbrook's persona as 'a rapidly-revolving vortex of charm and tempest'.[2]

Britain's challenges in the summer of 1940 were immense. The Air Ministry had foreseen American aircraft would be needed and from February 1939 one type, the Hudson Mk.I, was being delivered from Lockheed's plant in Burbank, California, for deployment in Coastal Command.

All American aircraft were shipped across the North Atlantic. This meant disassembling and crating them for ocean transport, then reassembling and testing before delivery to RAF squadrons. By mid-1940, a multitude of problems had emerged. Shipping was being attacked by the *Kriegsmarine*, especially U-boats. The increasing demand for war matériel from North America was restricting cargo space for aircraft at a time when demand for them was increasing dramatically. At the same time, with the fall of France, Air Ministry orders for American aircraft skyrocketed, 'representing about 12,000 machines in all, 2,500 of which were taken over from French contracts'.[3] Even if the ships completed the crossing, the process was expensive and took about three months from factory to operational squadron. Flying them across would reduce that to ten days.

The Air Ministry had dismissed flying the aircraft direct citing seemingly insurmountable challenges: the aircraft lacked the range; finding aircrew with the requisite experience and qualifications was impossible; and the North Atlantic was notorious for its weather, in particular in the winter months when it was the most changeable. Fog, storms, rain and winds were challenging at any time, but the key threat in winter was icing on aircraft.

As Don and his colleagues at Imperial Airways knew, there were also navigational challenges passing over a 1,900-mile stretch of water (the shortest route). There were no radio navigation aids unless ships were stationed across the Atlantic to provide position and meteorology updates, which meant relying on basic navigation techniques.

Don recalled someone at Coastal Command describing the idea of flying aircraft direct as 'absolute suicide'. Indeed, Don maintained one of the trenchant critics was Air Chief Marshal Sir Frederick 'Ginger' Bowhill, head of Coastal Command, despite being the principal beneficiary of any scheme should it succeed.[4] When Lockheed's London representative suggested fitting Hudsons with long-range kits, an RAF air marshal had commented it was too expensive and half the aircraft would be lost.[5]

Undoubtedly, the concerns were real and the challenges enormous, but the demand for aircraft, and more quickly, was paramount. What was needed was a small group prepared to address the challenges and bypass any opposition to get the job done.

When it was suggested to Beaverbrook that flying the Atlantic might be possible, he seized upon it.[6] The question was who could take it on. After various discussions, Air Commodore George Pirie, the British air attaché in Washington, contacted George Woods Humphery, the former managing director of Imperial Airways, now based in New York. He believed it was possible if he could 'get sufficient of his old Atlantic team' together and utilise the infrastructure of an existing organisation. He suggested the privately-owned Canadian Pacific Railways (CPR), with whom he had business interests.

CPR was not an unusual choice. Its president, Sir Edward Beatty, had been a part of aviation in Canada for many years. As recently as 1939, he had purchased Canadian Airways and all its affiliated companies and was intent on contributing to the war effort.

What was unusual was that negotiations were conducted without involving the Canadian or British governments, excluding, of course, the fact Beaverbrook was a minister of the crown. He entered discussions 'without seeking the approval of the War Cabinet or consulting with the Air Ministry'.[7] Engaging CPR was attractive because Beatty was an 'old and trusted friend'. For Beatty's part, he entered into negotiations without involving the Canadian Minister of Munitions and Supply, Clarence Howe, whom Don had met during his *Mercury* days.

As a key member of the 'old Atlantic team', Don was called into Beaverbrook's office in Millbank and asked his opinion, and he backed Woods Humphery's assessment. He recalled Trevor Westbrook, formerly works manager at Vickers-Armstrongs' aircraft factory, and Jack Bickell, one of Beaverbrook's key lieutenants, being in attendance. Not surprisingly, they were of a similar ilk to Beaverbrook, meaning Don was in his element. His focus fell on just two issues – range and aircraft icing. He felt both could be overcome.[8]

Beatty agreed to put CPR at the Ministry of Aircraft Production's (MAP) disposal, with an agreement reached on 30 July. It was signed on 16 August by Beatty, Morris Wilson (president of the Royal Bank of Canada and Beaverbrook's MAP representative in Canada and the US) and Woods Humphery. Beatty would create an operating company within CPR – the Air Services Department[9] – to provide the administrative framework. The agreement also dealt with issues of recruitment, salaries, expenses, etc.[10]

Under the agreement, Beatty was company chair, with Woods Humphery as vice-chair, overseeing operations from his New York home but leaving the day-to-day operations to his 'old Atlantic team' in Montreal. In essence, a privately-owned Canadian company was working for the British government. The Canadian government, despite not being a party to the agreement, committed to assist where possible.

'Air Services Department' was never going to catch on, so it quickly came to be referred to internally as the Atlantic Ferry, officially becoming the Atlantic Ferry Organisation (ATFERO) on 1 March 1941. ('CPR' is used to designate the organisation until that point.)

Even while the agreement was being finalised, the main players were already acting based on a 'gentlemen's handshake' arrangement.

In late July, CPR allocated Woods Humphery's 'old Atlantic team' office space above the Windsor Street Railway Station in Montreal. First to arrive from England, on 30 July, were Harold Burchall and Arthur Wilcockson. Burchall, formerly general manager at Imperial Airways, took on this role at CPR, getting the fledgling organisation and administration in place. Wilcockson, as superintendent of operations, oversaw everything regarding aircraft, aircrew and airports.

Two days later, Don, Humphrey Page and Ian Ross arrived from England by ship. Don's role was flying superintendent. Working alongside Wilcockson, he had three specific responsibilities: selecting, testing and preparing the aircraft; aircrew training, with a specific focus on navigation; and leading the first flight across the North Atlantic. Page and Ross, both experienced Imperial Airways pilots, were to assist him.

Added to the team was the vastly experienced pilot Griffith 'Taffy' Powell. He had conducted Don's 'B' Licence endorsement for taking command of *Cassiopeia* back in January 1937. Now a squadron leader and chief navigation officer of Eastern Air Command, Royal Canadian Air Force (RCAF), he was seconded to CPR with responsibility for overseeing the work at Gander airfield in Newfoundland, the launch base for the Atlantic flights.[11]

The final piece of the personnel matrix was John 'Joe' Gilmore. He had joined Imperial Airways as an engineer in 1933 before being appointed to the airline's Atlantic Division in 1938. He had been instrumental in establishing Imperial's flying boat bases at Botwood, Newfoundland, and Boucherville, Montreal. At the latter, he had managed to establish docking, refuelling and servicing arrangements in a matter of weeks before *Mercury*'s arrival in July 1938.[12] He now oversaw all such arrangements at Saint-Hubert airport, Montreal, the centre of operations, in an RCAF hangar shared with No 13 Service Flying Training School of the British

Commonwealth Air Training Plan.[13] Gilmore had a maintenance crew provided by the Canadian government through Trans-Canada Air Lines.

Don arrived in Montreal to find Gilmore already there and was very pleased to see him. '[Gilmore] turned out to be such a pillar of strength on the engineering side in that first year of the Atlantic Ferry ... He knew his job, and did it.'[14] The last sentence was the rarest of all Don's personal references. Their work together on the Hudsons was crucial.

After some rapid planning by the team, Don and Humphrey Page were dispatched to the Lockheed factory in Burbank, California, on 6 August to view the potential aircraft, in particular the Hudson.[15]

Lockheed had committed to deliver 50 of the new Hudson Mk.IIIs between August and October, with the first eight in August, then 20 to 30 in September. The duo arrived to find no aircraft would be ready for delivery until mid-September. Part of the delay was due to the Air Ministry's late decision to install American Bendix radio equipment, while Don and Page also requested the compasses be changed.[16]

Don took two test flights, one in a Hudson Mk.I, which he flew for an hour and 40 minutes, and a Model 18 Lodestar.[17] Don maintains he did an extensive fuel consumption test on the Hudson but, as his logbook shows, this occurred on his second trip to Burbank a month later.[18]

What Lockheed did provide were two test and training aircraft, both Hudson Mk.IIs (T9370 and T9365), to take back to Montreal. On 16 August, they flew out of Burbank, stopping first at Albuquerque, New Mexico, and then Omaha, Nebraska. They now got first-hand experience of the US government's progressive work at diluting the country's Neutrality Act due to the exigencies of the war in Europe.

The Neutrality Act, first passed in 1935, was a product of the American 'isolationist' movement, itself a response to a belief America's involvement in the First World War 'had been driven by bankers and munitions traders with business interests in Europe'.[19] It reflected a general principle in international law that neutral states should not supply war matériel to 'belligerent' nations, requiring US arms manufacturers to obtain an export licence.

It had been progressively amended over four years as war approached, opening up the possibility of sales to US allies with certain conditions – some material, some purely pretence – to preserve the notion of neutrality. One of these was that American pilots were prohibited from flying aircraft into the airspace of a 'belligerent' nation. With Canada classified as 'belligerent', American pilots could not fly the Hudsons across the border.

The solution was one of splitting diplomatic hairs. To meet this requirement, the Americans would land the aircraft on farmland on the north side of Pembina in North Dakota, taxiing up to the Canadian border. From there, the aircraft were hitched to draught horses and dragged across the border to Emerson, Manitoba, then flown to their designated Canadian destination by non-American pilots. One report suggests the manoeuvre went so far as Canadian horses standing on their side of the border with a line taken to the aircraft on the American side, meaning no Americans were hands-on in the handover.

By August 1940, due to the shortage of pilots, American pilots were being allowed, on the quiet, to fly in Canadian airspace, but to preserve the pretence, they did not fly across the border.

On arrival at Pembina, Don was bemused rather than irritated, referring to it as 'this most vital formality'.[20] Once across the border, the American pilots accompanying them hopped aboard and away they went.

Back at the Windsor Street Railway Station and nearby Saint-Hubert airport, things were advancing at pace. Nobody had attempted the North Atlantic after September, so it was preferable to have the first flight depart for Britain as soon as practicable to avoid the worst of the weather. It was now the last week of August.

The Montreal team tackled recruiting and training aircrew, sorting out the route details, and preparing the aircraft as they arrived from Burbank. As a civilian enterprise, they had to recruit civilian aircrew. Although the agreement specified MAP would provide up to 25 first-class pilots, navigators and wireless operators, CPR would need many more recruited from within Canada and the US. The team very quickly utilised existing networks, assisted by a very generous pay scheme authorised by Beaverbrook.[21] By September, they had managed to attract 197 pilots from a wide variety of aviation backgrounds,[22] of whom 44 would ultimately pass the team's qualifications and training thresholds for making Atlantic flights.[23]

Don quickly formed the view they were getting 'rejects', the 'throw-outs' from American airlines. He recounts one such pilot sitting on Wilcockson's desk and saying: 'I'll show you how to fly the Atlantic – I've been thrown out of every airline in America for drunkenness. I'll show you how to fly the Atlantic.'[24] That these recruits were to be paid significantly more than their experienced British counterparts rubbed salt in the wound.

It created a cultural clash for Don who undoubtedly struggled with an unashamed American tendency towards self-promotion tinged with bravado, particularly if he didn't think it was warranted. Carl Christie, Atlantic Ferry historian, recorded that towards the end of Don's time at the Ferry, '… some experienced American pilots had resigned because of the perfectionist Australian's sarcastic comments.'[25]

Attracting suitable navigators was problematic because American aircraft at the time generally flew without them. This left no alternative but to train recruits for a Hudson aircrew comprising just pilot, co-pilot and wireless operator. To get around the glaring deficiency in qualified navigators, the inaugural flight, very much a 'proof of concept' exercise, would comprise eight (ultimately seven) aircraft with Don piloting the lead aircraft and acting as navigator for the entire formation.[26]

Page and Ross did the bulk of the training, with Don handling the more difficult cases. He was across every detail, given the level of risk:

> … he impressed everyone he met with his masterful knowledge of aeroplanes and aviation and his skill as a pilot, and sometimes tried their patience with his obsession for detail. He left nothing to chance. He personally tested the pilots and radio operators selected and trained by his assistants, and drew up cruising cards for the Hudsons calculating the engine setting for each aircraft in fractions of horse-power.[27]

Fast Tracking the Ferry

In a public BBC broadcast 'Calling Australia' on 14 April 1942, Don offered a rare insight into how it had all integrated in his mind around the subject of range:

> The prime and fundamental consideration is the question of range ... [Aircraft] must be working efficiently; there must be no serious decrease of performance due to ice formation; the navigation must be reasonably good, and the head-winds not so severe as to seriously decrease the range ... We had to test the range of each type of aircraft taken over for delivery. We had to train the crews in the correct cruising procedure, so as to get high economy and had to see that the machines were so well serviced that their range was not affected by any inefficiency.[28]

The training regime had his fingerprints all over it. After having their flying credentials and logbooks checked, first priority for potential pilots was a week's training in the British system of navigation, a challenge in itself and barely enough for any navigational scenario, let alone flying the Atlantic. It was, however, the bare minimum back-up plan should the aircraft become separated in flight.

This was followed by a week of technical training, including having to draw the Hudson's fuel and oil systems, chart the hydraulic system, and, ultimately, know 'how to properly operate and make use of every bit of equipment in the aircraft'.[29] The knowledge was bedded down by hands-on instruction in the Saint-Hubert hangars.

Only then did they commence the flying training. They learned the essentials of flying the Hudson, negotiating its peculiarities, then practised flying on instruments alone, performing navigation tests and addressing potential scenarios like having to fly on one engine. They were then classified as pilot, co-pilot or unsuitable.

Wireless operators, many of whom had never flown before, were trained separately, again by experienced BOAC aircrew. They were given some limited instruction on the Hudson itself in case they might be called on to perform in-flight repairs.

By mid-September, Lockheed had the first of the Hudson Mk.IIIs ready and Don returned to Burbank around his 30th birthday. His principal issue was with the change of engine. The training aircraft had the 1,100-hp Wright Cyclone R-1820-G105A engine, running on 91-octane fuel. The Hudson Mk.IIIs had the 1,200-hp R-1820-G205A engine, running on 100-octane fuel.[30]

Don's extensive fuel tests with *Mercury* forewarned him about the impact of different octanes on fuel consumption and he was reticent to accept Lockheed's performance figures. He also now had considerable data on the fuel consumption of the Mk.IIs. So, he demanded a test flight. On 15 September, he took off for a flight lasting just over six hours, taking measurements showing a discrepancy of 8% between the manufacturer's performance figures and his, proving extra fuel would be needed to cross the North Atlantic.[31]

The episode is a quintessential illustration of Don's character and genius. On view is not Don as pilot or navigator, but engineer. He described sharing his post-flight analysis figures with Lockheed's engineers, citing the causes as 'partly air frame, partly engines and partly propellers'.[32] That, in itself, is telling, but what followed is extraordinary:

> The provision of torque meters was a vital factor in this analysis, which would not have been possible otherwise. It was the first occasion on which I had ever been provided with an aircraft to test in which torque meters had been fitted.[33]

Torque meters were being introduced for assessing engine performance by providing a real-time measure of torque.[34] As such, they helped accurately assess fuel consumption. Whether Don knew about them before arriving in Burbank is unclear, but he was able to demonstrate their application to assessing engine performance to a greater degree than Lockheed's engineers! And to the point where they were prepared to modify the aircraft to meet his requirements. He was delighted to find Lockheed had installed an additional fuel tank in the test aircraft by the following morning.

He promptly took it up for another seven hours to retest the figures, and again the day after for another hour. Satisfied, he headed back to

Fast Tracking the Ferry

Montreal, leaving behind the test aircraft (T9422) for final modifications; it would be his choice for the inaugural North Atlantic flight.

Due to ongoing delays at Burbank, it was not until 8 October that the first of the Hudson Mk.IIIs (T9416) arrived. Eight arrived in dribs and drabs with the last arriving as late as 5 November.

Don paid particular attention to icing:

> I took off with Hudsons into the worse icing conditions I could find to test them ... On the airframe icing we were perhaps a bit weak but we had on a chemical called Kilfrost which I had used extensively over Europe.[35]

As for ice accumulating on the propellers, the crews were instructed to over-rev the engines to clear it away.

Complicating matters, the aircraft did not arrive ready to fly the North Atlantic, extra fuel tanks notwithstanding. Gilmore's personal notes show the aircraft had a multitude of problems: 'shutters cracked', 'found metal in oil filter', 'antenna arm slack', 'fuel pressure failing', 'drift sight lined up not finished screws', 'requires metal work'.[36] As something got fixed, something else went wrong. Everything fixed needed testing.

'Taffy' Powell had arrived at Gander on 8 October to make preparations. He had minimal ground staff.[37]

Also there was Dr Patrick McTaggart-Cowan as the chief meteorological officer.[38] He had developed a reputation as the most authoritative person on North Atlantic weather and was known by all as 'MacFog'.[39] '[His] word at Gander was law above the word of air marshals, commanders-in-chief, Prime Ministers and Presidents.'[40] He had been collecting data since mid-1937 and his forecasts would be crucial. Don knew him from *Mercury* days. However, the war had created a specific data-collection problem:

> Before the war, regular reports had been received from the UK, Greenland, Iceland, Europe and a hundred and fifty or more ships

on the North Atlantic. The European reports and those from ships ceased with the outbreak of war. Those from Greenland and Iceland ceased soon after. When the first delivery flights were made, weather charts had no source of data between the United Kingdom and the north east coast of North America.[41]

It was decided to attempt the inaugural crossing on 10 November, but significant issues remained for the ten days prior. An engine on T9419 failed; it was replaced with an engine from T9417 which was removed from the flight, leaving seven aircraft.

As the aircraft arrived progressively at Gander from 29 October, the weather made its own contribution. An early winter snowstorm had left the runway in poor condition. One aircraft made a hard landing, damaging the flaps and elevators.

By 7 November, they had still not finished applying Kilfrost to the aircraft aerials, leading edges of the wings, etc.

The following day, Don took T9422 for a final flight early in the afternoon at Saint-Hubert, returning with carburettor trouble. It was one more thing for Gilmore to fix. His notes suggest some of the mechanics were making mistakes in those final days.

At this juncture, a rational observer may have concluded what Don himself had heard months earlier, that the mission was absolute suicide: the crews were a hotchpotch with little or no experience of oceanic flying, let alone the North Atlantic; training was drastically foreshortened; there was no dedicated navigator for each aircraft; Hudsons seemingly rushed off the production line with a plethora of last-minute problems; the weather deteriorating; and now attempting to cross the North Atlantic at a time of year well past previous limits.

Don would have scoffed at such thinking. When asked in 1976 whether they had any special concerns or worries about this first formation flight, he concluded characteristically:

> I should say there was no anxiety as there would be flying into a target area. It was a purely calculated risk and we knew exactly what we had to do to overcome it. But many of the American pilots just

didn't believe us when we told them that everything was under control.⁴²

It now fell to Beaverbrook to inject a sense of history and occasion. He sent a telegram on 8 November, reinforcing to the first formation of Hudson crews how vital this enterprise was for Britain's survival:

> Day to day we stand at our sentinel posts and at night we change our guard. The enemy may still invade and when the year brings around the longer days, our dangers will be greater. We have one resort only – to equip our men and women, all, to fight for a life without chains in a land without a tyrant. We have asked you for much, but we must ask for more. We will fight with pikes if we need, but we hope for guns and aircraft and ships and tanks.⁴³

Don flew to Gander on 9 November with the express intention of commencing the flight that evening. It was impossible because he arrived to find the Hudsons there heavily iced up. They immediately began to chip the ice off, which Don estimated as half-an-inch thick on the upper surfaces, using 'every type of tool', continuing into the morning of the 10th.

McTaggart-Cowan forecast a front in the second half of the crossing, but his involvement extended far beyond providing the forecast. He had been working very closely with Don in making preparations for the crews to navigate through weather fronts:⁴⁴

> What we did for the pilots that were to follow [Bennett] was a very detailed flight plan with compass headings and everything for each zone. We would try to pick a route that kept them south of any depression, so if the storm was deeper than we anticipated, they would drift south and then drift back north. They never knew they had been off course because it was self-correcting.⁴⁵

His candid admissions reveal senior leadership was confronting the sober realities:

> They were supposed to play follow the leader across the Atlantic. We knew it wouldn't work but the Air Ministry in Britain and Lord Beaverbrook didn't ... We had to get self-correcting conditions because, within the first hour, these were night flights too, they'd be off course; there was just no hope of following the leader.[46]

In effect, from the very first hour of a flight expected to take upwards of ten hours, Don and McTaggart-Cowan were anticipating some aircraft might be on their own, well before hitting any weather front. It all came down to the quality of Don's planning, which was now on view during the pre-flight briefing:

> We saw little of him until 7 pm, when we gathered for briefing in the control tower, then we realised the enormous knowledge of aviation this man had. In the few hours he had been there he had studied the weather maps, forecasts, etc., and made up flight plans for each plane. Cruising cards he had already made up from the hours he had spent on consumption tests. Radio facilities were at his fingertips. In fact there was not one question asked, so thoroughly had he covered every phase of this first flight.[47]

It is a telling observation. Amid the general nervous excitement at being part of a history-making adventure, Don, despite his own reservations, was able to instil confidence through a leadership style incorporating physical presence, breadth and depth of knowledge based in experience, and meticulous preparation. The whole enterprise was cavalier, a challenge to overwhelming conventional wisdom, carrying a litany of risks beyond Don's control, but he almost single-handedly produced in them a belief the naysayers had it wrong.

Recognising that maintaining visual contact might last barely longer than getting off the ground, Don would broadcast his position periodically, however, due to the possibility of transmissions being intercepted by the

Germans, all the aircraft would maintain radio silence east of longitude 20 degrees West (about 400 miles from the Irish coast).

Each crew was given the option of turning back up to four-and-a-half hours into the journey.

With McTaggart-Cowan's forecast to hand, the crews headed out to their Hudsons on the evening of 10 November.

Aside from Don's own account of the flight, three others emerged, one in May 1941 by an unidentified pilot going by the title 'Eye Witness', who captured perfectly the character of the enterprise:

> Comic relief was afforded by the sight of the pilots coming out to enplane. Our wool-lined suits had not yet arrived, so each man was dressed according to his own sartorial tastes. Some of the men wore business suits, others tweed sports jackets! One Australian wore a beaver hat, and many of the men had skiing outfits. We looked as though we were heading for a costume ball instead of a Transatlantic hop.[48]

The seven aircraft took off into the darkness, led by Don at 6:52 pm,[49] heading for Aldergrove, near Belfast, sent on their way by the pipe band of the Queen's Own Rifles of Canada. Don was want to comment: 'The whole navigation depended on me, and I was aware of the fact that with practically no radio aids available, my sextant and myself were rather vital to proceedings.'[50]

Radio operator CM 'Curly' Tripp, who was flying with Ralph Adams (pilot) and Dana Gentry (co-pilot) in T9418, highlighted McTaggart-Cowan's prediction about maintaining visual contact. Each aircraft had station-keeping lights but 'for the first hour there seemed to be planes all around us, and which one was the leader was the question'.[51] They ultimately found their place and the formation proceeded as planned.

The plethora of engineering issues Gilmore had been wrestling with for weeks took little time to surface in the air. On board T9418, they discovered a leak in the starboard oil tank, causing Adams to hold back while he decided whether to continue. When the oil flow slowed, they decided to press on, but by that time they were alone.

Tripp's radio now blew up and burst into flames. 'Ralph hollered to me, "Shut the _____ thing off", but I had beat him to the gun. With that load of gas it isn't pleasant to have fire skipping around the cockpit.'[52] Early the following morning, the indicator on the radio compass broke, leaving them with 'no transmitter or compass and out over the Atlantic Ocean'.[53]

All the crews had been provided with a rubber tube to insert in their mouths for taking in oxygen when at higher altitudes. This produced everything from a foul taste to nausea. 'The taste of the rubber was sickening, and on the first trip I vomited three times.'[54] It was also very cold.

Don said the weather front predicted by McTaggart-Cowan arrived later than expected. Icing became a problem and, while confident in his own capacity to address the issue, he worried for the other crews. He ascended to 22,000 feet in an attempt to climb over the front but failed. He gave the order to separate, in effect telling the other crews they were on their own. Given they were trying to navigate through a North Atlantic storm that would last much of the remainder of the trip, and in the early morning hours, the order was largely superfluous, but it did confirm there would be no attempt to re-establish the formation. They would, however, continue to pick up Don periodically reporting his position, at least until 20 degrees West.

V Edward Smith, the pilot of T9420, was also experiencing icing problems and had ascended to 18,000 feet which helped, but maintaining this height caused altitude sickness, incapacitating the second pilot. He measured the outside temperature at minus 23 degrees Celsius. Like Tripp back in T9418, he was experiencing equipment difficulties. His autopilot, on which he was now relying heavily, appeared to be operating sluggishly, but on disengaging it, he found the problem was low pressure in the hydraulic system which he attributed to 'faulty maintenance' allowing for 'some foreign matter' to get into the system.

Things got a little more fraught for Smith, who was now extremely tired, as T9420 neared the Irish coast on the morning of 11 November. They sighted an island but could not identify it. It was not helped by continuing poor weather. He knew he was low on fuel but the master gauge for his wing tanks was no longer working, so he didn't know just how low.

They spotted Don fly out of a cloudbank nearby only to disappear into another.

Smith decided to break radio silence, twice, resulting in receiving a bearing that got them on track to Aldergrove but produced 'quite a protest in British radio circles'.[55] Don was also unimpressed by Smith hitting the airwaves.

After a flight of ten hours and 17 minutes, Don arrived at Aldergrove first, with the remaining aircraft landing within the hour. Tripp recorded: 'Captain Bennett would not leave the window in the mess until the last plane was in, and I often wondered what his thoughts were as that seventh ship rolled to a stop.'[56]

It had been a resounding success. Given the need for aircraft, Sholto Watt maintains: 'Officials in the United Kingdom were prepared to give up the plan if fewer than three of the seven aircraft arrived safely.'[57] Don had ensured no such calculation was necessary.

Due to security surrounding the flight, the dishevelled crews, with accents from four countries and dressed to match, tried to explain to an undoubtedly quizzical Belfast hotel receptionist that they had just arrived from England. They flew across to the terminus at Speke the following day, where Lockheed had set up a factory to perform modifications on the aircraft and install armament.[58]

Most of the crews were sent to Liverpool to catch a ship back to Canada while Don drove down to London to report to Beaverbrook. While he found Beaverbrook 'was smiles from ear to ear. And how the Beaver can smile!',[59] he had created a problem for himself because he was convinced each aircraft should have a navigator. Beaverbrook was unconvinced; formation flights would continue in the manner just demonstrated.

Don sailed for Canada on 19 November with the five remaining crew members.[60] By his return, the second formation had already successfully crossed the Atlantic on 28/29 November, led by Page. Then Gordon Store, another BOAC pilot, led a third flight on 17/18 December. If the naysayers had been convinced Don's inaugural formation flight was a fluke, there were now 21 aircraft in England, with a 100% success rate. It did not last.

Don led the fourth formation flight in T9465, departing on 28 December. It was Christmas, and T9465, named 'The Spirit of Lockheed and Vega', was a gift to the British from the employees of Lockheed and Vega who paid for it by subscription. Of the seven aircraft, one crashed on take-off without loss of life, blocking another from getting off the ground. A third Hudson returned with engine trouble. T9465 experienced wireless transmitter issues. The flight reflected a certain inevitability; Don's level of planning drastically reduced the riskiness of the venture, but there would always be issues beyond his control.

Nonetheless, there were positives. T9465 had left the Burbank factory on 22 December and was in Britain a week later, a record that brought into stark contrast the speed of these delivery flights versus sending aircraft by sea.

Further, in his obligatory visit to London, Don had finally convinced Beaverbrook each aircraft needed a trained navigator and, with that decision, formation flying came to an end. Navigators trained in Canada as part of the British Commonwealth Air Training Plan would receive specialised training by BOAC navigators on flying the Atlantic.

21

Beaten by Bottlenecks and Bowhill

Don's return to Canada in mid-January 1941 after leading the fourth formation flight precipitated a breakdown of relationships in the original team in Montreal. The cohesion had lasted just five months. It centred on the delivery of the first Consolidated PBY-5 Catalinas, which Coastal Command were keen to acquire due to their long range.

He recalled Beaverbrook showing him a signal from Ottawa that 'Canadian Pacific Air Services were bungling the affairs of delivery, and had been stupid enough to have put a PBY flying boat into Halifax in November'.[1] Stupid, Don said, because seawater around parts of the Canadian coastline could freeze in winter, meaning 'to put a flying boat into Halifax at that time of the year was indeed asking for trouble'. He told Beaverbrook he had issued instructions that the Catalinas were to go via Bermuda, so if they had flown to Halifax, it was against those instructions.

Halifax/Dartmouth, Nova Scotia, was a deep-water port that very rarely froze over, so it made for a suitable staging point for flying the Catalinas across the Atlantic for much of the year. The question was at what point into winter did it become unsuitable.

The first of an order of seven PBY-5 Catalinas (AM264–270) designated for the RAF had started arriving at Boucherville, Montreal, in early December:[2]

> ... the first two Catalinas flew to Dartmouth, Nova Scotia, where they became coated with frozen spray. Their departure for Greenock on the Clyde in Scotland, the British reception base for flying boats, was delayed for more than two weeks while RCAF

[Royal Canadian Air Force] experts devised a way to keep them free from ice during engine run-up and take-off.³

Surprisingly, despite what Don said to Beaverbrook, he had flown a Catalina (AM265) from Port Washington, New York, to Halifax on 6 December, immediately after returning from the first formation flight in the Hudsons.⁴

Sholto Watt confirms that flight, adding it was the last: '... it was found necessary to abandon Halifax as a winter terminal owing to severe icing conditions, especially trying when the flying boats were taxiing on the water ... It had already been decided to deliver Catalinas from Bermuda ...'⁵ This suggests Don had checked conditions at Halifax for himself, resulting in his instructions to utilise Bermuda.

Records show that of the remaining five Catalinas, at least two (AM266 and AM269) flew to the UK from Bermuda, confirming Don's instructions had been obeyed after 6 December. Further, Don was in Canada throughout December until he led the fourth formation flight on 28 December, so it is impossible to see any more Catalinas flying to Halifax in contravention of his instructions.

This would have remained a purely operational matter had two Catalinas stranded in Halifax not become political. Beaverbrook was under pressure to get the aircraft and, when meeting with Don in late December/early January, they were still stuck in Halifax.⁶ It was potentially bad publicity. Don, taken by surprise by Beaverbrook, appears to have omitted to say the Catalinas had arrived there under his watch.

Things deteriorated on his return to Montreal. Sir Edward Beatty, who wanted an investigation, called in Don for an explanation which, when offered, led to Beatty getting into a contretemps with Woods Humphery and Burchall. According to Don, Burchall defended himself by blaming him, an argument which, on the facts, carried some merit. '[He] tried to infer I had let him down ... but nothing I could say could convince him otherwise.'⁷ The exact nature of the disagreement remains clouded.

Leaving aside the issues generated once again by Don's recollection of events in 1958 when writing his memoirs, it offers some insight into a contradiction in Don. His reputation was built on taking responsibility and leading from the front. Why not tell Beaverbrook the Catalinas had gone to Halifax, including one flown by him, only to experience icing difficulties that demanded routing the remaining Catalinas through Bermuda, and this was now underway. Once back in Canada, even if Don felt responsibility for the Halifax issue should be shared, it was easy enough to wear it to placate Beatty and move on – to 'take one for the team' – but he did not seem prepared to do that.

According to Don, the disagreement catalysed the departure of Woods Humphery and Burchall, and the 'old Atlantic team' broke up. It is unlikely, however, their disagreement was the only, even principal, issue. The initial successes had led immediately to a ramping up of pressure to deliver aircraft more frequently. The organisation was 'facing severe growing pains and becoming the target of national and international criticism':[8]

> … with an estimated 674 Hudsons, eighty-three Catalinas, fifty-eight Consolidated B-24 Liberators, and twenty Boeing B-17 Fortresses due to be received by June 1941, it was obvious that the three group leaders, thirty-five pilots, and eleven second pilots on the ferry roster in January 1941 would be hard-pressed to keep pace with the anticipated explosion in American production.[9]

The most acute issue was the difficulty in recruiting pilots, despite the salaries on offer. The great majority of potential civil pilots in Canada and the US were heading for the military or staying with their respective airlines. FJ Hatch, the Canadian military historian, avers Don was part of the problem: '… the majority of [American pilots] who showed an interest in the ferry service did not meet Bennett's minimum standards.'[10]

Sir Henry Self, chairman of the British Air Commission in Washington, made the point to the Ministry of Aircraft Production (MAP) that to overcome the problem 'the RAF must accept responsibility for provision of all aircrews either from home or Canada'.[11] Air Vice-Marshal John Slessor, head of planning at the Air Ministry, visited Canada and the US in December 1940, resulting in a letter to Air Chief Marshal Sir Charles Portal, Chief of the Air Staff: '… we shall want something of the order of at least 1,000 pilots on this job … I know this is a matter for MAP but it has many direct implications for the RAF.'[12]

The Air Ministry was shocked at the scale of the problem and the pace at which it had developed, coming just months after its persistent scepticism had been demolished by Don's initial 'proof of concept' flight. Portal and Sir Archibald Sinclair, Secretary of State for Air, agreed the Air Ministry would ultimately need to wrest control from MAP, but that required planning, so Canadian Pacific Railways (CPR) and MAP would be left in the short term to address the rapidly escalating delivery issues.[13]

At CPR, Beatty replaced Burchall with CH 'Punch' Dickens, who had been the general superintendent of Canadian Airways. Beatty himself then fell ill in March and resigned. Morris Wilson, Beaverbrook's personal representative in North America, took charge of the newly constituted Atlantic Ferry Organisation (ATFERO), allowing greater MAP control. Wilson appointed Harold Long effectively as Woods Humphery's replacement.[14]

While appreciating the organisation needed to expand, (by the end of March 1941 there were 263 'flying, ground and administrative civilian personnel'),[15] Don responded poorly to the changes, despite it signalling Beaverbrook was now largely running the show through Wilson. While some regarded Long, Don's immediate superior, as highly competent, Don thought he was a 'dead loss' and a 'catastrophe'.[16] It was no surprise; anyone perceived to be diluting his control or the centrality of his decision-making, or simply increasing the paperwork, would attract personal criticism. On the other hand, he and Dickens got along well because the latter exhibited 'common sense' and assured Don 'that he would let me get on with my part of the job without any interference'.[17]

Don's perpetual antipathy towards bureaucracy blinded him to the simple fact the ferry's existing administrative structures became obsolete as soon as the initial formation flights succeeded. With American factory output increasing and RAF demand rising, ATFERO was the obvious bottleneck. Don's leadership style – hands-on and, where necessary, micro-managing to surmount specific obstacles to flying the North Atlantic, leading from in the air and not behind a desk – so suited to the initial phase, was simply unsustainable. Overarching ATFERO's organisational issues were political ones involving the British, the Canadians and the Americans, conducted at inter-governmental level, or intra-governmental level when it came to internal British debates over who should control ATFERO. In that context, the flying superintendent had little influence.[18]

Discomfort with superiors and organisational change was one thing; Don was now having to confront the first fatalities. Back on 19 February, T9450 crashed on take-off at Saint-Hubert while on a training run, killing the crew of three.[19]

The day after produced a disaster of greater magnitude politically. T9449 took off on a delivery flight carrying the renowned Canadian scientist Sir Frederick Banting. Jointly awarded the Nobel Prize in 1923 for his work on discovering insulin, he had subsequently been involved in a range of military endeavours such as the first anti-gravity flying suit, understanding the physical effects of high-altitude flying, and even atomic and biological warfare.[20]

Joseph Mackey, T9449's pilot, recorded losing an engine 25 minutes after departing Gander and turning back. After the crash, Mackey regained consciousness to find his crew dead and Banting injured and delirious. Despite being injured himself, Mackey set off to get help but without success, returning to find Banting dead.

With T9449 missing, the RCAF performed a fruitless search for two days, hampered by poor weather and searching in the wrong area. The RCAF apparently prevented ATFERO crews from joining the search.

Don had returned to California, first to the Consolidated factory in San Diego where, on 13 February, he had his first test flight in an LB-30A Liberator B.Mk.I (AM258). This particular aircraft was the first production Liberator, having flown its initial flight just four weeks earlier on 17 January.[21] He was impressed. Five days later, he was back at Lockheed's Burbank factory performing consumption and carburettor tests in the new Hudson Mk.V. Satisfied, he proceeded to Washington for some business. It was there he learned of the accident.

He flew immediately to Saint-Hubert, called Mackey's wife,[22] then instructed all 11 Hudsons at Gander be on standby for conducting a search when he arrived. This he did the following morning (the 24th), only to find his crews hadn't waited and were aloft. While he was having discussions with the RCAF and airport authorities about where to take the search next, the wreckage was spotted with Mackey alive. Don flew to the crash site himself and, after a short period, Mackey was rescued.

The death of Banting was international news. There would be rumours of sabotage and a secret mission, but an inquiry found the reality was the more mundane engine failure.

While the high-level political skirmishes about ATFERO's performance, and who could or should do a better job of ferrying aircraft, continued through the first half of 1941, Don made two contributions of note, both in the Liberator. The first of six LB-30s designated for the RAF (AM258–263), began arriving in early March.

For some months the British, Canadians and Americans had been discussing the possibility of a northern route through Greenland (controlled by Denmark) and Iceland (in British hands), enabling the delivery of aircraft that lacked the range for the direct North Atlantic crossing while also opening up a new delivery channel. On 22 April, Don conducted the first survey flight to Greenland in AM263 to determine possible aerodrome locations. His account saw the 16-hour 20-minute flight, departing from Gander and returning direct to Montreal,[23] through two entirely different lenses.

One was the task at hand: flying up the south-west coast of Greenland from Cape Farewell photographing potential sites. At times they were hampered by the winds and low cloud. Don commented on the geography in terms of aerodrome suitability. Potential sites were marked on a map, but this, too, proved difficult 'with the compass swinging as badly as it does in that vicinity during manoeuvres'.[24] The map was then placed in a container and dropped to a Danish vessel off the coast as a precaution 'in the event of our not returning to base. Such optimism was a delightful feature of operations in those days.'[25] The flight would lead to an aerodrome being built at Narsarsuaq, codenamed Bluie West One, forming part of a route that ultimately took aircraft through Goose Bay in Canada, Greenland, and Reykjavik in Iceland.

The other lens was his immersing himself in the scenery – 'the dark rock mountains of Greenland, the ice and the icebergs … and the clear water'. The west coast 'rising steeply out of the sea' saw them flying in and out of the fjords, something his expression suggests was a pleasurable exercise not an endurance. Only a single episode with the engines interrupted the 'tranquillity'. New terrain in a new aircraft was an environment that made him happy and he remarked it was unforgettable.

So much so he admitted to breaking security by writing to his mother back in Brisbane to tell her what he had just been up to. He decided to use a most novel form of code, Reginald Heber's imperialist, evangelical hymn 'From Greenland's Icy Mountains', the first verse of which is:

From Greenland's icy mountains,
From India's coral strand,
Where Afric's sunny fountains
Roll down their golden sand;
From many an ancient river,
From many a palmy plain,
They call us to deliver
Their land from error's chain.

Knowing his mother was up on her hymns, he summarised his trip by telling her he 'had not been to India's coral strand', hoping she could work it out for herself! Even if she did, it was hardly informative. He never found out if she had decoded it accurately.

The other contribution was inaugurating the so-called Return Ferry Service: flying crew who had delivered aircraft to the UK back to Canada instead of sending them by sea; a vital component to increasing the pace of aircraft delivery.

On 4 May, Don flew AM258 from Saint-Hubert to Gander, departing late that evening for Prestwick in Scotland, which had replaced Aldergrove as ATFERO's eastern terminus in the UK, on the first west-to-east leg of the Return Ferry Service. Aboard was Air Chief Marshal Hugh 'Stuffy' Dowding, the chief of Fighter Command during the Battle of Britain, who had been sent by Beaverbrook to Canada to be a party to the Greenland discussions.[26] That same day, Alan Youell conducted the first east-to-west flight to Newfoundland carrying seven passengers, also in a Liberator.

Don's return flight on 8 May, from Squires Gate at Blackpool, was another milestone for him. He recalled carrying 24 pilots[27] on mattresses in the bomb bay. It was a most uncomfortable trip due to the aircraft's heating problems, especially as it lasted 16 hours and 55 minutes, but it carried echoes of the first Hudson formation flight just six months earlier. Back then they had aircraft across the North Atlantic in a day; now they had aircrew back across the North Atlantic in a day. Everything was now being conducted above the Atlantic.

They flew direct to Montreal, the first land plane to do so from the UK. It produced memories of flying *Mercury* non-stop to Montreal in July 1938, the first seaplane to do so. As was his custom, he recorded the achievement in his logbook: '1st Direct Flight Landplane UK – Montreal.'

On 27 May 1941, Wilson officially terminated the contract between MAP and CPR,[28] but by this stage the American administration was beyond frustrated with the delays. President Roosevelt wrote to Churchill on 29 May with a plan from General HH 'Hap' Arnold, commander of the US Army Air Corps: the Air Corps would take full responsibility for getting aircraft from factories 'to point of ultimate take-off' in Canada but, in return, Arnold recommended the Air Ministry take responsibility for ATFERO. The sweetener was that American civil pilots would be put at ATFERO's disposal for flying the North Atlantic.[29]

Churchill discussed it with Portal and agreed to the change. Beaverbrook had departed MAP on 30 April, removing resistance from that quarter; and the Canadian Prime Minister, Mackenzie King, assured Churchill of his cooperation 'to the fullest extent'.[30]

ATFERO became RAF Ferry Command on 20 July 1941 and the Air Ministry appointed Air Chief Marshal Bowhill as the new air officer commanding. Don now had to deal with an air chief marshal whose strong opinion against the viability of the enterprise he had emphatically proven wrong just eight months earlier.

It is unclear as to whether Don jumped or was pushed. He suggested Bowhill essentially invited him to jump using a bit of carrot and stick. On the one hand, Bowhill wanted a transition period, so Long was staying on, meaning Don had to deal with them both. If forced to choose, Bowhill wanted Long. On the other hand, Bowhill told Don he was needed in England and had apparently arranged for him to be made an acting group captain if he rejoined the RAF.[31]

Just three weeks after Bowhill's arrival, on 8 August, Don departed Saint-Hubert for Gander for the last time, flying out late that night for Prestwick in a Hudson III (T9157) on a delivery flight. Bowhill had also arranged for Ly and the two children, Noreen and Torix, who had been in Montreal, to return to England on a BOAC flight.

Don's time at Atlantic Ferry concluded with a significant tangential contribution to the war effort. Churchill was sailing in the battleship HMS *Prince of Wales* for Newfoundland for the Atlantic Charter Meeting with Roosevelt. He had summonsed Beaverbrook, now heading up the Ministry of Supply, to join him. When Don arrived in London on the morning of 10 August, Beaverbrook told him he was departing imminently for Canada on AM261, one of three Liberators recently arrived as part of the Return Ferry Service. Don told Beaverbrook AM915 had the best heaters and he should get switched to that aircraft, which he duly did. The two aircraft were set to depart within 30 minutes of each other that evening. AM261 took off first and crashed on the Isle of Arran, killing all 22 on board.[32]

The greatest emphasis of these Atlantic Ferry chapters has been on the lead up to, and conducting of, the first formation flight. The reason is Don led a paradigm change in thinking. He believed a collective view had evolved that surpassed reason: '… I think you could say that we succeeded in debunking the North Atlantic. Up to that time everyone had been overawed by the North Atlantic – it was like black magic.'[33]

What Don brought, uniquely, was his ability to assess the aircraft from an engineering perspective; establish an appropriate regime for selecting and training pilots; understand the navigation issues and associated training demands; ensure the radio operators were trained properly; work the meteorology issues with McTaggart-Cowan; set, as far as practicable, rigorous standards; apply a level of planning which he oversaw with the highest level of micro-management possible; and ultimately provide the necessary leadership by conducting the first flight. He believed he knew the panoply of risks; what was left after he'd worked on them were 'calculated risks'. All this was bolted on to that driving force to prove the collective opinion wrong, especially if contributors to that opinion were those for whom he held little regard.

Atlantic Bridge, the Air Ministry's official account of the ferry organisation in 1945, said the first flight made two noteworthy contributions:

> In the arrival of the seven Hudsons, so urgently needed by the hard-pressed RAF, and the inauguration of bomber deliveries from Canada to Britain by air, Atlantic flying progressed at least three years in the course of one ...
>
> The significance of that first Hudson flight was that the conceptions of aerial warfare were being extended once again. Reinforcement and supply by air are not new to the Air Force by any means: but trans-oceanic air delivery was an historic step forward.[34]

And it should not be forgotten this occurred at a time of great peril to Britain when the need for American aircraft in volume, and quickly, was most acute.

The magnitude of the change in thinking is illustrated by a single sentence in the *Atlantic Bridge* account: 'A hazardous adventure had to be reduced to a routine enterprise of war.'[35] And that is exactly what transpired.

Given the success of the whole operation was originally predicated on a 50% loss rate, the actual results for the first 11 months were staggering. During the CPR/ATFERO period up to 20 July 1941, when the organisation became Ferry Command, of 293 aircraft, 289 arrived in

Britain, with four aircraft and five lives lost.[36] This was the end result of the contributions of many, but the origins of the success lay with the 'old Atlantic team'. Ultimately, the ferry organisation in its various guises delivered more than 10,000 aircraft covering 17 different types built in American and Canadian factories.[37] It was not just aircraft; as the war progressed 'the North Atlantic airway was being used more and more to transport passengers, mail, and vital war supplies, including medicine, technical equipment, food, and sometimes even ammunition'.[38]

Carl Christie, the author of *Ocean Bridge*, the most comprehensive history of the ferry organisation, is afforded the final word about this phase of Don's life:

> There is no doubt that Bennett was the single most important individual in launching the transatlantic ferry service.[39]

22

A Very Short Hiatus

Arriving back in England on his last delivery flight on 9 August 1941, Don clearly believed he was to be offered a job back in the RAF as an acting group captain. Air Chief Marshal Bowhill had put him in the hands of Air Marshal Guy Garrod, Air Member for Training. Don clearly expected not just the rank but a command worthy of it. He was sadly disappointed on both fronts.

The critical issue from the RAF's perspective was that, despite his achievements on the international stage at Imperial Airways/BOAC and then his pioneering work at Atlantic Ferry, he had left the RAF in August 1935 as a flying officer. He had forsaken a military career to spread his wings in civil aviation and was now back without having risen through the ranks. He had never commanded a squadron. What to do with his undoubted expertise and experience?

Garrod, according to Don, sent him down the bureaucratic line to somewhere in the bowels of the Air Ministry at Holborn where he ended up with 'the junior of junior civil servants, to whom it was quite impossible to bring in an apparent civilian into anything like a senior position in uniform'.[1] It was unseemly treatment given his already considerable contribution to the war effort.

He was offered the acting rank of squadron leader which he challenged. It was Air Marshal Harris who claimed: 'I then got him back in the Air Force as a Wing Commander.'[2]

After several weeks, on 22 September 1941, Don received two letters signed by Charles Evans, Principal Assistant Secretary for Personnel in the Air Ministry.[3] Both were incorrectly dated 22 September 1940, which must have amused Don no end.

A Very Short Hiatus

The first dealt with rank. He was receiving a commission in the RAF Volunteer Reserve, 'appointed in the rank of Pilot Officer (with acting paid rank of Wing Commander)'. He was not even accorded the courtesy of retaining his 1935 substantive rank of flying officer.

The second letter revealed they had found him a job: '... you are required to report to the Officer Commanding, Navigation Training Unit, Eastbourne as soon as possible.' Enclosed was a railway warrant to get him there, and an instruction to change into uniform on arrival. Evidently, the best way to use one of the world's premier navigators was to appoint him as second-in-charge of a navigation training school. It carried all the echoes of his posting back to Calshot from Pembroke Dock as a lecturer in August 1933. After all he had achieved in the past six years in civil flying, including at Atlantic Ferry, the first thing the RAF did on his return was put him behind the proverbial desk. It was the start of a series of administrative mess-ups with his appointment that created a lingering bitter aftertaste.

He commenced at No 1 Elementary Air Observer School (1 EAOS) on 25 September.[4] Despite not getting along with the 1 EAOS commanding officer, Group Captain Dand, who Don thought had an unfortunate preoccupation with RAF decorations (of which Don had none), he knuckled down and within a few weeks had contributed to getting the school off the ground, including devising the syllabus. On the 12-week training course, trainees would have 60 to 70 hours practical experience, taking in aerial photography, reconnaissance and air navigation. The range of subjects in the classroom included basic navigation theory, signals, meteorology, armaments and aircraft recognition.

He calculated the staff required, by rank, for the different roles. He assessed numbers for the first intakes, due every fortnight commencing 1 November. He laid out the daily programme aside from the training, including various parades, drills, P/T and meal times. His notebook shows him dealing with a pet hate: 'No smoking in lectures.'[5]

It took no time for Don to confirm his gross underutilisation; he was hankering to fly operationally. Never one to delay once he had made a decision, barely three weeks into his appointment he went up to No 10 Operational Training Unit (10 OTU) at RAF Abingdon, outside Oxford, where he did some circuits and bumps in a Westland Lysander before taking to the air in an old Armstrong Whitworth Whitley III doing blind approach practice for an hour. The OTU supplied crews to No 4 Group, Bomber Command. It was Don's first time flying a British bomber, leaving aside the Westland Wapiti back at Point Cook over a decade earlier.

Back at Eastbourne, he stayed long enough to get the first course started on 1 November. Four days later, he wrote an official letter to Dand:

> Sir,
>
> I have the honour to apply for posting to an Operational Unit.
>
> In making this application, I would like to emphasise that I am perfectly willing to continue in my present appointment and to exert myself fully in it. But I feel that the shortage of experienced four-engine pilots caused by the advent of Stirlings, Halifaxes, Lanchesters [sic], Liberators, etc., into the Service indicates that I would be far more usefully employed with such a squadron than in my present work. I have about 5,000 hrs. [sic] flying time in command of 4-engined aircraft of many types and considerable operational experience and technical knowledge of them.
>
> I conscientiously believe that I would be more valuable to the Service in an Operational Squadron and I therefore beg your favourable consideration to this application.
>
> I have the honour to be,
>
> Sir,
>
> Your obedient Servant[6]

Accompanying the letter were his particulars in accordance with A.M.O. A.894/1940. Regarding the question about aircraft types on which he was proficient, he advised:

> 14 types of single-engined aircraft = about 400 hrs.
> 14 types of two-engined aircraft = about 600 hrs.
> 4 types of three-engined aircraft = about 400 hrs.
> 12 types of four-engined aircraft = about 4,100 hrs.

Under 'Civil licences held and dates', he noted:

B Pilots Licence No. 5230 issued 27.2.33.
2nd Class Navigators No. 171 issued 27.2.33.
1st Class Navigators No. 171 issued 20.7.34.
'A', 'C' & 'X' Ground Engineers No. 2727 issued 7.1.35.
W/T Air Operator's Licence No. 185 issued 17.7.35.

It is a fascinating letter. Its supplicatory tone is so at odds with Don's character. The language is soft, with no directness or demands; there is nothing combative or critical. In fact, there is a pleading, a tacit acknowledgement, perhaps, that he is trapped and in no position to free himself. While he says he's willing to stay on, he plainly isn't. He is throwing himself on Dand's mercy and considered judgement.

Notwithstanding, the letter is thoroughly strategic by his honing in on the growing presence of the four-engine bombers in Bomber Command. His pencil draft shows an initial response to the question 'Type on which proficient and number of hours on each' followed his usual approach when analysing his flying hours: he wrote 'Liberator, Hudson, Catalina, Short Empires & Atlantic flying boats, – and about …', then paused. He realised this typology was problematic; calculating hours in this way pointed towards a Coastal Command posting. It was a logical choice: his experience on flying boats, his time at Atlantic Ferry, his maritime air navigation skills and, if they were aware at the Air Ministry, his anti-submarine duties in *Mercury*. He did not want to go to Coastal Command.

Instead, his gaze was firmly fixed on Bomber Command. The challenges were of the highest order, demanding innovative thinking, technical development and drive; it was pioneering work of a new kind.[7]

He crossed out his initial nomenclature and calculated 'type' based on the number of engines. This marked him out from other pilots and commanders because few could boast as much experience on four-engine aircraft. Knowing he might be quizzed on specific aircraft, he listed them all, according to his new typology, though this does not appear in the typed version.[8] He typed up the letter: despite the vast majority of his hours on four-engine aircraft being on flying boats, none are named, just incoming Bomber Command four-engine aircraft and the Liberator, the only heavy bomber he had flown.

According to Don, Dand took it with good grace and sent him off to see Group Captain Henry 'Daddy' Dawes, Senior Personnel Staff Officer at Bomber Command Headquarters. Dawes, in turn, sent him off to 4 Group Headquarters at RAF Linton-on-Ouse in Yorkshire essentially with the promise of the command of a heavy bomber squadron:

> As I went in I noticed a figure leaning over the fireplace whose badges of rank were obscured by the way he stood. I spoke to the Squadron Leader Personnel and told him what I had come about, and to my delight Roddy Carr turned around from the fireplace and said, 'Hello, Bennett.'[9]

Air Vice-Marshal Charles Roderick 'Roddy' Carr had been appointed Air Officer Commanding (AOC), 4 Group, the previous July. He had come from being AOC, RAF Headquarters in Northern Ireland. He was well aware of Don's achievements at ATFERO and had, as Don said, 'been most helpful to me'.

Carr was also doubtless aware of Don's pre-war achievements and his peerless reputation for navigation. Given the issues Bomber Command was facing, Don's navigation expertise was just what he needed in a commander. And, given his own background, he was probably more attuned to what Don could bring than his fellow group commanders.

Another Colonial – a Kiwi from Feilding – Carr, like Don, had something of the pioneering and adventurous spirit, having been a part of Ernest Shackleton's final Antarctic expedition in 1921. On that trip, he had flown a modified seaplane version of the Avro Baby. Six years later, he had set a world distance record in a Hawker Horsley, only to lose it almost immediately to Charles Lindbergh. He understood Don's mindset and could recognise and identify with the magnitude of his achievements.

Don returned to Eastbourne having accepted Carr's offer of Officer Commanding No 77 Squadron at RAF Leeming. The squadron flew Whitleys, not four-engine bombers, but it was an operational squadron. With the paperwork done, he left Eastbourne just over two months after commencing the appointment to take command of 77 Squadron with effect from 2 December 1941.[10]

A Very Short Hiatus

Following his feats at Atlantic Ferry, Don was elected a full Fellow of the Royal Aeronautical Society (FRAeS) on 12 November, having only become an Associate Fellow on 30 November 1939. This was a significant development in its own right, being prestigious and coming at the age of just 31.

23

77 Squadron – Enter the 'Hot' War

Royal Air Force bombers had been undertaking operations from the early days of the war. Squadrons sent out single or small numbers of aircraft to attack a range of targets during daytime. The aircraft – primarily Blenheims, Wellingtons, Hampdens and Whitleys – all twin-engine medium bombers and small in numbers compared to later in the war, had their limitations, not least of which was the bomb tonnage they could carry. The heavy bombers – Halifaxes, Stirlings and Manchesters – began bombing operations from early 1941 but not in great numbers, and the early variants often suffered a variety of technical problems or, in the case of the Manchester, overwhelming underperformance.

Even before the Phoney War of late 1939 and early 1940, losses to German fighters forced a strategic shift from daytime bombing to operations at night. While that initially reduced losses, it also increased significantly the challenges for aircrew navigating to the target area, clearly identifying the target once there, and then accurately dropping what bombs they could carry.

Navigation was the sole preserve of each crew, deploying whatever training they had received in basic navigation techniques and using the rudimentary on-board equipment.[1] The more sophisticated navigation aids were yet to come, the first being Gee in early 1942. An experienced hand at dead reckoning might get an aircraft roughly within the vicinity of the target area but, even if you got close, cloud cover or haze posed diabolical problems for anyone trying to compare what they could see dimly on the ground with what was on the map. Seeing any ground features or landmarks with precision was impacted significantly by altitude, even on clear, moonlit nights. The Germans made things even more difficult by introducing blackouts, bolstering their anti-aircraft defences, including in the air, and an array of other countermeasures.

As early as 11 October 1939, Air Commodore Coningham, then in charge of No 4 Group, Bomber Command, neatly encapsulated the problem in a communication to Bomber Command Headquarters:

> The real constant battle is with the weather ... The constant struggle at night is to get light onto the target ... a never ending struggle to circumvent the law that we cannot see in the dark.[2]

Despite operations planning evolving through the first two years of the bombing campaign, and concerted attacks being mounted on a wide range of targets, some in the upper echelons of the Air Ministry and Bomber Command, particularly Air Marshal Sir Richard Peirse, head of the latter, retained a considerable blind spot: assessing accurately how effective these operations were in damaging or destroying targets. It was not that this area was given no attention; the problem was relying on the avenues of information available without subjecting them to objective, substantive, systematic scrutiny.

Predominant among these were the reports submitted by aircrew after operations and recorded in each squadron's Operations Record Book (ORB). These often over-optimistic reports were the bedrock of assessment in determining the success of the Bomber Command offensive. Unfortunately:

> The authors of the Bomber Command operations record books (ORBs) at a group, station and individual squadron level struck a consistently positive note of the damage done and the targets being reached. They appeared to suggest that not only were the bomber crews achieving an astonishing level of navigational accuracy but also an ability to distinguish a wealth of detail about precision targets. It was nonsense, of course.[3]

Independent on-ground reports, untainted by German propaganda, were filtering through from target areas indicating minimal, if any, damage. The RAF had established the first photographic reconnaissance unit (PRU) in mid-1940. The PRU aircraft delivered periodic reports containing photographs of targets confirming supposedly successful raids were failures, but they could not provide systematic coverage of all Bomber Command operations. Whatever reports they did produce failed to penetrate the self-delusion.

Bomber Command had also started mounting cameras in bombers, typically those flown by the better crews, enabling bombing efforts to be captured in real time, but these were not being systematically analysed. It was Professor Frederick Lindemann, Lord Cherwell, Churchill's scientific adviser, who commissioned just such an analysis. It led to the famous Butt Report, named after David Bensusan-Butt, Cherwell's private secretary, who came up with the idea and did the work.

Released on 18 August 1941,[4] just over a week after Don arrived back in the country from Atlantic Ferry, Butt's report contained an analysis of 'about 650' aircraft photographs taken over 100 night raids on 28 targets between 2 June and 25 July 1941 (summer months were the best bombing weather). Butt selected photographs showing 'enough ground detail for the position photographed to be plotted', while defining the 'target area' as 'within five miles of the aiming point' (equating to a target area of approximately 75 square miles).[5] His statistical conclusions are quoted here in full:

1. Of those aircraft recorded as attacking the target, only one in three got within five miles.
2. Over the French ports, the proportion was two in three; over Germany as a whole, the proportion was one in four; over the Ruhr, it was only one in ten.
3. In the Full Moon, the proportion was two in five; in the new moon it was only one in fifteen.
4. In the absence of haze, the proportion is over one half, whereas over thick haze it is only one in fifteen.
5. An increase in the intensity of A.A. [anti-aircraft] fire reduces the number of aircraft getting within five miles of their target in the ratio of three to two.

6. All these figures relate only to aircraft reported as <u>attacking</u> the target; the proportion of the <u>total sorties</u> which reached within five miles is less by one third.

Thus, for example, of the total sorties only one in five get within five miles of the target, i.e. within the 75 square miles surrounding the target.[6]

Among his recommendations was that suitably qualified staff be appointed to 'maintain statistical records of night photographs and any other evidence which may be available',[7] a clear message the self-delusion needed to stop. Part of Butt's reasoning lay in a paragraph highlighting why Don's expertise and experience was sorely needed. Analysing photographs, Butt said, was not just about damage to the target:

> The night photographs are, however, chiefly of value in measuring the success of navigation, which seems from the results already obtained to be a matter of equal if not greater importance. For this purpose it seems absolutely essential that they should be subjected to every possible form of analysis. Practically speaking, they are the only data about navigation brought back from Germany in which the human factor is slight.[8]

That he actually had to state this was indicative of the problem.

> The Butt Report, taken with other evidence available, seemed to prove with equal harshness that Bomber Command, with its present navigation equipment, standard of crew training and methods of tactical deployment, could not hit its targets with any accuracy, even on the best moonlit nights in summer weather.[9]

While it produced disbelief in some quarters and started a debate about the validity of Butt's analysis (an exercise in trying to shoot the messenger), it penetrated sufficiently to cause the future of Bomber Command's usefulness to be questioned, including by Churchill himself. Peirse was defensive. Air Chief Marshal Portal, Chief of the Air Staff, had to mount a rearguard action in defence of Bomber Command. For a number of squadron and station commanders, Butt's report posed few, if any, surprises.

En route from Eastbourne to RAF Leeming, Don headed back to No 10 Operational Training Unit (10 OTU) at RAF Abingdon where, from 2 to 8 December 1941, he did a course. There were several flights in Armstrong Whitworth Whitley Vs, including a 'Nickel' raid (dropping propaganda leaflets) on Paris and Orleans on the night of the 7th (the standard way to apply what had been learnt at the OTU). While good experience flying to the Continent, Don did not regard it as his first true operation, that is, dropping bombs; that would come shortly at 77 Squadron.

The course brought him up to speed on the Whitley V, but his principal focus was on reviewing the training for each role in the Whitley crew, and the practical implications of that training, in particular for navigators. He was unimpressed. Among his observations, he found the foundational skill of map reading was weak. One squadron leader commented new navigators were also weak on log keeping. Indeed, at a more general level, Don observed 'many [navigators] do not know why they do things, e.g. Air Plot'.[10]

The context of his observations about aircrew training was provided by those who would shortly loom large in his RAF career – Arthur Harris and Ralph Cochrane.

At the start of the war, Harris was appointed Air Officer Commanding, No 5 Group. One of his initiatives was 'the setting up of a comprehensive training organisation for aircrews, which was to become the blueprint for the later Bomber Command Operational Training Units and Heavy Conversion Units …', the result of what Dudley Saward, his biographer, describes as a 'fetish for training'.[11] In essence, aircrew, who had all undergone training in separate specialised training schools, would 'crew up' at OTUs and train intensively together under 'realistic conditions' until they reached a high degree of proficiency.

Harris's training regime was picked up by No 6 Group, the operational training group of Bomber Command, headquartered at RAF Abingdon. By pure coincidence, Group Captain Ralph Cochrane had been

Abingdon's station commander for a short period over the winter of 1939–40 before becoming the Senior Air Staff Officer to 6 Group's chief, Air Commodore MacNeese Foster. During his time as one of Harris's flight commanders at No 45 Squadron back in the early 1920s, Cochrane had also developed a passion for aircrew training. He now got the job of checking out 5 Group's training blueprint and their relationship was formally re-established.

So much so, Saward recounts an exchange between Harris and Air Chief Marshal Edgar Ludlow-Hewitt, then head of Bomber Command, when Harris heard Cochrane might be leaving 6 Group. On 5 March 1940, Harris requested Cochrane's services at 5 Group, citing a commonality of outlook, including 'some revolutionary ideas in regard to getting more work, and less useless paper maintenance, out of the men and aircraft we possess'. Ludlow-Hewitt responded in the negative, adding a statement that was prescient for Don's forthcoming relationship with the Harris–Cochrane axis:

> Isolated radicals are most valuable people, but if you get too many of them together they become dangerous!! And so I think for the moment things are better as they are.[12]

As it was, with 10 OTU at Abingdon up and running, Cochrane was appointed to command the newly formed No 7 (Operational Training) Group at Brampton Grange in July 1940, overseeing the development and expansion of training at multiple OTUs.

'Ralph was now developing a reputation for original thought among both his subordinates and his superiors.'[13] After just three months, Cochrane spoke to Portal during the latter's visit to 7 Group HQ:

> Peter Portal came to have a look around and having served with him I knew him well and I opened my mouth rather wide on the failings of Flying Training Command to train *pilots* [this author's emphasis] who were coming to us and to train them on the lines that were wanted, now there was a war on.[14]

Portal immediately appointed Cochrane the Director of Flying Training and he had been in the role for over a year when Don arrived at Abingdon for those few short days in December 1941. The critical issue was that OTUs were the final point of training prior to going on active duty. (Heavy Conversion Units would only be established with the widespread introduction of the four-engine bombers.) There was no better place than 10 OTU for Don to see the effectiveness of aircrew training generally, and navigation training in particular. If the training was turning out navigators so ill-equipped for operational duty, why were elements of the senior leadership surprised the bomber campaign was producing negligible results?

It is worth noting Harris, according to Saward, acknowledged major issues with the bombing campaign despite innovations in training:

> During this time, Harris's 5 Group was involved in all these activities, as well as in the somewhat abortive attacks on strategic targets in Germany, which he knew to be a failure because of the complete absence of adequate navigational aids and target finding devices, and because of the ridiculously small bombs available for attack.[15]

It was an implicit concession the training regime to that point was making no appreciable difference to Bomber Command's performance in hitting and destroying targets. Harris was sheeting home the failures to the absence of the necessary technology (and bombs) not the training.

Don's experience, however, was the training itself had significant shortcomings, at least when it came to navigation. And in that vein, he departed Abingdon expecting his highest priority at 77 Squadron was to raise navigation standards. That was an area within his control and where he could make an immediate impact.

The unit was a typical 4 Group squadron, comprising three flights with an operational strength during Don's time of 17 Whitley Vs. Up to 14 aircraft participated in operations on any one night.[16]

Don came to regard the Whitley V as a 'good work-horse' but, like other medium bombers, its limitations were exposed by Bomber Command's shift to nighttime operations:

> It soon became apparent that the Group's aircraft were inadequately equipped, both navigationally and operationally, for crews to regularly locate targets by any other means than visual map reading, which was only practical under conditions of bright moonlight. This gave a monthly window of opportunity for accurate target location which lasted no more than about seven days, weather permitting, and it seldom was in winter months.[17]

And it was the start of winter when Don took command.

To that point, 77 Squadron had flown on a wide range of bombing operations. Priorities were set by various Air Ministry directives issued to Bomber Command over the period. During the Phoney War up to May 1940, it primarily carried out daytime 'Nickel' raids, performed reconnaissance, and periodically bombed coastal targets. They had attacked the Germans in Norway, bombing ports and airfields, then the Channel Ports when Britain was facing invasion. Various French ports would continue to receive attention in an attempt to hit elements of the German surface fleet and disrupt the U-boat campaign.

In Germany proper, 77 Squadron's Whitleys had joined raids on ports such as Hamburg, Bremen and Wilhelmshaven, and a range of inland cities and centres, including Berlin. The Ruhr, Germany's heavily defended industrial heartland, was a regular operational target, coming to be known by aircrew as 'Happy Valley'. Some operations ranged as far as Poland and Italy. Disrupting and degrading the German war machine saw the Whitleys return time and again to synthetic oil plants, rail and marshalling yards, aircraft and engine plants, communication hubs, road and rail junctions, and canals and bridges.

These operations had come at a heavy cost. From the start of the war to when Don took command, 77 Squadron had lost 55 aircraft and 194 aircrew either killed, missing or captured:

> The number of aircraft and crews lost were almost twice the squadron's established strength.[18]

One pilot from the early days was future Group Captain TG 'Hamish' Mahaddie. He flew with the squadron from 1937, completing an operational tour on Whitleys by July 1940 before taking a training role at RAF Kinloss. He would participate in the very first Pathfinder operation against Flensburg on 18 August 1942 with No 7 Squadron, later becoming the infamous and celebrated 'Bennett's Horse Thief' as commanding officer of the Path Finder Force Navigation Training Unit.

Don arrived at 77 Squadron during a lull following the release of the Butt Report as high-level discussions took place about Bomber Command's future direction. The squadron would continue attacking a range of targets, though winter also contributed to a slower pace as the weather regularly prevented any action. For Don, it was, in effect, a 'nursery' command, one where he could 'learn the ropes' without the sorts of pressures he would experience from July 1942.

He scribbled down the names of his flight commanders and their deputies, featuring some other future luminaries:

A Flight: C – S/L Seymour-Price,[19] 2nd – A/F/L D.F.E.C. Dean[20]

B Flight: C – S/L L.H.W. Parkin, 2nd – F/L F.A. Drury

C Flight: C – F/O Oakes, Adj – Hall, Liston

To these he added:

Sqn Gunnery Ldr – A/F/L O'Rourke

Sqn Nav. & Bomb. – F/L H.H. Lawson

Signals Instructor – F/Sgt Jones & Sgt Sandford

Paratroops – Wheatley & Luff

77 Squadron – Enter the 'Hot' War

As was his custom, Don prepared detailed summary notes of the Whitley V's specifications to act as a ready reference, replete with a diagram of the fuel system. The Whitley V had two Rolls-Royce Merlin X engines. At the bottom of the page, Don wrote 'Mr Godfrey, Rolls-Royce', along with some data on valves, suggesting 'Mr Godfrey' was either going to get a phone call or they had already talked and Don wanted his name in the event of further questions.

Don's first operation of his so-called 'hot' war took place on the night of 16/17 December 1941. The pilot was Sergeant Grace in Z6641, one of two 'junior' pilots in the squadron. It came about because Group Captain Bill Staton,[21] the station commander, had advised Don do a couple of operations as second pilot before taking command of an aircraft himself. Grace had arrived in September and had seven operations under his belt, five in command. The target was Wilhelmshaven. The squadron contributed 12 Whitleys as part of a larger force of 83 aircraft.

The Z6641 account in the Operations Record Book (ORB) illustrates with succinct blandness the bombing issues Don was facing at 77 Squadron:

> SGT. GRACE experienced 7/10 cumulus cloud over the North Sea and 3/10 cloud over the target area, at 12,000 feet. EMS estuary was identified, but hazy conditions made accurate identification of the target impossible. Bombs were dropped from 10,000 feet, but no results were observed. Intense light flak was encountered in the target area. An enemy aircraft with a searchlight was seen 5335N. 0530E, but no attack was made.[22]

Despite aircrew reports of starting large fires, those below recorded 'slight damage' and no casualties.[23] This particular report exemplifies many aircrew accounts in 77 Squadron's ORB.

Don offered a much more detailed account of his second op six days later, this time with Sergeant Murphy in Z9231. Records suggest this was William A. Murphy, a Canadian pilot, who had completed just five operations with 77 Squadron. Once again, Wilhelmshaven was the target. As they crossed the Dutch coast, Don noticed the oil pressure gauge for one of the engines dropping and alerted Murphy, attributing it to 'symptoms ... typical of a failure of the capillary of the particular type of oil gauge in use', but Murphy 'felt that he knew better' and stopped the engine and feathered the propeller, leaving them to approach the target on one engine.[24]

They now encountered intense flak. The bomb aimer failed to locate the aiming point on the first two runs over the target. By this stage they were coned by searchlights and 'there was no chance of breaking free'. Murphy came around for a third run, with the bomb aimer again failing, having been blinded by the searchlights. Don became 'quite convinced' the bomb aimer was not up to the task and, with the Whitley down to 8,000 feet, prevented Murphy doing a fourth run by taking command. He unfeathered the propeller and handed control back to Murphy once they were out of the target area.[25]

The principal lesson of Don's first two operations was their offering 'a really great insight into the nature of young men on bomber operations ... The point was that our bomber operations were largely in the hands of lads like [Murphy]. He had little experience and little knowledge, but he had the courage of a thousand lions. That was typical of Bomber Command operations in those days.'[26] In other words, he had command of a squadron of brave but largely ineffective aircrew.

On 21 December, less than two hours before taking off on this second operation, Don had received a telegram at Leeming. It was from 10 OTU at Abingdon, passing on a message from the Air Ministry, who was still catching up with where he had been posted:

Following signal received from Air Ministry P2777 dated 20/12. Begins BOAC requests services of W/Cdr Bennett D.C.T. signal if willing to accept ends.

This was a potential inflection point. Having been exposed once again to all the difficulties of life in the RAF, Don was offered a choice. Shortly after, he scribbled a response:

77 Squadron – Enter the 'Hot' War

> Following from Wing Commander Bennett. Reference Air Ministry Signal P2777 20/12. After consultation with BOAC I do not repeat do not wish to return to civil aviation aaa DGCA informed. Bennett.[27]

What BOAC was offering remains unknown. What is clear is his decision was swift and emphatic: much bigger challenges lay ahead and there would be no turning back.

Close to Christmas, about three weeks into his command, Don took to his notebook and sketched out something of a 'to do' list.[28] Top of the list: 'Sqn Standing Orders.' He had evidently endured several weeks of inefficient, time-wasting nonsense and would conduct his own private war against the leading culprits.

Secondly, 'Training Instructions', with an added note, 'Arrival Procedure.' These needed to be streamlined and made fit-for-purpose for aircrew. The added note suggests all new crews to the squadron would know exactly where he stood on training priorities.

Thirdly, he honed in on the practicalities of bringing his navigators up to a standard he felt was acceptable, with a particular focus on core skills related to dead reckoning: '1499E incl. D.R. General Nav., Met., D.R. Plot., Astro Nav Practical, [*unreadable*], Recog P., Sigs P.'

Having experienced two operations, another priority was: 'List of Pilot Instructors – juniors.' The junior pilots would get special attention. Don doesn't identify the credentials he had in mind for his 'Pilot Instructors', but operational experience was almost certainly top of the list.

'Files on equipment' was another unsurprising priority because 'equipment' was crucial for his crews – air and ground – to get their job done.

'Photos models etc. Library Equipment' evidences his early appreciation the preceding priorities meant little if he and his crews were unable to measure progress in hitting the designated targets.

To these priorities he now brought his specific leadership style, taking the decision after the first two operations to continue flying the 'hot' war irrespective of the risks to him as the newly appointed commanding officer. His decision carried echoes of Laddie Clift back at RAF Calshot, and more recently Atlantic Ferry – he would lead the squadron as much as possible from the cockpit. He detested ignorance, especially when aircrew lives were on the line. How could a senior officer make pertinent decisions on behalf of his men if he didn't experience operational conditions to a range of targets alongside them?

To have the greatest impact, he made a decision increasing the risk to him personally:

> I always made it a point of flying with a different crew each night, leaving the captain concerned at home. I was thus able to check on the efficiency of each crew, and at the same time to give them what I could in the way of instruction on navigation and how best to do their work.[29]

It was perfectly logical: assess the capability of aircrews as they responded to operational conditions; instruct *in situ*, especially in navigation; and use his observations to improve tactics and training.

For his next 11 operations (giving a total of 13 while commanding the squadron), only three men flew with him twice, and none more than twice. Intentional or otherwise, it sent a message to all crews about his expectations and the standards he was setting. It probably didn't hurt to have all crews on notice that they better 'be at their best'; the boss didn't fly every op, and no crew knew if they would be picked the next time he decided to go.

It was also a welcome example to his crews to see his courage, knowledge and experience on display, even if he could be a little intimidating or demanding. He may not be loved, but he would be respected.

'The results of my navigational efforts soon began to be reflected in our bombing results.'[30]

Don immediately qualified this claim by referring to crews taking 'more and more successful night photos', which, as Butt had indicated, was the one real objective measure of improved navigation. He set out to drive a change in culture whereby what had been 'a considerable matter of pride' for a crew to return with such a photo became 'a matter of considerable disgrace' if they failed to get one, and not just of the target area but the aiming point.

Whether his crews were improving in bombing targets is nigh impossible to prove. Of the 32 operations flown by the squadron during Don's time in charge, except for one, where he flew alone on a 'Nickel' raid to Oslo, his Whitleys were part of raids of anywhere between 18 and 235 aircraft. Attributing damage done to targets by aircraft of any particular squadron was difficult without photographs. Also, some targets may have been hit, but there was insufficient or no independent corroborative evidence.

Even Don's claim about an increased number of crew photos over targets, as evidence of rising navigation standards, is difficult to prove, that is if 77 Squadron's ORB reports are any indication. Flight Lieutenant 'Dixie' Dean's crew scored one photo on each of the Wilhelmshaven raids on 16/17 and 22/23 December. Squadron Leader George Seymour-Price got two and Flight Lieutenant Drury got one on the night of 16 January, then Don scored four over Rotterdam one night at month's end. Come February, Don again got one on his 'Nickel' raid to Oslo, though that was of Kristiansand[31] as they crossed the coast south of Oslo. Nothing recorded in March. On the night of 5 April, he got three over Gennevilliers. In total, 13 recorded photographs by four crews over the entire period, with Don taking eight of them. Almost all were taken in 'excellent visibility', 'good visibility' or 'clear conditions'. A definitive conclusion is hampered by not knowing whether Don decreed crews had to record capturing a photo in their ORB reports.

It is evident, however, from 77 Squadron's ORB reports generally, that winter weather, industrial haze and German defences played havoc with target identification let alone affording a degree of visibility to enable a photo. Cloud cover is a regular feature of the reports. Taken as a whole, the operational results between 7/8 December 1941 and 12/13 April 1942 (Don's last operation with the unit) suggest accuracy in hitting targets had not improved significantly over previous periods even if the squadron's general level of navigational competence had risen.

During that time, the squadron lost 11 aircraft and 48 aircrew on six operations, with four of the aircraft crashing on return. That included the loss of Squadron Leader Leslie Parkin DFC, the B Flight commander. While 26 operations were conducted without loss, it was still a heavy price for what amounted to a trail of non-destruction.

Don's response to aircrew crashing on return should put paid to any portrayal of him as unfeeling, aloof and distant. Just as he had flown out from Gander to visit the crash site of Mackey's Hudson to do what he could, he now headed immediately for the crash site when one of his Whitley crews came down near Leeming.

He offers a particularly harrowing account of arriving at the crash site of Z6975 close to midnight on 13 March. It had stalled on approach, coming down well short of the runway after a raid on Boulogne with one other 77 Squadron aircraft. Both crews were 'nursery crews' – Don's 'juniors'. He recounted arriving by car before anyone else:

> I waded through the river and up the far bank, where I managed to get one of the crew out of the already blazing aircraft. He had already his face badly burned and his eyelashes stuck together. This was merciful, because he could not see his hands, what had been burnt off to the bare bone as far as just short of the thumb, a rather ghastly sight.[32]

It was Flight Sergeant Richard Lewis of the Royal Canadian Air Force, a wireless operator/air gunner.

At this point, an American RAF sergeant arrived, desperate to retrieve a friend from the aircraft, but Don managed to prevent him trying to enter the blaze, instead getting him to assist with Lewis. They got him across the Swale and into Don's car, who drove him to Sick Quarters, but he died of shock the following day. Don remarked that Lewis 'co-operated magnificently' despite his injuries, another form of bravery he witnessed in his crews. The remainder of the crew perished.³³ The pilot, Sergeant James Church, was on his third operation.

Don's lengthy account, written in the mid-1950s, has a shocking immediacy about it. There is no doubt he was reliving it in great detail. The experience would not deter him from continuing to act this way, even when at the head of the Path Finder Force. Crew losses under his command would never just be numbers. He shouldered the responsibility of command, but it weighed heavily on him when his boys died, perhaps more than anything else.

Don was to make much of his participation in the 'hot' war vis-à-vis other senior RAF commanders, so it is worth considering briefly his personal experience on operations at 77 Squadron.

There was a range of targets: four operations were to Germany (Wilhelmshaven twice, Düsseldorf and Mannheim); three to Norway (Stavanger – airfield, Oslo – 'Nickel' raid, and Herøya and Odda – aluminium factories); once each to Genoa and Rotterdam; and four to France (Saint-Nazaire, Boulogne-Billancourt, Poissy and Gennevilliers). Although he was to encounter German defences on most of these operations, only once does his crew's ORB report mention damage to the aircraft, the Gennevilliers raid on 5/6 April 1941: 'Aircraft was damaged by flak.'

On at least one other occasion, subsequent examination of his aircraft showed more than superficial damage. On the night of 1 April, 77 Squadron dispatched 12 Whitleys as part of a force of 41 aircraft to attack the Matford motor factory at Poissy, on the outskirts of Paris. It was Don's 11th operation with the squadron. He recalled an inordinate amount of time over the target due to the bomb aimer's poor night vision:

> I could see the target quite clearly myself and would begin a bombing run, but as soon as I was level the target disappeared under my nose and I could no longer see it. The bomb aimer, whose job it was to direct me would then fail to see the aiming point because of his poorer night vision, and was unable to drop accurately. During

the whole of this period a German night fighter was trying to get on our tail, and what with doing tight turns round behind him plus the poor bomb aimer, we gave plenty of time to the few gunners down below for their night's practice.[34]

The resultant damage to his Whitley (Z9438) – 'some fifty-odd holes' and a control cable nearly being severed – prompted the new Station Commander, Group Captain Strang Graham, to 'have a very fatherly talk' with him. This suggests Graham thought Don was riding his luck a bit and, as squadron commander, he should reconsider flying on operations. It was an appeal that failed in spectacular fashion within weeks.

Throughout Don's command of 77 Squadron, one operation alone stood out because of its significance within the context of Bomber Command as a whole, and as a foretaste of Don's future. This was the raid on the Renault factory at Boulogne-Billancourt on the night of 3 March, his tenth operation. The factory was producing German trucks.

Due to its proximity to the centre of Paris (less than four miles from the Eiffel Tower), 'the plan called for the mass use of flares and a very low bombing level so that crews could hit the factory without too many bombs falling in the surrounding town'.[35] Further, the first wave of aircraft would comprise experienced crews 'thus foreshadowing some of the "pathfinding" methods to be used later in the war'.[36]

It was also the largest operation of the war to date, with 235 aircraft of six types, signalling the arrival of Bomber Command's new leadership and a shift in strategy.

77 Squadron put 13 aircraft over the target, with one describing 'moderate visibility', another '5/10 cloud'. The first crews and the flares had done their job, however, and Don, exemplifying his crews as a whole, bombed from 1,800 feet.

The raid was an outstanding success for Bomber Command, with considerable damage to the factory, a loss of production for several weeks, and large numbers of trucks destroyed. The tragedy of the raid was that 367 French people died in the surrounding areas. This was tangible evidence the emerging tactics for precision bombing were only in their infancy, while also presaging what was about to be unleashed on the civilian populations of German cities.

Don himself remarked: 'It was, in my view, the first decent result achieved by Bomber Command up to that date.'[37]

Just weeks earlier the bomber offensive had reached a turning point, the culmination of a strategic shift that had evolved over the preceding year.

Back on 9 March 1941, the Air Ministry directive to Bomber Command had prioritised bombing targets in 'congested areas' to affect German civilian morale. Harris, then Deputy Chief of the Air Staff, had sent an amending directive nine days later regarding the targets of Mannheim and Stuttgart: 'Both are suited as area objectives and their attack should have high morale value.'[38] While it portrayed civilian casualties as being collateral to hitting specific designated targets, the undercurrent was clear: bombing cities to kill civilians to destroy morale and the will to fight was now recognised openly as a legitimate strategy.

The directive of 9 July 1941 reinforced this thinking: the choice of specific types of targets, in this instance the German transport system, but with the complementary objective of 'destroying the morale of the German population'.

Now, just over two weeks before the Renault factory raid, on 14 February 1942, after discussions in the War Cabinet and the Air Ministry about the future of Bomber Command, a new Air Ministry directive was explicit and direct:

> It has been decided that the primary objective of your operations should now be focussed on the morale of the enemy civilian population and in particular of the industrial workers.[39]

The justification for the so-called 'area bombing' was straightforward. Bomber Command was Britain's only offensive weapon and it needed to contribute substantively to the war effort. Contemplating the alternative, a slow, grinding land campaign, with fresh memories of the First World War stalemates – costly in military and civilian lives, and matériel – made for a simple and persuasive argument: it was far less 'costly' to bomb the German population into submission. With the arrival of the heavy bombers, in numbers and carrying far greater tonnages of bombs, the task of causing the 'German home front' to collapse could reasonably be sped up.

The strategic decision to bomb civilian targets, based on the prevalent concept of 'total war', remains controversial precisely because it touches issues of morality, the conceptualisation of 'good' and 'evil', and the relative 'costs' of war, in particular the loss of lives, especially non-combatants.

Given Don was about to play a major part in executing Bomber Command's strategy under this directive, his stance is of interest. The foundational military thinking to which he was exposed begins in the 'History and Strategy' lectures back at Point Cook in 1930–31. In a single lecture, he was imbued with the following:

> <u>Aim of War</u> & aim of Air Force in War – To bring another nation, as quickly and economically as possible to conform to your will or purpose. The Services are instruments of the nation's will. Aim of Air Force is to break down enemy resistance.

This is revealing in itself, but earlier in the lecture is this statement, acting as a form of umbrella:

> <u>Superiority</u> includes a moral, physical & material superiority.

'Moral' is not defined. If you believe – collectively and individually – that you have entered the war on the side of what you perceive to be just, right and good, it doesn't need to be; moral superiority is assumed. Even the word 'superiority' comes with connotations. War is a dirty business, but if you can fight it from the moral high ground, almost anything can be legitimised and rationalised, especially in comparative terms when considering the enemy you are up against.

Now, at 77 Squadron, Don found himself in something of a morality debate in a conversation with one of his pilots as he provided a justification for 'area bombing':

> Affairs of conscience are sometimes difficult. A young Canadian pilot came to me at one stage and asked for an interview. The

purport of this talk was simply that on a previous evening he had been asked to bomb an aiming point which was obviously a civilian-type target. He could not, in all conscience, do such a thing, even if they were Germans and even if they were killing us. I tried to explain to him that a modern war is a total war and that the whole population of Germany was in fact fighting against us, whether in making munitions at home or in shooting in a field-grey uniform in the front line. I think I helped him considerably, but there was no doubt about it that for a long period his conscience worried him whenever there was any target that was not glaringly military in nature.[40]

Given his adherence to the Christian beliefs of his upbringing, and his natural resistance to towing what one might term 'the party line', it is an argument surprisingly lacking in nuance. There is no discrimination between the German war machine and the civilian population – all Germans are complicit in the killing because 'modern war is total war'. In effect, all Germans are engaged in the Nazi war effort in some way and that is sufficient to legitimise them all as targets. The pilot's response shows he wasn't convinced.

Whether Don himself was entirely convinced by the 'total war' argument is a legitimate question because he implies it was an 'affair of conscience' for him, too. And, if it was, what impact did any intellectual dissonance have on him as the air war progressed?

Eight days after the 'area bombing' directive, Air Marshal Arthur Harris was appointed head of Bomber Command, charged with prosecuting the new bomber offensive. Whether or not Don was personally a full convert of Bomber Command's 'area bombing' offensive from a moral perspective at this juncture is immaterial, he was most definitely a convert of Arthur Harris, and therefore to the offensive by proxy. Don said Harris's appointment:

… cheered me up considerably … I had learnt [back at No 210 Squadron] that he was full of fire and dash, was not easily baulked, and was also remarkably intelligent without trying to show it. He was, I knew, a real man, and my hopes for the bomber offensive and its ultimate destruction of Germany were revitalised.[41]

Harris could be relied upon to cut through the Gordian knots of bureaucracy to get things done and achieve the overall objective. In Don's eyes, Harris was the most efficient senior commander he had ever met.

The appointment hadn't come a moment too soon. Don was ropable with the string of failures that had seen the battlecruisers *Scharnhorst* and *Gneisenau* escape from Brest to northern German ports in the so-called Channel Dash on 11–12 February. To Don, the 'disgraceful fiasco' exemplified the state of the air war. He believed Harris was the catalyst for change.

24

Bureaucracy Sows the Seeds of Bitterness

Throughout Don's time at No 77 Squadron, he was engaged in skirmishes with the bureaucracy over many issues, but none more compelling and enlightening than the battle over his rank and attendant pay.

He had arrived at Leeming already smarting from the treatment meted out to him on his return to England: the failure to acknowledge and recognise his expertise; his initial placement at No 1 EAOS; the delays in decisions being made; and the arguments over his rank. This latter issue now became a running sore; more than a supremely frustrating distraction, it exemplified the bureaucratic colossus that would continue to plague him.

He may have been appointed an acting wing commander, but the administration was lagging badly. In response to a query to No 4 Group Headquarters on 16 December, he received the commendably fast response the following day:

> From: Headquarters, No. 4 Group.
> To: R.A.F. Station, Leeming.
> Date: 17th. December 1941.
> Ref: 1007/P.2
>
> <u>CONFIDENTIAL</u>
>
> With reference to your LG/C.1003/P.2. dated 16.12.41., it is confirmed that the provisions of K.R. & A.C.I. para. 117 apply.
>
> 2. Since details regarding Wing Commander D.C.T. Bennett do not appear in the current issue of the "Air Force List", action has been taken to confirm his rank and seniority. A

further communication will be addressed to you in due course.

(sgd) E.A. Grant. S/LDR.,
For Air Vice Marshal,
Air Officer Commanding,
No. 4 Group, R.A.F.[1]

It had just the right amount of obfuscatory gobbledygook to have Don seeing red. Perhaps to his surprise, given the last sentence, he received that 'further communication' the very next day, suggesting Air Vice-Marshal 'Roddy' Carr, commander of 4 Group, himself had intervened, deciding it was inadvisable to antagonise his newest squadron commander too much:

With further reference to this Headquarters letter 1007/P.2 dated 17th. December 1941, information has now been received from the Air Ministry through Headquarters, Bomber Command, that the substantive rank of Acting Wing Commander D.C.T. Bennett is that of a Pilot Officer, he having been granted the acting rank of Wing Commander with effect from 25.9.41. Under the provisions of A.M.O. A.91340 he is eligible for the grant of the War Substantive rank of Squadron Leader with effect from 25.12.41.

2. Wing Commander J.A.H. Tuck is therefore the senior Squadron Commander at your Station.[2]

Somewhere in the ongoing bureaucratese was the simple message Carr had promoted him to the substantive rank of squadron leader effective from Christmas Day. This was welcome, and addressed one issue, but the more important one remained unresolved – his salary.

On 20 March, Don wrote to headquarters again. The response contained just the right amount of verbiage to distract him momentarily from bombing the Germans:

From: Headquarters, No. 4. Group.
To: O.C. R.A.F. Station, Leeming
Date: 24th. March 1942

Ref: 1001/1384/P.2

CONFIDENTIAL

A/W/Cdr. D.T.C. [sic] BENNETT

The matter of the substantive rank and seniority of the above-named Officer has been the subject of correspondence between this Headquarters and the Air Ministry.

2. Information has now been received that he was appointed to a Commission in the G.D. Branch of the R.A.F.V.R. in the rank of P/O with the paid acting rank of W/Cdr. w.e.f. 25.9.41.

3. Notice of his appointment to the basic rank of P/O appeared in the London Gazette of the 6th. January 1942 and an entry in respect of the grant of the acting rank of W/Cdr. will appear in A.M.P.L. No.197/1942.

4. P.O.R. action should be taken accordingly.[3]

Carr may have chosen to promote him to the substantive rank of squadron leader, but 1942 had dawned with official recognition still only as a pilot officer. That didn't matter so much because the salary issue was seemingly addressed: '… with the paid acting rank of W/Cdr w.e.f. 25.9.41.' It had taken the bureaucracy six months to refer to his original appointment letter regarding his pay rate. Consequently, on 2 April, 4 Group sent him three contracts to complete.

Nothing changed.

Don now corresponded directly with the Air Ministry. On 7 June, nearly nine months after returning to the RAF, he wrote to the Under Secretary of State. In the letter, the full impact of the bureaucratic failures were revealed:

> On the 28th April last I replied to your A157375/41/Acts.2.(a) and in my letter pointed out that a very considerable mistake appears to have been made in the amount of remuneration which has been paid to me. I

pointed out that the matter was one of considerable urgency to me and requested that the mistake be rectified as quickly as possible. It appears that nothing has yet to be done.

I rejoined the R.A.F. on the 25th September last as an acting paid Wing Commander, and as such I am due of the sum of 36/2d. per day less income tax from the 25th September 1941. It appears that the amount being paid into my account is 14/6d. per day. There has never been any doubt concerning the paid status of my rank which was authorised by the A.M.P. when I rejoined. The D.D.P.S. will confirm this if you require it.

I will be grateful if you can rectify the matter including the back pay as soon as possible.

[sgd] Wing Commander, Commanding
No. 10. Squadron, R.A.F.[4]

Don had served his entire time as commanding officer (CO) of No 77 Squadron and been eight weeks as CO of No 10 Squadron without resolution of his pay issue, despite it being clear in his re-appointment letter on 25 September the previous year. He was livid and it stoked his personal sense of grievance while concretising his view of the Air Ministry as thoroughly moribund and incompetent. Beyond the personal, it showed the ossified structures were unable to respond with any agility to anything, or anyone, new or out of the ordinary; the wheels were turning far too slowly in the face of a rapidly changing war landscape.

Even once the matter had been resolved, the fact it had occurred continued to be a great source of personal irritation.

Whatever Don had been expecting of his first operational command, and how much he felt he could achieve, it had been a chastening experience:

During this initial period with 77 Squadron my faith in the bomber offensive had been seriously shaken. Like many other Air Force officers, I had looked at this great offensive as the only means of victory … The offensive on which I had pinned such faith was clear to me as a relative failure …

> … if we were to destroy the industrial strength of Germany, our bombing not only had to be heavier but also far, far more accurate than it was. The necessity of a target-finding force became more evident than ever, and I hoped for the day when the 'powers-that-be' would see the sense of such a force.[5]

Publicly, however, he put on a brave face. With Australia now imperilled in the war against Japan, he took the opportunity to help boost morale back home. At 8:15 am on 14 April 1942, the day he took command of 10 Squadron, he recorded a broadcast on the BBC Pacific Service – 'Calling Australia' – for sending Down Under. Much of the broadcast was about his time with Atlantic Ferry but finished by referring to 'the unit' he was serving in. 'I'm glad to be having a direct crack at the Hun.' He spoke about the competitive spirit of the aircrew from across the Empire:

> … but the main theme is that all of us are British, and coupled with this feeling there's a tremendous enthusiasm and keenness combined with a very serious appreciation of our aim and object at all times – to do our part to destroy the international gangsters as soon as possible. This task we aim to carry out, not only as Britishers, but also because we believe that that is the duty of all decent human beings.
>
> Australia is threatened now. So also is Great Britain. Total tenacity in deed rather than word is the only way of achieving victory. True totality of effort must be achieved by every man, woman and child, great and small.[6]

True to form, he received a memo a week later from the Public Relations Officer at 4 Group Headquarters:

> Dear Wing Commander Bennett,
>
> Public Relations at Bomber Command are again worried over

procedure. They told me that the B.B.C. should have applied through them for you to take part in the gramophone record feature programme.

They ask me to ask you if you will be so kind as to write to whoever it was who made the request, pointing out that authority for the broadcast must be obtained from Wing Commander E.P. Beauman, at P.R.4., Air Ministry, Whitehall.

There is no objection of course to this broadcast, but arrangements should be made through Public Relations and not direct.

Sorry to put you to this trouble,

Yours sincerely

[signed]

W.V. Noble F/Lt

PUBLIC RELATIONS OFFICER[7]

Public relations officers, wherever they were, now joined Don's increasingly lengthy list of those he felt did nothing but constipate the bowels of the war effort. He kept the memo as a memento for the rest of his life.

25

Shot Down

Recently appointed as the commanding officer of No 10 Squadron, Don recalled the plan for crippling or even possibly sinking the German battleship *Tirpitz* as being pretty straightforward:

> The Admiralty worked it out that with the ship lying close to shore, it would be possible to roll some spherical mines down the shore side of the ship, which was only fifty feet from the steep sloping banks of the Aas Fjord [Åsenfjord], and that these spherical mines would, being fitted with appropriate depth fuses, then get under the comparatively vulnerable bottom of the hull and blow it in.[1]

On 16 January 1942, the *Tirpitz* arrived in Fættenfjord, an arm of Åsenfjord, about 17 miles east of Trondheim, Norway. The ship had sailed from Wilhelmshaven after completing sea trials in the Baltic. It took a week for the RAF and local intelligence sources to confirm it was there.

The *Tirpitz* was the sister ship of the more well-known *Bismarck*. They were the cream of Hitler's *Kriegsmarine*. Carrying a formidable armament,[2] with a maximum speed of 31 knots, they were the most powerful capital ships in Western waters. Consequently, they were greatly feared by the Allies.

It had almost been a matter of pure luck the Royal Navy had stopped the *Bismarck* in the Atlantic in May 1941 when a Fairey Swordfish had succeeded in hitting the battleship's rudder with a torpedo, sending the ship in circles and allowing naval firepower to sink it, although not before the *Bismarck* had blown up the famed battlecruiser HMS *Hood* with a direct hit and damaged the brand-new battleship HMS *Prince of Wales*. No one in the upper military or political echelons in Britain was under any illusions about what the *Tirpitz* was capable of should it sortie like the *Bismarck*.

Hitler had invaded Norway on 9 April 1940, taking control on 10 June. It was highly strategic militarily. From the *Kriegsmarine*'s perspective, the many deep fjords with steep cliffs, accompanied by regular poor weather, afforded safe havens for its ships because attack by air or sea was exceedingly difficult.

With Hitler's launch of Operation *Barbarossa* against the Russians on 22 June 1941, Stalin asked the Allies for supplies by sea. The first of the so-called Arctic convoys began in August 1941. The route took them from various Allied ports north through the Norwegian Sea, across the top of Norway into the Barents Sea, then down to the Russian ports of Murmansk and Archangel.

The convoys began with few merchant ships and a small number of Royal Navy escorts.[3] The new year would see a rapid increase in convoy sizes, and it was immediately obvious that, along with the U-boats, the *Tirpitz* posed an enormous threat to the Arctic and North Atlantic convoys. Whereas the Royal Navy might defend convoys against U-boats with destroyers and corvettes, capital ships were needed against the *Tirpitz*, ships that could be better deployed elsewhere.

The *Tirpitz*'s presence in Norwegian waters was also to counter a threat, at least a perceived one. Hitler appears to have developed the view the Allies were planning to invade Norway, posing a significant threat to his northern flank. He directed Admiral Raeder to locate his capital ships in Norwegian waters from early 1942.[4]

In London, the *Tirpitz*'s disappearance from Wilhelmshaven had caused alarm. Convoy PQ 9 was delayed. Once the battleship was located, Churchill wrote to Major-General Ismay, Secretary to the Chief of Staff Committee, on Sunday 25 January, not mincing his words:

1. The presence of *Tirpitz* at Trondheim has now been known for three days. The destruction or even crippling of this ship is the greatest event at sea at the present time. No other target is comparable to it. ... The entire naval situation throughout the world

would be altered and the naval command in the Pacific would be regained.

2. There must be no lack of co-operation between Bomber Command and the Fleet Air Arm and aircraft-carriers. A plan should be made to attack both with carrier-borne torpedo aircraft and with heavy bombers by daylight or at dawn. The whole strategy of the war turns at this period on this ship, which is holding four times the number of British capital ships paralysed, to say nothing of the two new American battleships retained in the Atlantic. I regard the matter of the highest urgency and importance. I shall mention it in the Cabinet tomorrow, and it must be considered in detail at the Defence Committee of Tuesday night.[5]

Cabinet duly discussed the situation the following day, recorded as follows:

The *Tirpitz*: In answer to a question by the Prime Minister, the First Sea Lord and the Chief of the Air Staff said that the possibility of attacking the *Tirpitz*, now in Trondheim Fjord, had been exhaustively examined and it was hoped that an attack would be made as soon as weather conditions were favourable.

The Prime Minister said that it was of the utmost importance strategically that the *Tirpitz* should, if possible, be disabled or sunk.[6]

Churchill's directive had created a flurry of activity rapidly engaging with a host of problems.

Fættenfjord, ran west to east, narrowing at the eastern end, with steep slopes. The *Tirpitz* was moored close to the northern bank towards the eastern end, rendering a torpedo attack impossible. The Fleet Air Arm was ruled out; it had to be an RAF job.

Adding to the natural defences was the weather. It was winter, presenting icing problems for aircraft getting to and from the target. Cloud and fog were common around the fjords, with the weather generally being fickle. Aside from the possibility of blanketing the target during a raid, fog made it difficult for reconnaissance to gather accurate weather information prior to any planned raid.

To this, the Germans added their own defences. The *Tirpitz* was camouflaged, an anti-torpedo boom erected on the rear starboard side, with light gun emplacements, multiple flak batteries and searchlights installed around the fjord on land and sea. The ship itself had formidable anti-aircraft defences. The Germans were also in the process of installing 'smoke pots', designed to fill the fjord with smoke in the absence of fog.

Even without those factors, the RAF faced an almost insurmountable list of problems. Bombing German ships had proven exceedingly difficult, even in harbours with favourable terrain. Further, even if you hit the *Tirpitz* with a 2,000-lb armour-piercing or 500-lb semi-armour-piercing bomb, it seemed generally understood these could not penetrate the armoured deck plating, though they would damage superstructure.

Carrying these weapons were the relatively new four-engine bombers, the Stirling and Halifax Mk.II. Some squadrons designated for the forthcoming raids were just taking delivery of these aircraft, meaning crews had few flying hours on the bombers, let alone the experience for special operations of this kind. Nor was Churchill offering any time to practise, unlike the 11 weeks No 617 Squadron aircrews had for the 'Dambuster' raids the following year. There were questions about the bombers' range. Bomb aimers were still using the non-stabilised bombsight, making precision targeting even more difficult.

Bombing at night carried its own set of risks, but this was mitigated to a degree by mounting raids around the time of the full moon.

The first raid, on the night of 29 January 1942, mere days after Churchill's edict, was a total failure. Stirlings from Nos 15 and 149 Squadrons of No 3 Group and Halifaxes from Nos 10 and 76 Squadrons of No 4 Group were assigned for the attack. Mechanical failures, bad icing on the journey out, fuel shortage problems and a weather forecast for Norway that proved wrong saw none of the Stirlings and just one Halifax reach the target. It did no damage. The Stirling was ruled out of future consideration, better forecasts were needed, and new weaponry required.

On 6 March, the *Tirpitz* made it to sea, heading for convoy PQ 12 and the return convoy QP 8. It had no success, returning to Fættenfjord a week later, but the sortie caused a considerable stir, exemplifying the very threat Churchill had pointed out. By late March, the pocket battleship *Admiral*

Scheer and heavy cruisers *Admiral Hipper* and *Prinz Eugen* were in the neighbouring Lofjord immediately to the north of the *Tirpitz*.

The second RAF raid was mounted on 30 March by Halifaxes from Nos 10, 35 and 76 Squadrons. Planning was for two waves. The first would drop the new 4,000-lb bombs (known as 'cookies'); the second would drop naval Mk.XIX anti-submarine mines. These small spherical mines carried 100 lbs of explosive and were fitted with hydrostatic fuses set to detonate at a 30-foot depth. The aiming point was the port side, between ship and shore, where it was hoped the mines would produce a concussive force that damaged the hull plating. Greater precision in dropping the mines close to the stern of the ship could potentially damage the rudder or the screws. The Halifax bomb bays had to be modified to carry four of these mines.[7] Even then, the doors only half closed.

Delivering the mines with such accuracy required a line of attack from west to east up the fjord at a suggested 600 feet, hugging the northern face, one aircraft at a time. Once the mines were dropped, each bomber would ascend rapidly and turn to port, heading north before turning west to return to the staging point in Scotland.

The raid was a complete failure. Some reached the target area only to find it blanketed in fog. A few dropped their 'cookies' but none found their target. Others jettisoned their load and returned to base. Six aircraft were lost.

Don assumed command of 10 Squadron at RAF Leeming on 14 April 1942, right in the middle of preparations for the third raid on the *Tirpitz*. The detailed plans for the late-night raid had been delivered to the participating squadrons just three days earlier.

Don, leading the squadron in the attack, had very few flying hours in the Halifax Mk.II. He had done a familiarisation flight in late February, then four short flights between 5 and 8 April, totalling just under three hours.

The squadron had only begun taking delivery of Halifax Mk.IIs in December and they now required modification to hold the mines. Don's first opportunity to fly his new kite, W1041 'B-Baker', was on 22 April, the day before they departed for RAF Lossiemouth in Scotland. It was not a practise bombing run but a flight to test W1041's fuel system. The flight produced a wonderful anecdote. Due to the mines not fitting entirely within the bomb bays, aircrew were told to hand-crank the bomb bay doors shut to the extent possible, avoiding use of the hydraulic system:

> Wing Commander Don Bennett warned of dire consequences should the doors be damaged. After taking his own aircraft W1041:B out over the bay, Bennett was alarmed that he had forgotten to carry out his own order. He went up to Corporal Saunders and quietly asked him to repair the doors with wire. This Saunders did, the aircraft flying out to attack the *Tirpitz* in this condition.[8]

Over at 35 Squadron, some aircrew recorded being given two days practice. 'They were to practise jinking to port and starboard at low levels ... fly as close as possible to any steep hillsides and cliffs that were within the practice area.'[9] In lieu of practical experience, they spent time examining three-dimensional models of the fjord.

On 23 April, Don led the 10 Squadron Halifaxes north to RAF Lossiemouth. On board was Group Captain Strang Graham, RAF Leeming's station commander, who was directing the operation.

The full briefing took place two days later. The overall approach followed the March plan. During phase one of the attack, Halifaxes from 76 Squadron, accompanied by Lancasters of Nos 44 and 97 Squadrons from No 5 Group (less than two months after the type's first operational sorties), would drop 'cookies' from between 6,000 and 8,000 feet, aiming to disrupt the defences. During phase two, 20 Halifaxes from Nos 35 and 10 Squadrons would perform the low-level attack with the mines. Don's unit would be last in. Diversionary raids by Coastal Command on aerodromes and shipping were planned to keep the *Luftwaffe* busy elsewhere.

There had been some significant 'developments' since the March raid. Commencing their timed run from Salt Island (Saltøya), less than two miles from the *Tirpitz*, the Halifaxes now aimed to release their mines at 200 feet. If they encountered a smokescreen, they were to fly into it, dropping their mines using dead reckoning.

The level of nuance didn't end there. The original plan of dropping the mines with precision in the water between the towering hillside and the battleship had undergone something of a reality check. Although not in the official battle plans, aircrew recall being briefed they could widen their horizon by dropping their mines on the steep 400-foot-high hillside, from where they would bounce down and into the water.

Local Norwegians who later learned of the plan pointed out that the contours of the hillside and the plethora of trees would have likely prevented any mines, so dropped, reaching the water. The aircrews, already on a flight path perilously close to the side of the fjord, and undertaking the attack with little or no practice, were unlikely to get close enough to give it a go anyway.

They were given one other option: if an attack was not possible, they were directed to attack the other capital ships in Lofjord just to the north.

The five squadrons, based at RAF Lossiemouth, RAF Kinloss and RAF Tain, all received high-level visits prior to the raid, reinforcing the stakes involved in military and political terms: Air Vice-Marshal Carr visited RAF Kinloss; Sir Archibald Sinclair, the Minister for Air, went to RAF Lossiemouth; and Churchill himself visited RAF Tain with Marshal of the RAF Trenchard. A message from King George VI was read out.

Don took off from Lossiemouth at 7:00 pm on 27 April. The raid had been delayed a couple of days due to fog over the target, but a late afternoon meteorology reconnaissance flight showed the night was set to be clear. The 10 Squadron crews, well aware of their wing commander's navigational reputation and pre-war feats, now witnessed something unusual. Acting Flight Lieutenant Miller, pilot of W1037, recorded:

> We flew low all the way to Norway led by Wing Commander Bennett who dropped flame floats all the way and as the night was calm, the sea dead flat, they were easy to follow.[10]

It was, indeed, perfectly clear over the target and as the 90-minute-long attack began, the Germans started filling the fjord with smoke. Phase one of the attack saw the 'cookies' dropped from between 6,000 and 8,000 feet as planned, with the German searchlights coning some of the aircraft for the flak batteries. One aircraft was lost. The *Tirpitz* wasn't hit.

The second phase did not begin for 50 minutes:

> For each Halifax going in on its bombing run, the *Tirpitz* seemed to be lying at the end of an avenue of flak that increased in intensity as the aircraft neared the target. The defences on both sides of the fjord were projecting a storm of cross-fire into the skies above the battleship.[11]

Despite that, seven of 35 Squadron's Halifaxes dropped their mines in the vicinity of the target.

By the time the Halifaxes of 10 Squadron commenced their attack, it was bordering on suicidal. Jack Watts, in W1038, offers a chilling account of being one of the first in:

> At that very moment, we flew into a smokescreen so dense it seemed to be solid. It was like flying in cotton wool. There was no sense of motion, no spatial relativity. We knew that we were thundering alongside a solid, rocky cliff wall, practically brushing it with our wingtips and speeding towards an even higher and equally solid cliff wall not far ahead – all sight unseen. ...
>
> We had reached the end of our timed run. The *Tirpitz* must be dead ahead of us, almost underneath us now. I pressed the bomb release, and we pulled up at full throttle, hoping against hope that we had not miscalculated ... No one would have wished to end their life like a fly squashed on the wall.[12]

Don's W1041 'B-Baker' was one of the last to attack. They passed over Salt Island and pressed the stopwatch to begin the run.

The flak was intense, as Flight Lieutenant HG 'Mick' How, the rear gunner recalled: '... we were being shot at from the ship itself, from the squadron of ships around the *Tirpitz*, and from the hills and downwards onto us as well.'[13]

They hit the smokescreen at 400 feet. Don said it came as a complete surprise, remarking that Intelligence had not warned them explicitly; they had only received a vague reference to it at the briefing.[14] 'A split second later the ship's superstructure passed beneath us.'[15]

Don determined on a second run, pulling up which, he admitted, rendered the Halifax 'a sitting bird'. They were repeatedly hit, including behind the starboard inner engine, setting the wing on fire.

How, who admitted to being 'shit scared' after rotating his gun turret in the rear and seeing the starboard wing on fire, now received a direct hit 'which blew all the perspex out and gave me a nasty cut on the face and a bit of shock. I immediately told the Captain that I had been hit, and he said, "Okay, keep quiet", which I did.'[16] (It transpires Don had asked if he was okay and when he replied that he was, he then told him to keep quiet.)

They climbed to port out of the fjord and turned west, attempting to gain height to come around for a second bombing run up the fjord. It soon became clear this was not going to happen. The wing was ablaze. Don turned off the fuel to the starboard inner engine and deployed the fire extinguisher, to no effect.

Not able to complete the second run, he turned south-easterly towards where he thought the *Tirpitz* was moored. 'I pointed towards the ship's position, and released the mines. I often wonder where they went!'[17] The Halifax had, in fact, travelled a lot further than Don estimated because the mines landed about half a mile south of Borås Farm, four-and-a-half miles to the south-east of the battleship.

Don issued the order to prepare to abandon the aircraft, but when the fire seemed to die down, turned west again, heading for the Shetlands. It was short-lived because the fire flared again and was encroaching on the fuel tank. The engine now cut, and with it the undercarriage hydraulics failed, dropping the starboard wheel and destabilising the aircraft. Don recalled the starboard flap had also begun to trail. They turned east again, looking for a place to put the aircraft down, but the entire area was forested, so Don gave the order to bale out:

> I regret to say that one member of the crew became a little melodramatic. He said, 'Cheerio, chaps; this is it, we've had it.' I told

him very peremptorily to shut up and not be a fool, that we were perfectly all right but that we would have to parachute.[18]

Don sent Flight Sergeant John Colgan, the flight engineer, to help How out of the rear turret and get his parachute on, but realised as soon as Colgan departed that he didn't have his own parachute on. Colgan, seeing the parachute on his way to How, returned to the cockpit with it and clipped it on to Don's harness, much to Don's relief, before disappearing once again.

Colgan got the rear turret doors open, allowing How to crawl out and put on his parachute before getting him out the escape hatch and jumping immediately after. How's boots were ripped off in the slipstream and shortly after he found himself 'hanging in a fir tree ... swinging backwards and forwards, thinking "Oh Christ, what am I doing here?!".'[19]

Up front, Sergeant Tom Eyles,[20] the navigator, was first out, followed by Sergeant Harry Walmsley, the second pilot, and the two wireless operator/air gunners, Sergeants Clive Forbes and John Murray. Walmsley records them being so low he hit the snow-covered ground within seconds.

Don was alone in the aircraft 'holding the wheel hard over to port' when the starboard wing exploded and collapsed at about 600 feet. As he headed for the escape hatch, his parachute got caught twice, delaying him precious seconds. By the time he made it out, the aircraft was at just 200 feet.

Despite the late hour, Olaf Horten, a local farmer, witnessed the crash: '... he saw a burning aircraft approaching over the wooded hills, and noted four parachutists in the sky, and immediately before the crash a fifth one jumped out of the blazing aeroplane.'[21]

Seconds later, at about 11:15 pm,[22] the Halifax hit the ground in the hills of Flornesvollen about a mile south of the railway station at Flornes, some 15-and-a-half miles south-east of the *Tirpitz*, and exploded in flames.

26

Escape

Don's parachute only had enough time to open before he hit the snow, just enough to break his fall. Quite remarkably, he was uninjured.

> My reaction, I suppose, should have been one of deep gratitude, but instead the only thing I thought of was the shock it would be to my wife; I said a few words out loud addressed to her, and suddenly realised that I would rather be getting on with the job of trying to get back to her.[1]

Don's crew, all safely on the ground, were spread over some distance in the Flornesvollen to the south of Flornes, contemplating, like Don, what to do. The Swedish border was about 25 miles to the east-south-east, but to get there they would have to walk over inhospitable terrain, staying away from the main road and avoiding German patrols.

Tom Eyles, the navigator, initially headed east through the snow before turning north and wading down through a stream after having seen wolf tracks and worrying they might follow his scent. Ending up soaking wet and cold, he knocked on the door of a house at a log-clearing station. The Kringberg family took him in but were well aware of the consequences of harbouring Allied aircrew and, having a Nazi sympathiser living nearby, decided to turn him in.

Mick How, the rear gunner, hanging in the tree, snapped his quick release and dropped to the ground. Not having his boots immediately became a problem in the snow. He had just buried his parachute when John Murray, one of the wireless operator/air gunners, appeared. They walked for much of the night and then the following day, consulting their silk handkerchief-sized maps. They had a small compass too, only to discover it didn't work. Mick found an old pair of ski-boots. After staying in an empty log cabin on the second night, they were awoken by a nervous Norwegian who showed them where they were on the map and left them another compass.

They walked all the following day in what they thought was the direction of Sweden only to discover they'd walked in a wide circle, ending up at the same log cabin. The second compass had failed them. Exhausted and starving, they fell asleep, only to be woken by some Norwegians who were sent to get them by the Germans. How was in such bad condition they stretchered him out. The Nielsen family in the village of Østkil provided them with food, rest and medical assistance before the Germans took them to Oslo and on to Berlin.

Murray ended up in *Stalag Luft VI* and Mick How in *Stalag Luft III*. Years later, in recounting those days, How remarked, 'If it hadn't been for the great airmanship and thoughtfulness of Wing Commander Bennett, I would not have written this account. He undoubtedly saved my life.'[2]

John Colgan, the flight engineer, and Harry Walmsley, the second pilot, landed close to one another, buried their parachutes and set out for Sweden.[3] After a short encounter with some nervous Norwegians, they turned north, heading for the railway line and the main local road connecting Norway and Sweden. By pure chance they happened upon Johan Øyan and his wife at Sona, who welcomed them in, fed them and treated a hand wound.

Johan took them to meet an English-speaking friend, Johan Schiefloe, in the telegraph station. Between them, they advised Colgan and Walmsley to follow the railway line east (which they traced on one of the silk maps), but only at night. They told them about the two towns on the line, Gudå and Meråker. The former had a railway bridge, where the rail line crossed to the north side of the river, guarded by German sentry posts.

They set off, walking much of the night, and successfully navigated through Gudå. On the north-western outskirts of Meråker, they had another piece of luck. Bunking down in the outhouse of a farm at around 4:30 am, they lit a fire which was spotted half an hour later by the early rising farmer, Herlof Juliussen. The family took them in but was acutely

Escape

aware of a Nazi sympathiser nearby, and that there were a lot of Germans in Meråker.

Having fed them, they sent them on their way. The route Johan Øyan had marked on their map took them north-east, away from the railway line and main road into the forested hills, aiming for Lake Hällsjøen, some 14 miles away, whose eastern tip was in Sweden. On reaching Lake Fjergen, they encountered a Norwegian couple who kindly marked the two-and-a-half-mile route to Lake Hällsjøen on their skis! There, a Swedish military patrol met them, ushering them across the border to Storlien. After questioning, they were taken to the internment camp at Falun by train.

Having decided to get back to Ly as soon as possible, Don also headed east, but on encountering a fast-moving mountain stream, turned south, hoping to reach a point where he could cross it. It was nearing dawn, but still quite dark, when:

> Quite suddenly, I almost ran into a man! In fact, I did not see him until I was within two yards of him, nor did he see me. We both stopped abruptly. I vaguely remembered I was carrying a small automatic which I always had with me on operations ...'[4]

Before he could reach it, the man raised his hands, and Don now recognised him as Clive Forbes, one of his wireless operator/air gunners.

Trekking north-east, they were often trudging deep in snow with the bright light eventually causing both to have vision problems. They waded through streams, filling their boots with water that then froze causing frostbite. With a ration of three barley sugars and five Horlicks tablets a day, which Don remarked was 'not a strengthening diet', they were rapidly becoming exhausted.

Late afternoon on the first day, the 28th, they arrived at Gudå.[5] Having looked at a railway official through the window of the railway station hut, they discovered the railway bridge and sentry posts Colgan and Walmsley had been warned about. Heading south and up the hill overlooking Gudå, they came across a house and knocked on the door.[6] When there was no answer, they went inside and called out. Still no answer, so they went to the shed where they bedded down.

Finding it too cold to sleep, they set off again at 1:00 am. After walking much of the next day, they were exhausted and knocked on a farmhouse door. While Don stood on the doorstep explaining their exhaustion and hunger by means of words and hand gestures, the occupants just stared back in suspicion, saying nothing:

> I stopped talking to the Norwegians, therefore, and without taking my eyes off them I said to Forbes, 'Get ready to run. I don't like the look of this; when I give you the word to go, run as fast as you can.' No sooner had I said this than a grin broke out on the face of the man in front of me, and he said in perfect English, 'It's quite all right – come in.' I almost collapsed. He explained to me that his unfriendly reception was because he thought we might be German *provocateurs*. Apparently the police were in the district looking for us, and he was frightened that the Germans were trying to trick him.[7]

It turned out to be Bjørnås Farm in Torsbjørkdalen, and the man was John Dalamo. The family were the only English speakers in the area, John having been to America and Canada with the timber trade.[8]

While they were receiving treatment for frostbite, a young girl, Sofrid, came in and 'saw them taking off the airmen's boots, then peeling down their long, thick socks and oh, oh, terrible, their skin came off with their socks. Horrible.' John gave her a note to take to her father requesting they 'meet us at the usual place at 1:00 am'.[9]

After Don and Forbes had slept in front of the kitchen stove for a few hours, John took them to rendezvous with Sofrid's father, who led them up into the mountains to a small farm, Mannsæterbakken, where they slept some more.

Close to dawn, the owner, Trygve Dalanes, whom Don knew only as 'J.G.M.', set out on skis with them trudging along behind. On arriving at a high plateau above the tree line he pointed out 'some distant ridges' they understood to be Sweden and departed.

After walking for much of the day they came upon ski tracks running north-east to south-west. At regular intervals were colour-marked 'hummocks of stone'. Taking this to be the border, they were concerned the ski tracks might be evidence of German patrols, who would not hesitate to chase them across the porous border, so they kept going east, now downhill, as fast as they could go.

Around 10:30 pm, they saw a light and headed for it. They came across a fully lit building and saw people dancing inside, some of whom were in uniforms that looked quite German. They decided to knock on the back door, which was then opened and slammed shut in their face by a young girl. Turning to go, they were stopped in their tracks by two men.

> To my immediate relief and surprise, however, they once again greeted us – as had the Norwegians – in perfect English, saying, 'Welcome to Sweden, come inside.'[10]

They had reached Storvallen, a ski resort about 1.8 miles from the border town of Storlien. The very friendly welcome was shortly interrupted by the arrival of Captain Skoogh of the Swedish Army, who announced they were under arrest and locked them up for the night.

By the time Skoogh had escorted them to Storlien the following day, his officiousness had softened somewhat and Don persuaded him to send a cable to England to let Ly know he was alive. Don and Ly had agreed on a pre-arranged signal for such an eventuality during his time at Imperial Airways, which was the single word 'Love'. Don found it humorous because, due to wartime regulations, Skoogh had to sign it himself.

Ly duly received the signal 'Love, from Captain Skoogh, Swedish Army, Storlien' and shared it with Strang Graham, the Station Commander at Leeming, who promptly joined Ly and some officers to search an atlas to find Storlien.

From Storlien, Don and Clive Forbes were dispatched by train to the Falun internment camp.

On arrival, they were reunited with Harry Walmsley and John Colgan. They had already reported Don killed because, in seeing the Halifax go down, they were convinced he had not made it out. This simply magnified the importance of Don's early decision to get Skoogh to send his 'love' to Ly.

Relentless Skies – The Most Efficient Airman

The escape from Norway had been a close-run thing. The morning after the crash, eight Germans arrived at Olaf Horten's place and, on hearing he had seen the crash and the parachutes, made him lead them to the wreck. 'On approaching the smouldering Halifax, some of the sparks triggered one of the machine guns and shot one of the Germans in the lower abdomen.'[11]

Having searched the wreck, they mobilised rapidly to track down the crew members. In Don and Forbes's case, John Dalamo recorded '20 steel helmets with fixed bayonets' arrived the very morning the airmen had departed. The Germans had been following their tracks in the snow.[12] Dalamo didn't explain how they managed to bluff their way through an interrogation by the German soldiers, whom he found 'not very tender hearted', but the Germans left the farm to continue following the tracks.

Dalanes also felt the heat. After returning from ushering Don and Colgan towards the Swedish border, he too received a visit.

> When I got back I had a rest, then in the evening when we were eating, the dog began to bark and we saw a troop of Germans. In all, there were thirteen of them. The leader spoke good Norwegian and said they had followed the tracks of the two English pilots to the house. I denied there had been any pilots there. The Germans searched the shed and grounds, but not the house. I opened the window in the room where the pilots had stayed because there were bandages that smelt of petrol.
>
> Fortunately for us, the Germans were satisfied with the search, and carried on in the direction we had gone.[13]

Apparently Dalanes believed the German officer in charge wasn't in favour of shooting Norwegian civilians who helped Allied airmen to escape and this accounted for a lack of thoroughness in the search.

Escape

The internment camp at Falun was a fairly relaxed affair, with Allied airmen having the run of the town. Despite the attractions of the place, Don wanted his time there to be as brief as possible, maintaining a single-minded determination to get back to Ly and the squadron.

On the first day, he and Forbes received tickets to the local baths in Falun. He detoured to a local menswear shop where he ordered 'one of everything'. The only piece of paper extant from Don's time in Sweden, one on which he wrote names and made lists, reveals 'one of everything' was: a shirt, two collars, six handkerchiefs, a vest, a pair of underpants, two pairs of socks, pyjamas, a razor and blades, a brush, toothbrush, soap and a case.[14] His memoirs indicate that on arrival in the shop he added a suit, a tie and a pair of shoes. He also wanted a towel, but they didn't have one. He invited Forbes to shop likewise and then told the owner to 'charge it to the British Legation in Stockholm'.[15]

The names of the British Legation were also on that sheet of paper. Group Captain Richard Maycock[16] had been appointed Air Attaché in Stockholm in August 1941. His assistant, Squadron Leader LDH Fleet, took Don's call on that first day and, on hearing Don's demand he be released immediately, and why, offered a response one could describe as lukewarm and typical of official diplomatic channels. Don found the response unsatisfactory.

He was contemplating what to do next when he received some unexpected visitors. Åke Sundell, his wife Stella, and their friend Lena Bensenitz had driven up from Stockholm after reading in the papers there was an Australian wing commander being held at Falun. Apparently, the two women had been born in Australia.

Sundell is one of several intriguing characters on Don's sheet of paper. Don recalled in his memoirs Sundell was the general manager of the Carbon Gas Company, though he wrote '[*unreadable*]'[17] Gengas Aktiebolaget' next to Sundell's name. It was likely one of many companies working under the direction of the Wood Gas Board providing gasifiers to enable cars to run on wood.[18]

Sundell was well connected in Stockholm circles. He advised Don to get parole and head to Stockholm. Don applied and, while waiting, made another unexpected entry on his piece of paper: 'The Lady Constance

Malleson, Ramsnäs, Sundborn.' Malleson was a well-known British–Irish aristocrat, stage actor, activist and pacifist, known for having a life-long relationship with British philosopher Bertrand Russell, with whom she corresponded regularly. She was also a travel writer, especially about her beloved Nordic countries. In mid-1938, she had moved to Ramsnäs with her friend Dorothy Mallows, living there throughout the war:

> Malleson was a well-connected aristocrat and an actress (she signed 'Colette', her stage name) who worked for the British Council in Sweden. She had connections in the British Legation in Stockholm and seems to have operated in a kind of unofficial capacity when it came to airmen wishing to return to the war.[19]

She wrote about her day trips into Falun to shop and the Allied airmen she met there, including regular correspondence with Bertrand Russell.[20] On 12 May 1942, she wrote to him, concluding as follows:

> Sometimes, when I get in from shopping in the church-village, I find a scribbled note stuck in the door, signed by quite unknown R.A.F. men from Canada, Australia, Poland and England (sent to me by my Korsnäs hostess). They've all been either winged on their way here from Berlin, or else have come over the Norwegian mountains on foot – in which case they are sometimes damaged by frostbite, and have passed through the hands of my Lapland doctor friend, Einar Wallquist, of the cottage hospital at Arjeplog (where I stayed with him before the war).

> I was impressed by an Australian, Bennett by name: just about the most decisive character I've ever met – and with the broad outlook of having mixed with all sorts of people of different nationalities. (He was a commercial airways pilot before the war.) I took him to buy silk stockings for his Swiss wife, and the way he managed that transaction – quite without my help as a Swedish translator – was a wonder to behold. It's clear he'll manage our Legation and Swedish Authorities with the same decision – and will get himself flown to England without the slightest delay. A most refreshing character.[21]

On receiving his parole, Don headed for the British Legation in Stockholm, only to find an in-person interview appeared to produce no greater sense of urgency in repatriating him than he had achieved over the phone. He visited the Sundells, who arranged a meeting with Count Folke Bernadotte, who worked with the Swedish Red Cross. (Don believed he worked at the Foreign Office.) Bernadotte gave Don a good hearing as he mounted a spirited legal argument for immediate repatriation. It resulted in his being allowed to stay in Stockholm while his case was considered.

There were two possible paths for returning to England, and Don was now introduced to the first – Carl Florman, the managing director of AB Aerotransport (ABA), Sweden's international airline. ABA had commenced regular flights between Stockholm and RAF Leuchars in March 1942 using DC-3s painted with a *Sweden/Schweden* in large black lettering to identify them as neutral. Safe passage was not guaranteed because the Germans had placed demands on knowing passenger and cargo manifests, which the Swedish refused to give. The weather was variable and all flights took place in daytime.

The service was still in its infancy when Don arrived in Stockholm demanding a flight home. His referral to Florman meant ABA was under consideration. But there was another option: the British had their own service. The war, particularly with Hitler's invasion of Norway, had thrust Sweden into the spotlight. The British needed Swedish ball bearings. They were concerned about Sweden's ongoing relationship with Germany and were expanding their intelligence apparatus in Scandinavia. And Allied airmen, escaping the Germans, were heading for neutral Sweden.

British Airways had established a covert operation in late 1939 from RAF Leuchars to Oslo, Stockholm and Helsinki using Junkers Ju 52/3m trimotors and Lockheed 14 Super Electras (the airliner from which the Hudson was developed). The service had transferred to BOAC after the British Airways and Imperial Airways merger in 1940. Some of the pilots were Norwegians. The planes carried the company's livery, were slow, unarmed and struggled with the weather, making them prime targets for the *Luftwaffe*. In April 1940, one was attacked by a German seaplane, delaying flights for a year. By late May 1942, it was a night service.

It appears the British Legation had been more active than Don supposed because he and Walmsley, being pilots, were now offered a flight on a Lockheed 14. Don was expecting a night flight, but the Norwegian pilot decided, based on a faulty forecast about heavy cloud in the Kattegat, to take off in the middle of the day, resulting in their flying across the danger zone in clear skies. Having just been shot down by the Germans, Don was unnerved, but they arrived safely at RAF Leuchars. Don flew back to Leeming to be reunited with Ly and his children, Noreen and Torix, arriving just one month after being shot down.

On 29 May, Don and Harry Walmsley presented for interrogation by Intelligence in London. The report contains two curious entries regarding what happened after the crew baled out:

> 20. Shortly after landing, W/Cdr. Bennett was taken prisoner by a single German on skis. Soon after he had surrendered, the skier went away to round up some of the others, thus leaving him alone. He took full advantage and made off in the direction of Sweden. ...
>
> 21. Sgt Walmsley, after landing, heard shouts, and met F/Sgt Colgan but they were very soon taken prisoner by three German soldiers. Before long two of them went away to try and capture some of the other members of the crew ... thus leaving one man with a pistol to guard them.[22]

Walmsley goes on to say Colgan tackled the German, they knocked him out and escaped.

Both accounts were a fiction.

Don was aware of the Swedish law with respect to repatriating Allied airmen who had been downed in Norway and made it to Sweden. It drew a distinction between airmen captured by the Germans but who escaped ('escapers') and those who evaded capture ('evaders'). The former were 'technically "escaped prisoners of war" and, since they had lost their "belligerent" status, they were entitled to immediate repatriation'.[23] Not so the 'evaders'.

On arrival at Falun, Don had been interrogated by Löjtnant NO Hansson. Realising being an 'evader' placed in jeopardy an early return to

England, he manufactured the story of his capture. Evidently, Forbes, Walmsley and Colgan were all similarly aware of this legal technicality, with the latter two devising their own 'capture' story. This undoubtedly lay at the heart of the lengthy legal discussion Don refers to having with Count Bernadotte once he made it to Stockholm.[24]

For the sake of consistent paperwork and to avoid any further questions, Don and Walmsley now repeated their concocted stories for Intelligence back in London.

Don's story of escaping after being captured did create an amusing complication. He understood it was customary for Intelligence to recommend escapers not be permitted to fly on operations again. To counter this, he assured Intelligence he hadn't killed any Germans during his escape. True enough. In fact, he hadn't actually seen any Germans, except perhaps some sentries on duty at the bridge at Gudå.

While in London, Don was summonsed to Bomber Command Headquarters at High Wycombe, where he met with Harris for the first time since his appointment as head of Bomber Command. Harris told him he had been awarded an immediate Distinguished Service Order (DSO) for his contribution to the raid. Walmsley would receive the Distinguished Flying Medal. There was no hint of what was to come. The DSO citation appeared in *The London Gazette* of 16 June 1942:

> One night in April, 1942, Wing Commander Bennett and Sergeant Walmsley were the captain and second pilot respectively of an aircraft which attacked the German naval base in the Trondheim fjord. In spite of a fierce defensive barrage, the attack was carried out at an extremely low level. The aircraft was hit by shell-fire and, later, burst into flames. Wing Commander Bennett and Sergeant Walmsley were forced to escape by parachute but both landed safely in occupied territory. Both Wing Commander Bennett and Sergeant Walmsley displayed excellent resource and, after escaping from German soldiers and police, they eventually reached Swedish territory after a most arduous and trying journey across snow-clad

mountains. Throughout, both Wing Commander Bennett and Sergeant Walmsley displayed courage, initiative and devotion to duty of the highest order.

News of Don's DSO prompted an immediate response from Australia. On 16 June, a letter was penned in the Brisbane suburb of Ascot:

> The President, Vice-Presidents, Council and Members of the Brisbane Grammar School Old Boys' Association extend hearty congratulations to Wing Commander Donald Clifford Tyndall Bennett of the Royal Air Force on his elevation by His Majesty, the King, to the high honour of the Distinguished Service Order.
>
> We have watched with pride and satisfaction his brilliant career both before and during the War and especially his latest splendid achievement in Norway and Sweden which has been, as it were, a glorious addition and crowning achievement to his many previous illustrious deeds. We wish him many more years of success and of happiness.
>
> For and on behalf of the Council,
>
> J.G. Nowlan
> Honorary Secretary[25]

Don may not have matriculated at Brisbane Grammar, but they would have him as one of their own.

Don resumed command of 10 Squadron, but within a month was called back to Bomber Command Headquarters, where Harris broke the news of his personal decision to appoint Don commander of the newly created Path Finder Force. The appointment came with a promotion to Group Captain. It was almost exactly 12 years since he had commenced as a cadet at RAAF Point Cook in July 1930. Less than half those years had been in the Air Force, and even fewer in the RAF itself. He was 31.

Appendix 1
Chronology

14 Sep 1910	Born at 'Fairthorpe', the family home, Toowoomba, Queensland
Late 1919	Moves to Auchenflower, Brisbane, Queensland
1926	Achieves Junior Matriculation, Brisbane Grammar School (BGS)
Oct 1927	Father pulls him out of BGS
Late 1927/Early '28	Decides on career in aviation
27 Feb 1928	Bert Hinkler flies into Brisbane after record flight from England
9 Jun 1928	Charles Kingsford Smith and Charles Ulm fly into Brisbane after first trans-Pacific flight
1 Sep 1928	Joins the Citizen Forces of the Australian Army
Mid 1929	Accepted by Royal Australian Air Force for pilot training
24 May 1930	Amy Johnson arrives in Brisbane after record flight from England
15 Jul 1930	Commences as cadet at RAAF Point Cook
12 Jun 1931	Gains his wings as pilot officer
Aug 1931	No 5 Flying Training School, RAF Sealand
Sep 1931 – Jul 1932	No 29 (Fighter) Squadron, North Weald – fighter pilot training • Promoted to flying officer
Jul 1932 – Dec 1932	School of Naval Co-operation and Aerial Navigation, RAF Calshot – Flying Boat Pilot's Course
Jan 1933 – Jul 1933	No 210 (Flying Boat) Squadron, Pembroke Dock • Achieves civil Pilot's 'B' Licence (Feb 1933) • Achieves civil Second Class Air Navigator's Licence (Feb 1933)
Aug 1933 – Aug 1935	School of Naval Co-operation and Aerial Navigation, RAF Calshot – lecturer in navigation and flying boat instructor • Purchases DH.53 Humming Bird (Oct 1933)

	- Achieves First Class Air Navigator's Licence (Mar 1934)
- Flies with civil airline, Jersey Airways (June–Jul 1934)
- Competes in 1934 Centenary Air Race (Oct 1934)
- Achieves 'A', 'C' & 'X' Ground Engineer's Licences (Jan 1935)
- Achieves W/T Air Operator's Licence (Jul 1935)
9 Aug 1935	Departs Royal Air Force
21 Aug 1935	Marries Elsa 'Ly' Gubler
Oct – Dec 1935	In Australia on honeymoon
- Contemplates job with New England Airways
- Writes *The Complete Air Navigator*, published 1936
Jan 1936 – Jul 1940	Imperial Airways Limited
- Commences as first officer
- Writes *The Air Mariner* (Mar 1936), published 1938
- Daughter Noreen is born (14 Nov 1936)
- Takes command of Empire flying boat *Cassiopeia* (Feb 1937)
- Achieves first Egypt to UK flight in one day (May 1937)
- Son Torix is born (28 May 1938)
- First commercial transatlantic flight; first non-stop east to west flight from UK to Montreal; fastest east to west Atlantic crossing (Jul 1938)
- Awarded Oswald Watt Gold Medal
- Awarded Johnston Memorial Trophy
- Achieves world seaplane distance record – Dundee to South Africa (Oct 1938)
- Achieves first non-stop flight from UK to Egypt (Nov 1938)
- Part of in-flight refuelling trials and inaugurating commercial in-flight refuelling transatlantic services (Jul – Aug 1939)
- Takes distress call from torpedoed SS *Athenia* during transatlantic flight (Sep 1939)
- Elected Associate Fellow of Royal Aeronautical Society (Nov 1939)
- Rescues Polish General Staff from France (June 1940)

Aug 1940 – Aug 1941	Atlantic Ferry – flying superintendent • Leads first transatlantic flight of Lockheed Hudsons (Nov 1940) • Performs first survey flight of Greenland (April 1941)
25 Sep 1941	Rejoins RAF as acting wing commander
Sep 1941 – Dec 1941	No 1 Elementary Air Observer School, Eastbourne – 2IC • Establishes syllabus and gets school up and running • Elected Fellow of the Royal Aeronautical Society
Dec 1941 – Apr 1942	No 77 Squadron, RAF Leeming – Officer Commanding • Flies 13 operations
Apr – Jul 1942	No 10 Squadron, RAF Leeming – Officer Commanding • Shot down attacking the battleship *Tirpitz* • Awarded the Distinguished Service Order

Appendix 2
Aircraft Don Bennett flew to July 1942

Armstrong Whitworth Siskin III & IIIA	Lockheed Hudson II, III & V
Armstrong Whitworth Whitley III & V	Lockheed Model 18 Lodestar
Avro 504N	Lockheed Vega DL-1A
Avro 618 Ten	Miles M.2 Hawk
Avro Anson	Saunders-Roe A.27 London
Avro Tutor	Saunders-Roe A.29 Cloud
Boulton & Paul P.71A	Short Bros L.17 Scylla
Bristol Bulldog II	Short Bros S.8 Calcutta
Consolidated LB-30A Liberator B.Mk.I	Short Bros S.8/8 Rangoon
Consolidated PBY-5 Catalina	Short Bros S.14 Sarafand (passenger)
de Havilland DH.53 Humming Bird	Short Bros S.17 Kent
de Havilland DH.60G III Gipsy Moth	Short Bros S.20 *Mercury* (Mayo Composite)
de Havilland DH.60M Metal Moth	Short Bros S.23 Empire
de Havilland DH.60X Cirrus Moth	Short Bros S.30 Empire
de Havilland DH.84 Dragon	Short Bros S.33 Empire
de Havilland DH.86 Express	Spartan Three Seater
Fairey IIIF	Stinson (model not recorded)
Handley Page H.P.42	Supermarine Scapa
Handley Page Halifax II	Supermarine Southampton II
Hawker Hart	Westland Lysander
Klemm Kl32-V	Westland Wapiti Mk.IA & IIA

About the Author

Ian Campbell has had a lifelong interest in military history. Growing up behind the Australian War Memorial in Canberra, he spent many hours wandering the halls thoroughly absorbed in the displays. His wise mother fostered his love of reading and history by introducing him very early to the works of Paul Brickhill.

Ian has qualifications in History and Politics from the Australian National University. After a varied career in the government, private and not-for-profit sectors, he retired to research and write. His first book, *Thinks He's a Bird*, about Australian Pathfinder and Lancaster pilot, Flight Lieutenant Keith Watson, was released in 2022.

https://www.bigskypublishing.com.au/books/thinks-hes-a-bird/.

Ian is currently curator of the Bennett/Vial Archive at the Queensland Air Museum. This archive comprises three collections, of which AVM Don Bennett's private collection is one.

Ian and his wife, Kathy, have lived in many places in Australia, as well as New Zealand and Canada. They now live in Brisbane. When circumstances permit, they love to travel, but he has vowed never to move house again.

Bibliography

Air Vice-Marshal D.C.T. Bennett Personal Papers and Collection, Bennett/Vial Archive, Queensland Air Museum

Air Ministry, Paperwork related to First Class Air Navigator's Licence.

Air Ministry, Paperwork related to Seaplane Instructor's Certificate.

Air Ministry, *Pilots' Licences (Class 'B'): Extension of Licences to Cover Further Types of Flying Machines*.

Air Ministry, *Syllabus of Examination for Licence to Operate Radiotelephony Apparatus on board British Civil Aircraft*.

Bennett, D.C.T., *Code for Reporting State of Flight*, for *Mercury's* world seaplane record, 6-8 October 1938.

Bennett, D.C.T., *Flying Logbook*, 8 volumes, facsimile.

Bennett, D.C.T., 'Investigation into the fuel consumption of the Rapier VI engine in connection with long range operation', undated.

Bennett, D.C.T., Letter to the Commanding Officer, No 1 E.A.O.S., Eastbourne, requesting transfer to an operational squadron, 5 November 1941.

Bennett, D.C.T., Letter to the Under Secretary of State, Air Ministry, Stroud, Glos., Reference No A157375/41/Accts.2(a), 7 June 1942.

Bennett, D.C.T., 'Long Range Routes', memo to Imperial Airways management, 2 January 1939.

Bennett, D.C.T., Map collection.

Bennett, D.C.T., *Pathfinder*, Frederick Muller Ltd, London, 1958.

Bennett, D.C.T., *Personal Notebook – 1 EAOS to Formation of the Path Finder Force*.

Bennett, D.C.T., *Personal Notebook – Early Years*.

Bennett, D.C.T., *Personal Notebook – Imperial Airways*.

Bennett, D.C.T., *Personal Notebooks – RAAF Point Cook and RAF Calshot*.

Bennett, D.C.T., Personal notes during internment in Sweden, 1942.

Bennett, D.C.T., Personal notes for preparation of *Pathfinder* memoirs.

Bennett, D.C.T., Personal notes for writing about Atlantic Ferry.

Bennett, D.C.T., Personal papers regarding Imperial Airways flights, including transatlantic and *Maia/Mercury* flights.

Bennett, D.C.T., 'Proposal for Long Range Test Flight in Mercury using 100 Octane Fuel', memo to Imperial Airways management, March 1939.

Bennett, D.C.T., 'Proposed Route for Long Distance Test Flight', memo to Imperial Airways management, 7 March 1939.

Bennett, D.C.T., *The Air Mariner*, 2nd edition, Sir Isaac Pitman & Sons Ltd, London, 1943.

Bennett, D.C.T., *The Complete Air Navigator*, 6th edition, Sir Isaac Pitman & Son, London, 1954.

Bennett, Ly, 'My Nostalgic and Sentimental Journey to Norway, July 1989', unpublished account.

Canadian Pacific Railway Company & Ministry of Aircraft Production, Flying times for Captain D.C.T. Bennett, November 1940 to May 1941.

Evans, S/L, 4 Group HQ, Letter to Officer Commanding, RAF Station Leeming, Reference No 1001/1384/P.2, 24 March 1942.

Fédération Aéronautique Internationale, *Records Officiels*, Paris, 1 January 1964.

Grant, S/L E.A., 4 Group HQ, Letters to Officer Commanding, RAF Station Leeming, Reference No 1007/P.2, 17 and 18 December 1941.

HM Signal School, *Admiralty Handbook of Wireless Telegraphy, 1931*, His Majesty's Stationery Office, London, 1932.

Imperial Airways, *Handbook of Instructions and General Information*, Copy No 184, with amendments, Don Bennett's personal copy.

Nowlan, J.G., Honorary Secretary, Brisbane Grammar School Old Boys' Association, letter to Don Bennett on awarding of his DSO, 16 June 1942.

Vial, A.J.V., Record of phone conversation with Jack Kogler, 18 May 2006.

Interviews with Don Bennett

Imperial War Museum, 'Bennett, Donald Clifford Tyndall', IWM Sound Archive 9378, 11 August 1986, 5 reels, 130 minutes.

McGrath, Amy, 'Interview with Air Vice-Marshal Donald Bennett', National Library of Australia, Canberra, 1981.

Royal Air Force Centre for Air Power Studies (RAF CASPS Historic Interview), 'Maverick Leader and Navigator *Par Excellence*, Air Vice-Marshal Donald 'Pathfinder' Bennett CB CBE DSO', December 1980.

Books & Manuscripts

Air Ministry, *Atlantic Bridge: The Official Account of R.A.F. Transport Command's Ocean Ferry*, HMSO, London, 1945.

Air Ministry, *Merchant Airman: The Air Ministry Account of British Civil Aviation, 1939–1944*, HMSO, London, 1946.

Anderson, W., *Pathfinders*, Jarrolds Publishers Ltd, London, 1946.

Barnes, C.H., *Shorts Aircraft Since 1900*, Putnam Aeronautical Books, Great Britain, 1989.

Blanchett, C., *From Hull, Hell and Halifax: An Illustrated History of No 4 Group 1937–1948*, Midland Counties Publications, England, 1992.

Bluffield, R., *Over Empires and Oceans*, Tattered Flag Press, Great Britain, 2014, e-book edition.

Bramson, A., *Master Airman*, Airlife Publishing Ltd, Shrewsbury, 1985.

Campbell-Wright, S., *An Interesting Point: A History of Military Aviation at Point Cook 1914–2014*, Air Power Development Centre, Department of Defence, Canberra, 2014.

Carey, P., 'A Brief Account of the Career of Group Captain Alban Majendie Carey, CBE, compiled by his son, Peter, from various sources and discussions, for private circulation to family and friends', unpublished.

Casey, L.S. and Batchelor, J., *Seaplanes and Flying Boats*, Phoebus Publishing, Great Britain, 1980.

Cassidy, B., *Flying Empires*, Queen's Parade Press, Bath, UK, 1996.

Christie, C.A., *Ocean Bridge: The History of RAF Ferry Command*, University of Toronto Press, 1995.

Cookson, B., *The Historic Civil Aircraft Register of Australia (Pre War)*, The Print Approach, Narangba, 1996.

Coverdale, C.J., *Celebrating Bennett the Man 1910–1986*, self-published, Great Britain, 2018.

Feast, S., *Pathfinders: The Definitive Story*, Key Publishing Ltd, Great Britain, 2022.

Fysh, Sir Hudson, *Qantas Rising*, Rigby Limited, Sydney, 1965.

Goodrum, A., *School of Aces: The RAF Training School That Won the Battle of Britain*, Amberley Publishing, 2019.

Gula, J., *The Roman Catholic Church in the History of the Polish Exiled Community in Great Britain*, University of London, Great Britain, 1993.

Gunn, J., *The Defeat of Distance: Qantas 1919–1939*, University of Qld Press, 1985.

Harris, Sir Arthur, *Bomber Offensive*, Collins, London, 1947.

Harrison, J., 'Queensland Methodism until 1902', in Glen O'Brien & Hilary Carey, eds., *Methodism in Australia: A History*, Routledge Methodist Studies Series, Routledge, 2018.

Jackson, A.S., *Pathfinder Bennett*, Terence Dalton Limited, Suffolk, 1992.

Jefford, C.G., *Observers and Navigators and Other Non-Pilot Aircrew in the RFC, RFAS and RAF*, e-book edition, Grub Street, London, 2014.

Kieza, G., *Bert Hinkler: The Most Daring Man in the World*, e-book edition, Harper Collins, Sydney, 2012.

Lewis, J., *Jimmy Woods: Flying Pioneer*, Fremantle Arts Centre Press, Western Australia, 1989.

McKay, G., *Dreams, Deeds and Dedication: A History of the Real Estate Institute of Queensland*, REIQ, Brisbane, 2008.

Maynard, J., *Bennett and the Pathfinders*, Arms and Armour Press, London, 1996.

Mead, R., *Dambuster-in-Chief: The Life of Air Chief Marshal Sir Ralph Cochrane*, Pen & Sword, Great Britain, 2020.

Middlebrook, M. & Everitt, C., *The Bomber Command War Diaries: An Operational Reference Book 1939–1945*, e-book edition, Pen & Sword Aviation, Great Britain, 1985.

Morris, R., *Cheshire*, Penguin, England, 2000.

Neumann, C., *Australia's Citizen Soldiers, 1919–1939: A Study of Organization, Command, Recruiting, Training and Equipment*, MA Thesis, University of NSW (Duntroon), 1978.

Parnell, N. & Boughton, T., *Flypast: A Record of Aviation in Australia*, Australian Government Publishing Service, Canberra, 1988.

Penrose, D., 'De Havilland D.H.53 Humming Bird', in ed. David Ogilvy, *From Blériot to Spitfire*, Airlife Publications, England, 1977.

Probert, H., *Bomber Harris: His Life and Times*, Greenhill Books, London, 2001.

Retinger, J., *Memoirs of an Eminence Grise*, ed. John Pomian, Butler & Tanner, Great Britain, 1972.

Saward, D., *'Bomber' Harris: The Authorised Biography*, Cassell Limited, London, 1984.

Shinkfield, H., *Esse Potius Quam Videri: A Brief History of No 77 (Bomber) Squadron, R.A.F. during the War Years of 1939–1945*, privately published, Great Britain, 2000.

Sims, P.E., *Adventurous Empires: The Story of the Short Empire Flying Boats*, e-book edition, Pen & Sword, Great Britain, 2013.

Smith, N., *Tirpitz: The Halifax Raids*, Air Research Publications, Great Britain, 1994.

Ward, C., *4 Group Bomber Command: An Operational Record*, Pen & Sword Aviation, Great Britain, 2012.

Watt, S., *I'll Take the High Road: A History of the Beginning of the Atlantic Air Ferry in Wartime*, Brunswick Press, Fredericton, 1960.

Wilcockson, A.S., *The Reminiscences of A.S. Wilcockson*, Oral History Research Office, Columbia University, 1960.

Wilkinson, P., 'Sikorski's Journey to England, June 1940', in Keith Sword, *Sikorski: Soldier and Statesman – A Collection of Essays*, Appendix 1, Orbis Books, London, 1990.

Willey, K., *The First Hundred Years: The Story of Brisbane Grammar School 1868–1968*, Macmillan Company of Australia Pty Ltd, South Australia, 1968.

Military, Government & School – Archival Records & Publications

Australian War Memorial, AWM 65, 292, 'Donald Clifford Tyndall Bennett (32065)'.

Bensusan-Butt, D., *Night Photographs in June–July 1941: A Statistical Analysis*, AIR 14/1218.

Brisbane Grammar School, *General Register, Volume II, 1911–1927*.

Brisbane Grammar School, *The B.G.S. Magazine, Vol 30, June 1927*.

Brisbane Grammar School, *The B.G.S. Magazine*, Vol 159, December 1986.

Bennett, Captain D.C.T., 'Preliminary Report on Atlantic Crossing in "Mercury" – 20th to 27th July 1938, 29 July 1938', Air Ministry, Trans-Atlantic Experimental Flights – Official Reports of Flights by First Pilots, Part II, Reference 692151/37.

Bennett, Captain D.C.T., 'Amplifying Voyage Report on Atlantic Crossing in "Mercury" – 20th to 27th July 1938, 29 July 1938', Air Ministry, Trans-Atlantic Experimental Flights – Official Reports of Flights by First Pilots, Part II, Reference 692151/37.

Commonwealth of Australia Gazette, No 225, 20 August 1942, reference 'No 57184 Clyde Kinsey Bennett (5965)'.

Daily Commercial News & Shipping List, Industrial Section Commonwealth Applications section, 5 December 1928.

Gander Airport's Watch Log, excerpts, supplied by Sandra Seaward, Executive Director, North Atlantic Aviation Museum, Gander.

Hatch, F.J., 'ATFERO and Ferry Command', Norman Hillmer and W.A.B. Douglas collection (89/97), Box 3, File 13, Directorate of History and Heritage, Department of National Defence, Government of Canada, 1974. This is the draft of what would appear as the much-shorter 'Ferry Command' in the official publication.

Hatch, F.J., 'Ferry Command', Appendix D, in Hillmer, N., Editor-in-Chief, *The Creation of a National Air Force: The Official History of the Royal Canadian Air Force, Volume II*, University of Toronto Press, 1986.

Hatch, F.J., 'Ferrying and Air Transport on the North Atlantic Routes', Norman Hillmer and W.A.B. Douglas collection (89/97), Box 3, File 11, Directorate of History and Heritage, Department of National Defence, Government of Canada, 1974.

Hatch, F.J., 'Interview with Air Vice-Marshal D.C.T. Bennett C.B., C.B.E., D.S.O. regarding the inauguration of the North Atlantic Ferry Organisation and the formation of the Pathfinder Force in Bomber Command', Biographical File: Bennett, D.C.T., Directorate of History and Heritage, Department of National Defence, Government of Canada, 1976.

Hillmer, N., et. al, 'Interviews with retired Air Marshals and Air Vice-Marshals', Norman Hillmer and W.A.B. Douglas collection (89/97), Box 1, File 7, Directorate of History and Heritage, Department of National Defence, Government of Canada, 1973, p. 174.

Hixson, Lt. Col. J. & Cooling, Dr B.F., 'Combined Operations in Peace and War', Appendix B: Anglo-French Liaison, 1939–1940, US Army Military History Institute Publication, Pennsylvania, Revised edition 1982.

Les Hourtiquets Base logbook, excerpts, supplied by Alice Vivancos, Musée de l'Hydraviation, Biscarrosse, France.

National Archives of Australia, 'Australian Military Forces: Attestation Form for Persons Voluntarily Enlisted in the Citizen Forces', Record: Bennett, Donald Clifford, Army No 120622, NAA: B4747 Item 9263682.

Office of the Historian, Foreign Service Institute, United States Department of State, 'The Neutrality Acts, 1930s', at https://history.state.gov/milestones/1921-1936/neutrality-acts.

'Rescue of Polish Cabinet from France, June 1940', provided by Musée de l'Hydraviation, Biscarrosse, France.

Sikorski, General W., 'Record of Daily Events', Polish Institute and Sikorski Museum, Archive Ref KOL 1/1 DCZ.NW, London.

Toowoomba Grammar School, *The Toowoomba Grammar School Magazine and Old Boys' Register*, selected entries from November 1914 to November 1917 on Clyde Kinsey Bennett and Aubrey George Bennett.

UK National Archives, *Operations Record Book, 77 Squadron*, December 1941 – April 1942, AIR 27/655 and AIR 27/656. Available from the 77 Squadron Association website.

UK National Archives, Statement by W/Cdr D.C.T. Bennett and Sgt H. Walmsley, No. 10 Squadron – No 4 Group, regarding 27 April 1942 attack on the *Tirpitz*, taken by W/Cdr S.D. Felkin, 29 May 1942, A.I.1. (k) No E62/1942, AIR 40/258.

Woods, J., *Personal Diaries 1934*, State Library of Western Australia, ACC 4638A/39-50.

Woods, J., *Correspondence*, State Library of Western Australia, ACC 4638A/24.

Woods, J., *England to Australia air-race. Official papers*, State Library of Western Australia, ACC 4638A/66.

Conference Presentations, Theses & Journal Articles

Ayliffe, Alec, *RAF Navigation Between the Wars*, A History of Navigation in the Royal Air Force, RAF Historical Society Seminar, RAF Museum, Hendon, 21 October 1996.

Chugh, S.R.S., 'Tides in Hooghly River', *Hydrological Sciences Journal*, 1961; 6(2):10–26.

Davis, Jeffrey, 'ATFERO: The Atlantic Ferry Organisation', *Journal of Contemporary History*, Vol 20, No 1 (January 1985).

Freer, Paul, *Circumventing the Law that Humans cannot see in the Dark: an Assessment of the Development of Target Marking Techniques to the Prosecution of the Bombing Offensive during the Second World War*, Doctor of Philosophy in History, University of Exeter, 2017.

Holman, B., 'The Meaning of Hendon: the Royal Air Force Display, aerial theatre and the technological sublime, 1920–1937', *Historical Research* 93, 259 (2020).

Holman, B., 'Ending Hendon – VI: 1935–1937', *Airminded*, 2 December 2011, at https://airminded.org/2011/12/02/ending-hendon-vi-1935-1937/.

Jefford, C.G., *The Epic Flights*, A History of Navigation in the Royal Air Force, RAF Historical Society Seminar, RAF Museum, Hendon, 21 October 1996.

Jefford, Jeff, 'Corps Reconnaissance 1914–18', Royal Air Force Historical Society, *Journal 54*, UK, 2013.

McCarthy, J., *An Air Force Fit for Air Displays? The RAAF 1921–1939*, conference paper for 90 Years of RAAF History, Canberra, 2012.

Male, C., 'A Short History of the Royal Aeronautical Society', *Aerospace*, Journal of the Royal Aeronautical Society, January 2016, available on Royal Aeronautical Society website.

Metcalf, Dr W.J., 'Dr Thomas Pennington Lucas: Queensland Scientist, Author, Doctor, Dreamer and Inventor', *Royal Historical Society of Queensland Journal*, Vol 19, No 5, February 2006.

Page, D., 'The Early Years', A History of Navigation in the Royal Air Force, RAF Historical Society Seminar, RAF Museum, Hendon, 21 October 1996.

Pirie, G., 'Incidental Tourism: British Imperial air travel in the 1930s', *Journal of Tourism History*, Vol 1, No 1, March 2009.

Richards, Clive, 'All Undiluted Nonsense? The Royal Air Force in the Army Co-operation Role, 1919–1940', Royal Air Force Historical Society, *Journal 54*, UK, 2013.

Royal Queensland Aero Club, 'The History of the Royal Queensland Aero Club, Part V', Joystick Jottings, Vol 6, No 1, July/Aug 1984.

Salvemini, G., 'Ciano's Diaries', *The Atlantic*, March 1946, at www.theatlantic.com/magazine/archive/1946/03/ciano-diaries/656062.

Vernon, J.E., 'Horses on the Payroll', *Royal Canadian Air Force Journal*, Vol 5, No 2, Spring 2016.

Vié-Klaze, Marie-Paule, 'En Juin 1940, Les Hourtiquets Sont le Theatre D'un Fait Historique'.

Watson, D., 'British Overseas Airways Corporation 1940–1950 and its legacy', *Journal of Aeronautical History*, Paper No 2013/03.

Non-Bennett Correspondence

The Lady Constance Malleson, Letter to Bertrand Russell, 12 May 1942, supplied by David Gray, University of Dalarna, Sweden.

Select Online Portals, Posts and Blogs

77 Squadron Association, '77 Squadron History 1937–1945', at https://77squadron.org.uk/.

ADF-Serials: Australian & New Zealand Military Aircraft Serials & History, www.adf-serials.com.au/raaf1.htm.

Anson, J., 'World War II Bordeaux: Life Under Occupation', at http://janeanson.com/world-war-ii-bordeaux-life-under-occupation.

'British Imperial Airways Handley Page H.P.42/45 Aircraft', at barthworks.com/aviation/aviation_historic_artifacts/handleypage/handleypage.htm.

Bliss, Philip (words), 'Dare to be a Daniel', hymn.

Bloise-Brooke, M., 'The Schneider Story', at www.supermarineseaplane.co.uk/the-schneider-story.

British Pathé Archive, '"Mayo" Composite Plane (1938)', at www.britishpathe.com/asset/45168/.

British Pathé Archive, 'Planes Across the Border (1940)', at www.britishpathe.com/asset/46881/.

Burgess, C., 'Air Vice Marshal D.C.T. Bennett: The man who formed the Pathfinders', *Transit*, Qantas Airways, 1983.

Cuskelly, R., 'Vega VH-UVK', Adastron Aviation History, at www.adastron.com.

Cuskelly, R., 'Wicko Cabin Sports VH-UPW', Queensland Air Museum website, at qam.com.au/collection/wicko-cabin-sports-vh-upw/.

Davies, I., 'A Casualty of War & Fate', *Aeroplane*, September 2001, pp. 36–41.

Dougherty, K., 'The Story of Australia's First Airmail – Part 5', Powerhouse Museum, at https://collection.powerhouse.com.au/.

East Anglian Film Archive, 'A newsreel-style amateur film celebrating the opening of Luton Airport in July 1938', at http://eafa.org.uk/work/?id=1049755.

'Eye Witness', 'I Did the Transatlantic 'Hop' with American Bombers', *The War Illustrated*, No 89, Vol 4, May 16th 1941.

Fagan, D., 'RAF Calshot', at www.hampshireairfields.co.uk/airfields/cal.htm.

Forces War Records, 'Unit History: RAF Calshot', at www.forces-war-records.co.uk/units/2035/raf-calshot.

'Glossary of Terms used in the Australian Flying Corps', School of Literature, Language and Linguistics, ANU, at slll.cass.anu.edu.au/centres/andc/glossary-terms-used-australian-flying-corps.

Goodall, G., Aviation History Site articles: 'Australian de Havilland D.H.84 Dragons', 'MacRobertson Miller Airlines (MMA)', 'The Australian GENAIRO Series', at www.goodall.com.au.

Gray, D., 'Constance Malleson's journey in Dalarna', *Dalarna Women's History Association*, at www.dalarnas-kvinnohistoriska.se/constance-mallesons-resa-i-dalarna/.

Gray, J.M., 'Harris, Bennett and Flying Boats', RAF Pathfinders Archive, at https://rafpathfinders.com/harris-bennett-flying-boats/.

Hampshire Aeroplane Club, 'About Us', at www.flyhac.co.uk.

'Hampshire Aeroplane Club', *Grace's Guide to British Industrial History*, at www.gracesguide.co.uk/Hampshire_Aeroplane_Club.

Kaijser, A., 'Driving on wood: the Swedish transition to wood gas during World War Two', Taylor & Francis Online, 31 March 2022, at www.tandfonline.com/doi/full/10.1080/07341512.2022.2033387.

Lewis, V., 'They Bridged the Atlantic', *Daily Sketch*, 13 February 1945.

Lockwood Family Collection, 'Royal Naval Review, July 16th 1935', Screen Archive South East, University of Brighton, at https://screenarchive.brighton.ac.uk/detail/8968.

McCallum, P., 'Stephen James Patch (1881–1943)', at wikitree.com/wiki/Patch-772.

Movietone News, 'The Short Mayo Composite', Bomberguy Aviation History, at www.youtube.com/watch?v=bYtazEBQ1K8.

Pearce, W., 'Napier H-16 Rapier Aircraft Engine', *Old Machine Press*, at https://oldmachinepress.com/2020/08/20/napier-h-16-rapier-aircraft-engine/.

Pelley, R.G., 'John Joseph Gilmore MBE', *Bob's Gander History*, at http://bobsganderhistory.com.

Pett, G., 'Blind Landings – the audio', *White Water Landings*, at http://whitewaterlandings.co.uk/voice-clips/blind-landings-the-audio.

Prophet, M., 'The PBY Catalina (the early history)', at www.michaelprophet.com/News_articles/PBY_CAT.html.

RCAF Association, 'Frederick Joseph Mawdesley', at www.rcafassociation.ca/heritage/search-awards.

Royal Aeronautical Society website, at www.aerosociety.com.

The Signals Corps, USA, *Catalogue of Official A.E.F. Photographs, Volume 2*, War Department Document No 903, Washington, 1919.

'The Do 18 Dornier's Whale Calf', *Air International*, Vol 18, No 4, April 1980.

Tibbo, F., 'Bennett', 'McTaggart-Cowan', 'The First Ferry With Tripp', 'Ferryman', *Gander Airport Historical Society*, at www.ganderairporthistoricalsociety.org/ferry_command.htm.

Toowoomba Region, 'The First Flights on the Darling Downs', under *Toowoomba Aerodrome*, at www.tr.qld.gov.au/our-region/history/historic-locations/117-historic-toowoomba-region-locations.

Wright Aeronautical Division & Curtiss-Wright Corporation, *Historical Engine Summary (beginning 1930)*, at www.enginehistory.org/Piston/Wright/C-WSpecsAfter1930.pdf.

Newspaper and Magazine Articles

'£10,000 Centenary Air Race', *The Herald*, Melbourne, 29 March 1933.

Australian Associated Press London correspondent, 'Another Flight Across the Atlantic – German Catapult Aeroplane', *The Courier Mail*, 25 July 1938.

'A New Agency Firm', *The Telegraph*, Brisbane, 12 April 1919.

'Air Cadets – Successful Trainees for England', *The Herald*, Melbourne, Saturday 20 June 1931.

'Air Marshal Who "Grew Up Early"', *Sunday Mail*, Brisbane, 11 June 1944.

'Amy Johnson Crashes at Eagle Farm', *Daily Standard*, Brisbane, 30 May 1930.

'Australian Air Force: Chief Answers Criticisms', *The Argus*, Melbourne, 10 October 1928.

'Aviator's Feat', *The Telegraph*, Brisbane, 30 June 1913.

Australian Associated Press London correspondent, 'Brisbane Pilot Flies Atlantic – Pick-a-Back Record', *The Courier Mail*, Friday 22 July 1938.

Australian Associated Press correspondent, 'Atlantic Flown Twice – The Mercury Returns to England', *The West Australian*, Friday 29 July 1938.

Bennett, Arnold, 'Learning to Fly at 70', *Sunday Mail*, Brisbane, 26 April 1981.

'Brisbane Airman's Rapid Rise', *The Courier-Mail*, Brisbane, 10 January 1944.

British Official Wireless, '"Pick-a-Back" Plane. Two Records. Eire to Canada Non-stop.', *Sydney Morning Herald*, Saturday 23 July 1938.

'Coming of Age', *The Brisbane Courier*, 26 August 1931.

Deiley, R., 'The Ultimate! When 'boats' ruled the air', *Gold Coast Bulletin*, Saturday, July 23 1988.

'Downs Aviator. Mr A.W. Jones. Sketch of his Career', *Darling Downs Gazette*, 13 June 1913.

'Egypt–England in 15 Hours', *The Argus*, Melbourne, Tuesday 25 May 1937.

Evans, J, recorded by Bruce Sinclair, 'When "Bomber" Harris flew at Pembroke Dock', Tivy-*Side Advertiser*, 4 July 2020, www.tivy-sideadvertiser.co.uk/news/pembrokeshire_news/18548431.bomber-harris-flew-pembroke-dock/.

'G.T. Bennett and Co.', *The Brisbane Courier*, 12 April 1919.

'In the Air. Mr A.W. Jones Makes Successful Flight', *Darling Downs Gazette*, 16 June 1913.

McPhie & Co., Auctioneers, Toowoomba and Dalby, 'For Auction at Our Office, on Friday, June 15th, at 11 a.m., under instructions from Mr Geo. T. Bennett, 'Kanimbla", *The Dalby Herald (Qld)*, Friday 1 June 1928, p. 2.

'Personalities in Real Estate – Mr G.T. Bennett', *The Courier-Mail*, Brisbane, 24 November 1936.

Phillips, L., 'Tributes flow as air ace makes his last journey', *The Advertiser*, 24 September 1986.

'President to Get Saturday's Grouse', *The Brooklyn Daily Eagle*, Monday 14 August 1939.

QTR, 'Atlantic Ferry Command – A Wartime Triumph', *The Argus*, Melbourne, 15 May 1943, p. 1.

Rollason, K., 'Emerson at War', *Winnipeg Free Press*, 8 November 2019, at www.winnipegfreepress.com/featured/2019/11/08/field-of-schemes-emerson-at-war.

Stuart, Frank S., 'This Flier Will Make Air History', unidentified newspaper article, in Australian War Memorial, AWM 65, 292, Donald Clifford Tyndall Bennett (32065).

'Successful Trials, New Farman Aeroplane', *Sunday* Mail, Brisbane, 17 November 1929.

'The end of an era – 'Woodsie' dies', *The West Australian*, Saturday 10 May 1975.

Endnotes

Introduction

[1] Sholto Watt, *I'll Take the High Road*, Unipress, Fredericton, NB, Canada, 1960, p. v.
[2] D.C.T. Bennett, *Pathfinder*, Frederick Muller Ltd, London, 1958.
[3] British Broadcasting Organisation, 'The Reith Lectures 2017', at www.bbc.co.uk/mediacentre/proginfo/2017/24/the-reith-lectures-2017. The quote was part of the first lecture in the series.
[4] Those making the selection were Robin Prior, Associate Professor of History, University College at the Australian Defence Force Academy, and Trevor Watson, Emeritus Professor of History, University of Adelaide.

Chapter 1: Decision Time at 'Kanimbla'

[1] Bennett, *Pathfinder*, p. 17.
[2] Personal notes for Pathfinder under heading 'The Author's Life Summarized', Bennett/Vial Archive, Queensland Air Museum.
[3] Bennett, *Pathfinder*, p. 21.
[4] Ibid.

Chapter 2: A Glimpse of the Future

[1] Bennett, *Pathfinder*, p. 21. The Wright brothers never came to Australia. Don's brother's memory of a biplane enabled a search identifying the only one landing at Toowoomba racecourse in 1913.
[2] Arnold Bennett, 'Learning to Fly at 70', *Sunday Mail*, Brisbane, 26 April 1981, p. 2.
[3] 'Aviator's Feat', *The Telegraph*, Brisbane, 30 June 1913, p. 4.
[4] 'Downs Aviator. Mr A.W. Jones. Sketch of His Career', *Darling Downs Gazette*, 13 June 1913, p. 7.
[5] Arnold's recollection that it was the first flight into Toowoomba was partly correct. It was the first flight onto, and from, the Toowoomba racecourse. The first flight to Toowoomba had been by another barnstormer, the famous Arthur Burr 'Wizard' Stone on 27 July 1912. Toowoomba Region local government website, 'The First Flights on the Darling Downs', under *Toowoomba Aerodrome*, at www.tr.qld.gov.au/our-region/history/historic-locations/117-historic-toowoomba-region-locations.
[6] 'In the Air. Mr A.W. Jones. Makes Successful Flight', *Darling Downs Gazette*, 16 June 1913, p. 5.
[7] Ibid.
[8] 'Downs Aviator', p. 7.
[9] Ibid.

Chapter 3: The Last Hope of the Bennetts

[1] 'A New Agency Firm', *The Telegraph*, Brisbane, 12 April 1919, p. 10.
[2] McPhie & Co., Auctioneers, Toowoomba and Dalby, 'For Auction at Our Office, on Friday, June 15th, at 11 a.m., under instructions from Mr Geo. T. Bennett, 'Kanimbla'', *The Dalby Herald (Qld)*, Friday 1 June 1928, p. 2.
[3] Ibid. The sale advertisement mentions 300 cattle. Pat McCallum, 'Stephen James Patch (1881–1943)', mentions numbers between 390 and 460, wikitree.com/wiki/Patch-772.
[4] For a biography of Steve Patch see Pat McCallum, 'Stephen James Patch (1881–1943)', at wikitree.com/wiki/Patch-772.
[5] 1861 census in England.
[6] Imperial War Museum, 'Bennett, Donald Clifford Tyndall', IWM Sound Archive 9378, 11 August 1986, at www.iwm.org.uk/collections/item/object/80009167.
[7] Toowoomba Grammar School, *Toowoomba Grammar School Magazine and Old Boys' Register*, November 1914, p. 19.
[8] Toowoomba Grammar School, *Toowoomba Grammar School Magazine and Old Boys' Register*, November 1917, p. 13.
[9] Bennett, *Pathfinder*, p. 15.
[10] Dr William J. Metcalf, 'Dr Thomas Pennington Lucas: Queensland Scientist, Author, Doctor, Dreamer and Inventor', *Royal Historical Society of Queensland Journal*, Vol 19, No 5, February 2006, pp. 788–804.
[11] I am grateful to Dr Bill Metcalf, a Brisbane historian, for supplying me with his research on Dr TP Lucas as well as taking me on a walking tour of South Brisbane where the Lucas family lived, including lunch at the Ship Inn, opposite where the Lucas house used to stand.
[12] Metcalf, 'Dr Thomas Pennington Lucas', p. 793.
[13] Ibid., p. 797.
[14] Bennett, *Pathfinder*, p. 16. Interestingly, over 20 years later, he heard this quote repeated by his squadron commander at No 29 (Fighter) Squadron, North Weald, in 1931.
[15] In a phone call by Jack Kogler, who knew Don in his youth, to Allan Vial, President of the PFF Association (Qld), on 18 May 2006.
[16] Bennett, *Pathfinder*, p. 15. An excellent newspaper article marking the church's 21st birthday in 1931 describes its various activities, including 'Mr Bennett' running the boys' club. 'Coming of Age', *The Brisbane Courier*, 26 August 1931, p. 13.
[17] John Harrison, 'Queensland Methodism to 1902', in G. O'Brien & H.M. Carey, *Methodism in Australia: A History*, Routledge, London, 2016, p. 77.
[18] Bennett, *Pathfinder*, p. 94.
[19] Ibid., p. 16.
[20] Don Bennett, *Personal Notebook from Early Years*, pp. 1–2. Bennett/Vial Archive, Queensland Air Museum.

Chapter 4: Nil Sine Labore – the Brisbane Years

[1] 'Personalities in Real Estate. Mr G.T. Bennett', *The Courier-Mail*, 24 November 1936, p. 22. See also G. McKay, *Dreams, Deeds and Dedication: A History of the Real Estate Institute of Queensland*, REIQ, Brisbane, 2008, p. 35.
[2] 'Air Marshal Who "Grew Up Early"', *Sunday Mail*, Brisbane, 11 June 1944, p. 8.
[3] Ibid.
[4] Bennett, *Pathfinder*, p. 18.
[5] 'Air Marshal Who "Grew Up Early"', p. 8.
[6] Department of Air, Directorate of Public Relations, an episode of 'These Men Make RAAF History', broadcast over Station 2KO Newcastle, 19 October 1944, in Australian War Memorial, AWM 65, 292, 'Donald Clifford Tyndall Bennett (32065)', p. 52.
[7] *Daily Commercial News & Shipping List*, Industrial Section Commonwealth Applications section, 5 December 1928, p. 8.
[8] Bennett, 'Learning to Fly at 70', p. 3.
[9] Ibid.
[10] Bennett, *Pathfinder*, p. 22.
[11] Brisbane Grammar School, *General Register, Volume II, 1911–1927*, p. 234.
[12] Keith Willey, *The First Hundred Years: The Story of Brisbane Grammar School 1868–1968*, Macmillan Company of Australia Pty Ltd, South Australia, 1968, p. 379.
[13] Bennett, *Pathfinder*, pp. 20–1.

Chapter 5: The Hinkler Effect

[1] The timing of Don's decision is fascinating. Kanimbla's manager, Steve Patch, was convicted on a charge of 'ringing in' over two appearances in court on 23 May and 27 June 1928. He had been caught droving a larger number of cattle to the Dalby cattle sales than he had a permit for. See McCallum, 'Stephen James Patch (1881–1943)', at wikitree.com/wiki/Patch-772; also '£90 in Fines. Stock Prosecutions', *The Dalby Herald (Qld)*, 25 May 1928, p. 5. There is no suggestion Patch had done this at George Bennett's direction or that Don himself was involved. However, Patch's conviction and Don's decision not to become the manager of the property coincided with George's decision to put 'Kanimbla' up for sale on 1 June 1928. If George had hoped Don would succeed Patch, whose position was now untenable, he would have been sorely disappointed.
[2] Bramson recounts Celia being more opposed than George to Don's decision to pursue an aviation career, arguing it was dangerous and not regarded as a 'profession' because it did not require a university degree. '[George] held similar views but nevertheless adopted a generous position. "It's your life – you do what you want."' Alan Bramson, *Master Airman*, Airlife Publishing Ltd, Shrewsbury, 1985, p. 8. In Australian terms, the quote offered can be read two ways – 'a generous position' or, as I've chosen to see it given the context, thorough exasperation, an acknowledgement Don was going to do whatever he wanted.
[3] Bennett, *Pathfinder*, p. 21. The precursor to the Royal Flying Doctor Service was the Aerial Medical Service, established by the Rev John Flynn of the Australian Inland

Mission in league with Qantas on 27 March 1928 at Cloncurry. This suggests Aubrey was flying on medical missions with Qantas pilots in outback Queensland prior to that time.

[4] Imperial War Museum, 'Bennett, Donald Clifford Tyndall', IWM Sound Archive 9378, 11 August 1986, at www.iwm.org.uk/collections/item/object/80009167. See also *The Medical Journal of Australia*, 13 November 1926, p. 677, for registration of Aubrey George Bennett at Camooweal.

[5] Bennett, *Pathfinder*, p. 21.

[6] Amy McGrath, 'Interview with Air Vice-Marshal Donald Bennett', National Library of Australia, Canberra, 1981, at nla.gov.au/nla.obj-215265933/listen.

[7] A thoroughly entertaining biography of Hinkler is by Grantlee Kieza, *Bert Hinkler: The Most Daring Man in the World*, Harper Collins, Australia, 2012.

[8] Ibid., Chapter 22.

[9] Bennett, *Pathfinder*, p. 22.

[10] National Archives of Australia, 'Australian Military Forces: Attestation Form for Persons Voluntarily Enlisted in the Citizen Forces', Record: Bennett, Donald Clifford, Army No 120622, NAA: B4747 Item 9263682. For details see Claude Neumann, *Australia's Citizen Soldiers, 1919–1939: A Study of Organization, Command, Recruiting, Training and Equipment*, MA Thesis, University of NSW (Duntroon), 1978.

[11] Bennett, *Pathfinder*, pp. 22–3.

[12] Maynard notes the RAAF Selection Board had a rule that candidates wait six months for re-assessment following any form of surgery. John Maynard, *Bennett and the Pathfinders*, Arms and Armour Press, London, 1996, p. 31.

[13] Ibid., p. 23.

[14] Ibid.

[15] Geoffrey N. Wikner was six years older than Bennett and a cousin of Captain Edgar Percival. See Ron Cuskelly, 'Wicko Cabin Sports VH-UPW', Queensland Air Museum website, at www.qldairmuseum.au/qam-content/aircraft/wicko/VH-UPW.htm. The aeroplane highlighted in this article is Wikner's successor to VH-UHM.

[16] 'Successful Trials, New Farman Aeroplane', *Sunday Mail*, Brisbane, 17 November 1929, p. 4.

[17] Bert Cookson, *The Historic Civil Aircraft Register of Australia (Pre War)*, The Print Approach, Narangba, 1996, entry for G-AUHM Farman Sport. With the change of registration serials in March 1929 it became VH-UHM. Following the accident at Dayboro, it would be officially SOR – struck off register – on 27 May 1930.

[18] 'Amy Johnson Crashes at Eagle Farm', *Daily Standard*, Brisbane, 30 May 1930, p. 7.

[19] Bennett, 'Learning to Fly at 70', p. 2. Don recalled it in a 1983 interview with Col Burgess, editor, 'Air Vice Marshal D.C.T. Bennett: The man who formed the Pathfinders', *Transit Magazine*, Qantas Airways, p. 4.

[20] 'Flying Training Course: List of Trainees', The Age, Melbourne, 9 July 1930, p. 11.

Chapter 6: Entering the New World

[1] Bennett, *Pathfinder*, p. 24.
[2] Ibid.
[3] 'Aerial Defence: New Commonwealth Force', *The Argus*, Melbourne, 21 April 1919, p. 4, quoted in Steve Campbell-Wright, *An Interesting Point: A History of Military Aviation at Point Cook 1914–2014*, Air Power Development Centre, Department of Defence, Canberra, 2014, p. 64.
[4] Ibid., p. 101.
[5] Australia, as part of the empire, had taken its military preparation cues from the British, who had formulated the so-called 'Ten Year Rule' at the end of the First World War. This suggested the British Chiefs of Staff should plan on the assumption there would be 'no major war for ten years'. This had led to an atrophying of the military machine, stifled development across all three Services, and solidified First World War thinking around strategy and tactics. Such thinking persisted in political and military circles through to the start of the Second World War, contributing to Britain and empire being unprepared for war. Air Chief Marshal Sir Arthur Harris was particularly scathing of this in his 1947 memoirs, *Bomber Offensive*, Collins, London, 1947, pp. 11ff.
[6] Public Record Office [now National Archives UK], London, Air 8-69, 'Notes for the Minister on the Fleet Air Arm', May 1926, quoted in John McCarthy, *An Air Force Fit for Air Displays? The RAAF 1921–1939*, conference paper for 90 Years of RAAF History, Canberra, 2012.
[7] 'Australian Air Force: Chief Answers Criticisms', *The Argus*, Melbourne, 10 October 1928, p. 8.
[8] Ibid.
[9] Campbell-Wright, *An Interesting Point*, p. 43.
[10] Ibid., p. 47.
[11] Ibid., p. 88.
[12] McGrath, 'Interview with Air Vice-Marshal Donald Bennett'.
[13] Don Bennett, *Personal Notebooks for RAAF Point Cook and RAF Calshot, Volume 2*, Bennett/Vial Archive, Queensland Air Museum.
[14] Sir Frederick Scherger KBE CB DSO AFC would rise to become the first Australian to hold the rank of air chief marshal. He was then Chief of the Air Staff (1957–61) and Chairman of the Chiefs of Staff Committee (1961–66), a role which now carries the title Chief of the Defence Force.
[15] Bennett, *Personal Notebooks for RAAF Point Cook and RAF Calshot, Volume 2*.
[16] These notes appear to have some loose affinity with the RAF's Air Publication AP1176, *Royal Air Force Manual of Army Co-operation*, cited in Clive Richards, 'All Undiluted Nonsense? The Royal Air Force in the Army Co-operation Role, 1919–1940', in Royal Air Force Historical Society, *Journal 54*, UK, 2013, pp. 40–3.
[17] Bennett, *Personal Notebooks for RAAF Point Cook and RAF Calshot, Volume 2*.
[18] On the emergence of the Royal Flying Corp's role in support of the Army on the ground in the First World War, see Jeff Jefford, 'Corps Reconnaissance 1914–18', in Royal Air Force Historical Society, *Journal 54*, UK, 2013, pp. 8–35.

19 See ADF-Serials: Australian & New Zealand Military Aircraft Serials & History, at www.adf-serials.com.au/raaf1.htm.
20 Bennett, *Pathfinder*, p. 27.
21 Ibid., p. 26.
22 Don Bennett, *Flying Logbook, Volume 1*, entry for 17 October 1930, facsimile in Bennett/Vial Archive, Queensland Air Museum.
23 Keith Watson, a future Pathfinder pilot at No 97 Squadron, did his Tiger Moth training at No 5 Elementary Flying Training School, RAAF Narromine, in just under seven weeks in 1942, clocking up just over 62 hours flying time. See Ian Campbell, *Thinks He's a Bird*, Big Sky Publishing, NSW, 2021, chapter 5.
24 Bennett, *Pathfinder*, p. 26.
25 Ibid., p. 32.
26 Campbell-Wright, *An Interesting Point*, p. 115.
27 The Signals Corps, USA, *Catalogue of Official A.E.F. Photographs, Volume 2*, War Department Document No 903, Washington, 1919, p. 239. Also Jefford, 'Corps Reconnaissance 1914–18', pp. 30–2.
28 Bennett, *Pathfinder*, p. 32.
29 For a fascinating overview of the evolution of military photography and gridded maps of the Western Front in the First World War, see Jefford, 'Corps Reconnaissance 1914–18', pp 11–18.
30 Bennett, *Personal Notebooks for RAAF Point Cook and RAF Calshot, Volume 1*.
31 'Glossary of terms used in Australian Flying Corps', School of Literature, Language and Linguistics, ANU, at slll.cass.anu.edu.au/centres/andc/glossary-terms-used-australian-flying-corps.
32 Bennett, *Pathfinder*, p. 32.
33 Ibid., p. 25.
34 Ibid., p. 32.
35 'Air Cadets – Successful Trainees for England', *The Herald*, Melbourne, Saturday 20 June 1931, p. 10.
36 Sir Hudson Fysh, *Qantas Rising*, Rigby Limited, Sydney, 1965, p. 163.
37 Ibid., p. 198.
38 Jack Peterson, *Joystick Jottings: Official Journal of The Royal Queensland Aero Club*, Vol 6, No 1, July/Aug. 1984, p. 9. Also Fysh, *Qantas Rising*, p. 193.
39 McGrath, 'Interview with Air Vice-Marshal Donald Bennett'.
40 Don did not fully complete his logbook for these flights. Research by Robert Livingstone, Aviation Historical Society of Australia (Queensland), and Geoff Goodall (see Geoff Goodall's Aviation History Site, www.goodall.com.au), has reduced the options to one of three Qantas DH.60 Moths from the Qantas flying school – VH-UGH, VH-UGW, VH-UOK. Refer John Gunn, *The Defeat of Distance: Qantas 1919–1939*, University of Qld Press, 1985, Appendix A, pp. 359–61. Local newspapers contain no records of the flights.

Chapter 7: Fighter Boy

[1] Alastair Goodrum, *School of Aces: The RAF Training School That Won the Battle of Britain*, Amberley Publishing, 2019, p. 20.
[2] Bennett, *Pathfinder*, p. 35.
[3] Royal Air Force Centre for Air Power Studies (RAF CASPS Historic Interview), 'Maverick Leader and Navigator *Par Excellence*, Air Vice-Marshal Donald 'Pathfinder' Bennett CB CBE DSO', December 1980.
[4] 'In Wireless Telegraphy or Telephony we deal with the transmission, propagation and reception of Electromagnetic or Ether waves.' 'Radio Telephony (R/T) consists in the transmission of speech, music or any audible sound to a distance by means of electro-magnetic waves.' HM Signal School, *Admiralty Handbook of Wireless Telegraphy, 1931*, His Majesty's Stationery Office, London, 1932, pp. 1 and 675 respectively.
[5] A photo of Don at the time suggests he was wearing the RAF 1930 Pattern flying helmet, which was progressively adapted for telephone receivers and other fittings. See RAF Museum, Hendon, 'RAF Stories: Headgear', p. 33, at www.rafmuseum.org.uk/documents/LargePrintGuides/Headgear%20-.Pembroke Dockf.
[6] Bennett, *Pathfinder*, p. 37.
[7] Ibid., p. 41.
[8] In final section of Bennett, *Flying Logbook, Volume 2*, facsimile. Bennett/Vial Archive, Queensland Air Museum.
[9] M.B. Barrass, 'Air Commodore H.D. O'Neill', Air of Authority – A History of RAF Organisation, www.rafweb.org/biographies/O'Neill.htm.
[10] Bennett, *Pathfinder*, p. 42.
[11] Ibid., p. 41.
[12] Ibid.
[13] Bennett, *Personal Notebooks for RAAF Point Cook and RAF Calshot, Volume 1*.

Chapter 8: Boats That Fly

[1] David Fagan, 'RAF Calshot', at www.hampshireairfields.co.uk/airfields/cal.htm.
[2] Forces War Records, 'Unit History: RAF Calshot', at www.forces-war-records.co.uk/units/2035/raf-calshot.
[3] For an informative look at the early days of flying boats, from initial designs through to their use in war and domestic travel, see Louis S. Casey and John Batchelor, *Seaplanes and Flying Boats*, Phoebus Publishing, Great Britain, 1980.
[4] Bennett, *Personal Notebooks for RAAF Point Cook and RAF Calshot, Volume 1*.
[5] Bennett, *Pathfinder*, p. 46.
[6] Christopher Coverdale, *Celebrating Bennett the Man 1910–1986*, self-published, Great Britain, 2018, p. 18.
[7] Alban's son, Peter, wrote about his father's time at Calshot: 'During the course, my father's enthusiasm was such that he so frequently used the term "Let's get cracking" that it ensured him the lasting nickname "Crackers".' In Peter Carey, 'A Brief Account of the Career of Group Captain Alban Majendie Carey, CBE, compiled by his son, Peter, from various sources and discussions, for private

circulation to family and friends', p. 8, in the possession of John Evans, Pembroke Dock historian, used with permission.
[8] Coverdale, *Celebrating Bennett the Man*, p. 18.
[9] Ibid. Coverdale provides no detailed attribution for the Clift quote.
[10] McGrath, 'Interview with Air Vice-Marshal Donald Bennett'.
[11] Ibid.
[12] Royal Air Force Centre for Air Power Studies, 'Maverick Leader and Navigator *Par Excellence*'.
[13] Flying Officer K.F.T. 'Percy' Pickles went on to become a squadron leader at the Marine Aircraft Experimental Establishment, No 16 (Reconnaissance) Group at Felixstowe. He was there when Don turned up with Mayo to gain certification for the Mayo Composite. The website 'Cranwellians' shows him to be a contemporary of CDC Boyce, graduating together in 1927.
[14] Bennett, *Pathfinder*, p. 48.
[15] Ibid., p. 46.

Chapter 9: Under Harris's Wing

[1] 'Harris, with his flight commanders Saundby and Cochrane, had laid the keel of the long-range heavy bomber of the future …' Dudley Saward, *'Bomber' Harris*, Cassell Limited, London, 1984, pp. 29–31.
[2] Henry Probert, *Bomber Harris: His Life and Times*, Greenhill Books, London, 2001, p. 62. Saward (ibid., p. 36) says the position was Senior Staff Officer.
[3] Ibid., p. 63.
[4] Harris, *Bomber Offensive*, p. 15.
[5] Saward, *'Bomber' Harris*, p. 5.
[6] Ibid., p. 60.
[7] Bennett, *Pathfinder*, p. 47.
[8] Dr Jennie Mack Gray, 'Harris, Bennett and Flying Boats', RAF Pathfinders Archive, at https://rafpathfinders.com/harris-bennett-flying-boats/.
[9] No 480 (Flying Boat) Flight would be expanded to become No 201 (Flying Boat) Squadron on 1 January 1929.
[10] Harris, *Bomber Offensive*, p. 25.
[11] Ibid.
[12] Ibid.
[13] Coverdale (*Celebrating Bennett the Man*, p.28) quotes Don as saying, 'I felt incredibly honoured to have been put in charge of night flying by Harris. To be in charge of overseeing the night flying routines, making sure that everyone was up to speed to perform the right night flying procedure, take-off and landings from Milford Haven estuary.' Unfortunately, this quote is not sourced. Don's logbook shows that after returning to 210 Squadron from the night flying at Calshot, he only flew three times at night at Pembroke Dock, two in preparation for the fishery protection exercise on the third of those nights. All remaining night flying was at Dover as part of the ADEE exercises in June and July.
[14] Probert, *Bomber Harris*, pp. 86–7.
[15] Ibid., p. 90.

16 Interview with John Evans, the Pembroke Dock aviation historian, recorded by Bruce Sinclair, 'When "Bomber" Harris flew at Pembroke Dock', *Tivy-Side Advertiser*, 4 July 2020, at www.tivy-sideadvertiser.co.uk/news/pembrokeshire_news/18548431.bomber-harris-flew-pembroke-dock/.
17 William Anderson, *Pathfinders*, Jarrolds Publishers Limited, London, 1946, p. 51.
18 Item in the Bennett/Vial Archive, Queensland Air Museum.
19 "Efficient", Google English Dictionary, provided by Oxford Languages.
20 Central Flying School, 'Training in Instrument Flying', RAF Wittering, October 1931. In Bennett, *Personal Notebooks for RAAF Point Cook and RAF Calshot, Volume 1*.
21 Bennett, *Pathfinder*, p. 47.
22 Ibid.
23 Flight Lieutenant Fred Mawdesley was a Canadian pilot on exchange duties at RAF Pembroke Dock when Don was there. He had arrived on 23 March 1933, about the same time as Harris. He had considerable experience in flying boats, having commenced seaplane training in Vancouver in 1925. For further background, see 'Frederick Joseph Mawdesley', at www.rcafassociation.ca/heritage/search-awards. A photo of Mawdesley can be found in *Canadian Military History*, Vol 22 [2013], Issue 1, Article 7, p. 12, at https://scholars.wlu.ca/cmh/vol22/iss1/7.
24 The identification of Lang comes after an extensive review comparing Don's flying logbook for the period with Don's account in his memoirs. The account offers several crucial pieces of information: (i) 'a more senior officer'; (ii) Don as second pilot; (iii) the officer being an 'old hand'; and (iv) the flight was to RAF Calshot. During the ADEE exercises, there were four occasions when Don flew back to Calshot at night, once with Lang and three times with Mawdesley. Strictly speaking, Mawdesley was 'a more senior officer' as a flight lieutenant. However, Don always recorded in his logbook whether he was flying as first or second pilot, and on each of the three occasions with Mawdesley, Don records himself as the first pilot. Only the account of flying with Lang shows him flying as second pilot.
25 Don noted in his memoirs that during his time at Pembroke Dock, they began on Southampton IIs but 'we soon got Singapores'. His logbook is clear, however, that he never flew the Singapore. They did not enter service with No 210 Squadron until January 1935, eighteen months after Don had left.

Chapter 10: Back to the Future

1 Bennett, *Pathfinder*, pp. 47–8.
2 Don's paperwork in the Bennett/Vial Archive, Queensland Air Museum, includes the following: Letter from the Deputy Director of Civil Aviation; a certificate to be completed by his instructor; 'Requirements for Seaplane Instructor's Certificate'; 'Particulars of Examination for Flying Instructor's Certificate Issued by the Guild of Air Pilots and Air Navigators of the British Empire'; and Don's application form.
3 Bennett, *Pathfinder*, p. 48. Don posted a photo of the three of them together in his logbook.

[4] Desmond Penrose, 'De Havilland D.H.53 Humming Bird', in ed. David Ogilvy, *From Blériot to Spitfire*, Airlife Publications, England, 1977, p. 107. My personal thanks to Carmel Attard, Malta, for bringing this article to my attention.
[5] 'De Havilland D.H.53 Humming Bird', at www.baesystems.com/en/heritage/-de-havilland-dh53-humming-bird. Alan Cobham became the first aviator to fly from England to Australia and return in a de Havilland DH.50J between 30 June and 1 October 1926, a feat for which he was knighted, at www.rafmuseum.org.uk/research/online-exhibitions/sir-alan-cobham-pioneering-aviator/australia-flight.
[6] Penrose, 'De Havilland D.H.53 Humming Bird', pp. 111–12.
[7] Bennett, *Pathfinder*, p. 54.
[8] Penrose, 'De Havilland D.H.53 Humming Bird', p. 112. A video of a DH.53's first engine run in 85 years can be found at www.youtube.com/watch?v=DeIi9o9C5tA.

Chapter 11: The Invisible Barrier Over the Horizon

[1] '£10,000 Centenary Air Race', *The Herald*, Melbourne, 29 March 1933, p. 1.
[2] Ibid.
[3] Ibid.
[4] Imperial War Museum, 'Bennett, Donald Clifford Tyndall', IWM Sound Archive 9378, 11 August 1986, at www.iwm.org.uk/collections/item/object/80009167.
[5] Bennett, *Pathfinder*, p. 55. He repeated this point many years later. 'And also one little thing was that I decided that the Air Force in peace time was not for me ultimately and I was determined to get into civil aviation.' McGrath, 'Interview with Air Vice-Marshal Donald Bennett'.
[6] Bennett, *Pathfinder*, p. 50.
[7] C.G. Jefford, *Observers and Navigators*, 2nd edition, e-book, Grub Street, London, 2014, Annex J.
[8] Ibid.
[9] Recorded in letter from Acting Wing Commander D.C.T. Bennett to the Commanding Officer, No 1 EAOS, Eastbourne, 5 November 1941, requesting posting to an operational unit. Bennett/Vial Archive, Queensland Air Museum. The 2nd Class Air Navigator's Certificate was also awarded to those who completed the navigation course run at Andover, the 'Long N' navigation course at RAF Calshot or other specific navigation courses.
[10] Peter Carey sheds some light on Don holding both the military certificate and the civil licence. In writing about his father 'Crackers' Carey, he recorded: 'He was pleased to get this enviable posting [to Calshot] and in addition to passing the course with a good rating, he obtained commercial flying and navigation licences issued by the Ministry of Civil Aviation. He thinks that he and his friend Don Bennett were the only serving officers in the RAF to hold them.' Peter Carey, 'A Brief Account of the Career of Group Captain Alban Majendie Carey, CBE, compiled by his son, Peter, from various sources and discussions, for private circulation to family and friends', p. 8.
[11] Jefford, *Observers and Navigators*, Annex J.

[12] Captain Arthur Wilcockson, who flew with Don at Imperial Airways, observed that the key difference between the 2nd and 1st Class Air Navigation Licences was the emphasis on celestial navigation. 'The first class ticket was very difficult to get – well it is comparable to the captain's ticket at sea.' A.S. Wilcockson, *The Reminiscences of A.S. Wilcockson*, Aviation Project, Oral History Research Office, Columbia University, 1960, p. 15.
[13] Air Ministry, 'Civil Air Navigators' Licences, First Class, General Information and Conditions of Examination', p. 1. Date of issue not in documentation. Bennett/Vial Archive, Queensland Air Museum.
[14] Bennett, *Pathfinder*, p. 50.
[15] Air Ministry, 'Civil Air Navigators' Licences, First Class, General Information and Conditions of Examination', p. 1.
[16] Ibid., p. 2.
[17] Ibid., p. 6.
[18] Ibid., p. 7.
[19] Bennett, *Pathfinder*, p. 50.
[20] McGrath, 'Interview with Air Vice-Marshal Donald Bennett'.
[21] Royal Air Force Centre for Air Power Studies, 'Maverick Leader and Navigator *Par Excellence*'.

Chapter 12: The Unshackling Begins

[1] Bennett, *Pathfinder*, p. 55.
[2] Coverdale, *Celebrating the Man*, p. 43.
[3] Alec Ayliffe, 'RAF Navigation Between the Wars', *A History of Navigation in the Royal Air Force*, RAF Historical Society Seminar, RAF Museum, Hendon, 21 October 1996, p. 24.
[4] David Page, 'The Early Years', *A History of Navigation in the Royal Air Force*, RAF Historical Society Seminar, RAF Museum, Hendon, 21 October 1996, p. 10.
[5] See Hampshire Aeroplane Club, 'About Us', at www.flyhac.co.uk.
[6] 'Hampshire Aeroplane Club', *Grace's Guide to British Industrial History*, at www.gracesguide.co.uk/Hampshire_Aeroplane_Club.
[7] Air Ministry, *Pilots' Licences (Class 'B'): Extension of Licences to Cover Further Types of Flying Machines*, Bennett/Vial Archive, Queensland Air Museum.
[8] Don's Pilot's 'B' Licence was awarded on 27 February 1933 at the completion of his Flying Boat Pilot's Course. This fact recorded in a letter from Acting Wing Commander D.C.T. Bennett to the Commanding Officer, No 1 EAOS, Eastbourne, 5 November 1941 requesting posting to an operational unit, Bennett/Vial Archive, Queensland Air Museum.
[9] Coverdale, *Celebrating Bennett the Man*, p. 43. See also, Jerripedia, 'Jersey Airways', at www.theislandwiki.org/index.php/Jersey_Airways.
[10] Bennett, *Pathfinder*, p. 55.
[11] In one of those great coincidences, G-ACMO was imported into Australia in March 1938 and registered with South Queensland Airways Pty Ltd in June of that year. In October, it was registered as VH-ABK and named 'City of Toowoomba', at www.goodall.com.au/australian-aviation/dh84-pt1/dh84-dragon-pt1.htm.

12 Bennett, *Flying Logbook, Volume 3*, entries for 23 and 24 June 1934, facsimile in Bennett/Vial Archive, Queensland Air Museum.
13 Bennett, *Pathfinder*, p. 55.
14 Ibid., p. 55.

Chapter 13: The Vega's Leg

1 Bennett, *Pathfinder*, p. 51. Remarkably, Biard's name is spelt correctly in Don's logbook.
2 Mark Bloise-Brooke, 'The Schneider Story', at www.supermarineseaplane.co.uk/the-schneider-story.
3 Frank S. Stuart, 'This Flier Will Make Air History', unidentified magazine or newspaper article, in Australian War Memorial, AWM 65, 292, Donald Clifford Tyndall Bennett (32065), p. 82. Date of article unknown, but the reference to Don's rank as air commodore suggests sometime in 1943.
4 Ibid., p. 82.
5 Bennett, *Pathfinder*, p. 51.
6 Julie Lewis, *Jimmy Woods: Flying Pioneer*, Fremantle Arts Centre Press, Western Australia, 1989, p. 19. The employer was either the Mayor of Auckland or a Member of Parliament.
7 Ibid., p. 24. 'The term was based on physical fact. Pilots learned to respond to the way in which muscles of the buttocks reacted to changes in height and direction.'
8 Ibid., p. 21.
9 Ibid., p. 57. On Woods's death, Norman Brierley commented, 'Sometimes I had to curb his enthusiasm in the interests of safety and sometimes I had to draw on it in the interests of progress.' In 'The end of an era – 'Woodsie' dies', *The West Australian*, Saturday 10 May 1975.
10 In a masterful branding exercise, the DH.61 was called 'Old Gold', with its first flight freighting chocolates from Melbourne to Adelaide. The full story of Horrie Miller's early years and the start of MacRobertson-Miller Aviation Company Ltd can be found at Geoff Goodall, 'MacRobertson Miller Airlines (MMA)', at www.goodall.com.au/australian-aviation/mma-1/mma.html.
11 The full scale of Woods's preparation deficiencies is outlined in Lewis, *Jimmy Woods*, pp. 70–1.
12 See Ron Cuskelly, 'Vega VH-UVK', Adastron Aviation History, at www.adastron.com/lockheed/vega/vega.htm. This site offers full chronological details of the life of G-ABGK.
13 Goodall, 'MacRobertson Miller Airlines (MMA)'.
14 Lewis, *Jimmy Woods*, pp. 91–2.
15 Jimmy Woods, *Personal Diaries 1934*, entry for Monday 17 September, State Library of Western Australia, ACC 4638A/39-50.
16 Letter from B. Burt, HM Customs & Excise, to Mr Jimmy Woods Esq., 27 September 1934, Jimmy Woods, *Correspondence*, State Library of Western Australia, ACC 4638A/24.
17 Ibid.
18 Bennett, *Pathfinder*, p. 51.

19 Woods, *Personal Diaries 1934*, entry for Saturday 13 October 1934.
20 R.D. Bedinger, US Department of Commerce Inspector, 'Notice to Operators, Model: Vega DL-1, Serial No: 155'. This undated card also specified maximum fuel and oil loads, along with a list of standard and special equipment and the weight of each. Woods, *Correspondence*.
21 James Gerschler, Project Engineer, Lockheed Aircraft Corporation, letter to Mr James Woods, 20 July 1934, Woods, *Correspondence*.
22 Horace Miller, letter to Jimmy Woods, 31 July 1934, Jimmy Woods, *Correspondence*.
23 The *Fédération Aéronautique Internationale* is the world governing body for air sports. It maintains world records for aeronautical activities.
24 The Automobile Association, Aviation Department, Invoice 4557, State Library of Western Australia, ACC 4638A 66 – 'England to Australia air-race. Official papers'. That £20 was paid in cash and that passport changes were made indicates Jimmy picked up the maps that day.
25 Bennett, *Pathfinder*, p. 51. Don's original navigation flight plan is in the Bennett/Vial Archive, Queensland Air Museum.
26 Bennett, *Pathfinder*, p. 51. Woods's diary entry confirms he went off to London.
27 Glendale News Press article, Woods, *Correspondence*.
28 State Library of Western Australia, ACC 4638A 66 – 'England to Australia air-race. Official papers'.
29 Woods, *Personal Diaries 1934*, entry for Tuesday 16 October 1934.
30 Ibid., entry for Wednesday 17 October 1934.
31 Ibid., entry for Thursday 18 October 1934.
32 Lewis, *Jimmy Woods*, p. 94.
33 Woods, *Personal Diaries 1934*, entry for Friday 19 October 1934.
34 Bennett, *Pathfinder*, p. 52.
35 Ibid.
36 Bennett, *Flying Logbook, Volume 3*, entry for 20 October 1934.
37 Letter from Woods to David Robertson, Aleppo, Syria, 21 October 1934, Woods, *Correspondence*.
38 Bennett, *Pathfinder*, pp. 52–3.
39 Ibid., p. 53.
40 Ibid.

Chapter 14: Ly and her Complete Air Navigator

1 Coverdale, *Celebrating Bennett the Man*, p. 52.
2 Ibid.
3 Bennett, *Pathfinder*, p. 53.
4 Wing Commander William 'Andy' Anderson, who worked alongside Don during the war, wrote of seeing him with Ly. 'Mrs Bennett is very fair, very attractive and most charming, with a delightful accent, for she comes from Switzerland. And it is with her that you can see Donald Bennett as he really is; not a machine, but a man.' William Anderson, *Pathfinders*, p. 51.
5 Coverdale, *Celebrating Bennett the Man*, p. 52.
6 Bennett, *Pathfinder*, p. 49.

[7] Ibid., pp. 49–50.
[8] Ibid., p. 50.
[9] Recorded in letter from Acting Wing Commander D.C.T. Bennett to the Commanding Officer, No 1 EAOS, Eastbourne, 5 November 1941, requesting posting to an operational unit, Bennett/Vial Archive, Queensland Air Museum.
[10] In 1935, Barry Littlejohn sold the Klemm and it was shipped to Australia, where it became the property of Maude Rose 'Lores' Bonney, who registered it as VH-UVE in December 1935. She flew it from Darwin to Cape Town between 13 April and 18 August 1937, the very first flight between Australia and South Africa in a light aircraft. The Klemm would meet its end in a fire in the Qantas hangar at Archerfield on 28 June 1939.
[11] The *Challenge International de Tourisme* was the *Fédération Aéronautique Internationale*'s Tourist Plane Competition. The 1932 Challenge was the third of these, taking place between 12 and 28 August, in Berlin, Germany. The Challenges, from 1929 to 1934, were considered major aviation events.
[12] Coverdale, *Celebrating Bennett the Man*, p. 59.
[13] Ibid., p. 52.
[14] Brett Holman, 'The Meaning of Hendon: the Royal Air Force Display, aerial theatre and the technological sublime, 1920–1937', *Historical Research* 93, 259 (2020), pp. 131–52. Holman's paper provides a fascinating and comprehensive assessment of the Hendon Air Displays, especially how they evolved in the inter-war period to meet different objectives. See also Brett Holman, 'Ending Hendon – VI: 1935-1937', *Airminded*, 2 December 2011, at https://airminded.org/2011/12/02/ending-hendon-vi-1935-1937/.
[15] Richard Morris, *Cheshire*, Penguin, England, 2000, p. 19.
[16] Holman, 'The Meaning of Hendon', ibid.
[17] For amateur video footage of the Southamptons, see Lockwood Family Collection, 'Royal Naval Review, July 16th 1935', Screen Archive South East, University of Brighton, at https://screenarchive.brighton.ac.uk/detail/8968.
[18] Air Ministry, *Syllabus of Examination for Licence to Operate Radiotelephony Apparatus on board British Civil Aircraft*, Bennett/Vial Archive, Queensland Air Museum.
[19] As stated by Air Marshal Sir Patrick Dunn at the passing of Don in September 1986. Laura Phillips, 'Tributes flow as air ace makes his last journey', *The Advertiser*, 24 September 1986. See also Anderson, *Pathfinders*, p. 50.
[20] Bennett, *Pathfinder*, p. 54.
[21] Don has 'J. Gill' in his logbook.
[22] Bennett, *Flying Logbook, Volume 3*, entry for 8 August 1935.
[23] Bennett, *Pathfinder*, p. 55.
[24] Ibid., pp. 55–6.
[25] Ibid., p. 56.
[26] Ibid., p. 57.
[27] The origins of Don initially connecting with Imperial Airways are unclear. Sir Arthur Harris claims he made the introduction (*Bomber Offensive*, p. 129). Coverdale quotes Don as saying he had been advised by his 'Air Force masters' to make the contact (*Celebrating Bennett the Man*, p. 58). In his memoirs, Don says he made the contact, but offers no reason why (*Pathfinder*, p. 58). In the 1980 Centre for Air

Power Studies interview, ('Maverick Leader and Navigator *Par Excellence*'), Don says it was Air Vice-Marshal Sir Thomas Webb-Bowen, now at Imperial Airways, who approached him because Imperial was preparing for the Atlantic crossings. The difficulty with this recollection is Don adding that Webb-Bowen had told him to get all his licences, which does not square with his memoirs or the chronology, with Don aiming for his First Class Air Navigator's Licence in January 1934, almost two years before joining Imperial. A possible resolution is it was Harris who contacted Webb-Bowen recommending Don sometime in 1935 when it was clear to all Don was not remaining in the RAF. This suggests, however, Harris had maintained an interest in Don's career after Harris departed Pembroke Dock in August 1933.

[28] Letter from E.C. Johnson, Controller of Civil Aviation, Department of Defence, dated 18 October 1935. In the back of Don's *Flying Logbook, Volume 3*.

[29] Bennett, *Pathfinder*, p. 57.

[30] Queensland Times (Ipswich), Saturday 26 October 1935, p.6.

[31] Don maintains in his memoirs that Young was the chief pilot. He wasn't. This position was held by Keith Virtue, one of the founders of the company. See Goodall, 'The Australian GENAIRCO Series', at www.goodall.com.au/australian-aviation/genairco/genairco.html. Tommy and Keith had accompanied Charles Kingsford Smith in a formation flight over Sydney Harbour for the opening of the Harbour Bridge in March 1932. Photo of VH-UPI at www.airhistory.net/photo/127265/VH-UPI.

[32] Photo of the aircraft at www.airhistory.net/photo/140500/VH-URL.

[33] D.C.T. Bennett, *The Complete Air Navigator*, Sir Isaac Pitman & Sons Ltd, 6th edition, Great Britain, 1954, p. viii. Prefaces for previous editions published in the front of this edition.

[34] Ibid.

[35] Ibid.

Chapter 15: Imperial, Empire and 'Empires'

[1] Original in the Bennett/Vial Archive, Queensland Air Museum.

[2] Don Bennett, *Personal Notebook – Imperial Airways*, Bennett/Vial Archive, Queensland Air Museum.

[3] Bennett, *Pathfinder*, p. 60.

[4] It should be noted Imperial and Qantas operated a common fleet.

[5] In 1922, two years before the formation of Imperial Airways, Sir Samuel Hoare, Secretary of State for Air, expressed the view that establishing air routes could contribute to uniting the empire countries. So, when Imperial Airways was formed as a merger of four British companies, it was 'privately financed but government backed'. P.E. Sims, *Adventurous Empires: The Story of the Short Empire Flying Boats*, e-book, Pen & Sword, UK, 2013, chapter 1.

[6] Gordon Pirie, 'Incidental Tourism: British Imperial air travel in the 1930s', *Journal of Tourism History*, Vol 1, No 1, March 2009, pp. 52–4. This research paper provides an extraordinary insight into passenger travel with Imperial Airways (and, in part, other airlines) during the 1930s.

[7] Sims, *Adventurous Empires*, chapter 2.
[8] The Postal Museum, www.postalmuseum.org/collections/airmail/, p. 7.
[9] Sims, *Adventurous Empires*, chapter 2.
[10] Imperial's third Kent at Alex had been lost on 9 November 1935 when it caught fire refuelling at Brindisi.
[11] I am indebted to Paul Sheehan for his extraordinary work on the flight history of every Empire flying boat. He was able to point out the only instance during Don's time at Imperial Airways when this episode could have happened and why, namely a weather event at Alex. In doing so, he corroborated Don's recollection in *Pathfinder*, p. 62. It should be noted that while Don arrived in Brindisi on 18 March and left on the 22nd, 'days' are measured midnight-to-midnight, hence Don's reference to 'five days'.
[12] Bennett, *Pathfinder*, p. 60.
[13] Ibid.
[14] Ibid., p. 62.
[15] For an excellent map of an early African route see Bryan Swopes, '20 January 1932', *This Day in Aviation*, at www.thisdayinaviation.com/tag/handley-page.
[16] 'British Imperial Airways Handley Page H.P.42/45 Aircraft', at barthworks.com/aviation/aviation_historic_artifacts/handleypage/handleypage.htm
[17] On what became the 'Kangaroo Route' because it involved many hops between England and Australia, Imperial would fly England to Karachi, then a joint venture of Imperial and Indian Trans-Continental Airways flew the legs from Karachi to Singapore, with Qantas operating the final legs from Singapore to Australia.
[18] Russell Deiley, 'The Ultimate! When 'boats' ruled the air', *Gold Coast Bulletin*, Saturday, July 23 1988, p. 47. Interview with Hudson 'Hoddie' Howse. For a full description of flying with Imperial Airways, and the flying boats in particular, see Sims, *Adventurous Empires*.
[19] Bennett, *Pathfinder*, p. 63.
[20] Ibid.
[21] Records often indicate there were 16 original pilots, but this number was taken from Brackley's diary. Information supplied by Paul Sheehan. Another list can be found in Robert Bluffield, *Imperial Airways: The Birth of the British Airline Industry 1914–1940*, Ian Allan Publishing, UK, 2009, p. 46.
[22] Brian Cassidy, *Flying Empires*, Queen's Parade Press, Bath, UK, 1996, p. 111.
[23] Bluffield, *Imperial Airways*, p. 170.
[24] Imperial Airways, *Handbook of Instructions and General Information*, Sheet No G 18, 'General Instructions – Captains and Crew', amended 15 February 1936, Bennett/Vial Archive, Queensland Air Museum.
[25] Bennett, *Pathfinder*, p. 64.
[26] Bennett, *Flying Logbook, Volume 3*, entry for 23 May 1937.
[27] 'Egypt–England in 15 Hours', *The Argus*, Melbourne, Tuesday 25 May 1937, p. 10.
[28] Bennett, *Pathfinder*, pp. 63–4.
[29] Pirie, 'Incidental Tourism', p. 62.
[30] Bennett, *Pathfinder*, p. 65.
[31] A remarkable audio piece exists where Geoffrey Pett, who worked for Imperial Airways in ground control operations, including Africa, recorded discussions he had

about how to get Empires onto the water at Lindi in bad weather at http://whitewaterlandings.co.uk/voice-clips/blind-landings-the-audio.
[32] My thanks to Paul Sheehan for making this discovery; it speaks volumes about Don's character.
[33] Imperial Airways, *Handbook of Instructions and General Information*, Appendix I to Sheet No G 18, 'Instructions to Captains by Air Superintendent', amended 15 February 1936, Bennett/Vial Archive, Queensland Air Museum.
[34] Don was not the only Imperial captain engaged in such activities. On 11 October 1937, John Burgess flew Khartoum to Brindisi in a day, totalling 16 hours.

Chapter 16: *Mercury* and the Atlantic

[1] For an excellent discussion of the history of flying the North Atlantic for the period 1919–1940, covering politics and technology, see Carl A. Christie, *Ocean Bridge: The History of RAF Ferry Command*, University of Toronto Press, Canada, 1995, Chapter 1 'Atlantic Pioneers', pp. 3–24.
[2] Sims, *Adventurous Empires*, chapter 5.
[3] Bennett, *Pathfinder*, p. 65.
[4] For a full discussion on the design aspects of the Short Mayo Composite, see C.H. Barnes, *Shorts Aircraft Since 1900*, Putnam Aeronautical Books, Great Britain, 1989, pp. 302–7.
[5] For a good collection of Movietone News footage of the Mayo Composite from its launch in 1937 to Don's seaplane record flight, see Bomberguy Aviation History, 'The Short Mayo Composite', at www.youtube.com/watch?v=bYtazEBQ1K8.
[6] Arthur Wilcockson provides some wonderful descriptions of his first flights across the Atlantic. See Wilcockson, 'The Reminiscences of A.S. Wilcockson', pp. 10–12.
[7] Don Bennett, *Personal Notebook – Imperial Airways*, Bennett/Vial Archive, Queensland Air Museum.
[8] Bennett, *Pathfinder*, pp. 66–7.
[9] Ibid., p. 68.
[10] Ibid.
[11] William Pearce, 'Napier H-16 Rapier Aircraft Engine', Old Machine Press, at https://oldmachinepress.com/2020/08/20/napier-h-16-rapier-aircraft-engine/.
[12] Don's original map is in the Bennett/Vial Archive, Queensland Air Museum. It contains the following notation in his handwriting: 'My first Atlantic Crossing 20th – 21st July 1938. BMA "Mercury" G-ADHJ'. Foynes – Montreal direct – N. York. Return flight N. York – Botwood – Horta – Lisbon.'
[13] Footage of this separation can be seen at: www.britishpathe.com/asset/45168/.
[14] East Anglian Film Archive, 'A newsreel-style amateur film celebrating the opening of Luton Airport in July 1938', at http://eafa.org.uk/work/?id=1049755.
[15] Captain D.C.T. Bennett, 'Preliminary Report on Atlantic Crossing in "Mercury" – 20th to 27th July 1938, 29 July 1938', Air Ministry, Trans-Atlantic Experimental Flights – Official Reports of Flights by First Pilots, Part II, Reference 692151/37.
[16] Captain D.C.T. Bennett, 'Amplifying Voyage Report on Atlantic Crossing in "Mercury" – 20th to 27th July 1938, 29 July 1938', Air Ministry, Trans-Atlantic

Experimental Flights – Official Reports of Flights by First Pilots, Part II, Reference 692151/37.
[17] Bennett, *Pathfinder*, p. 69.
[18] Don's navigation chart indicates he had planned for this eventuality.
[19] Captain D.C.T. Bennett, 'Preliminary Report on Atlantic Crossing in "Mercury".' The engineer was John Joseph 'Joe' Gilmore, who becomes an integral part of the Atlantic Ferry story.
[20] Sims, *Adventurous Empires*, chapter 5.
[21] Bennett, *Pathfinder*, p. 70.
[22] Australian Associated Press London correspondent, 'Brisbane Pilot Flies Atlantic – Pick-a-Back Record', *The Courier Mail*, Friday 22 July 1938, p. 1.
[23] British Official Wireless, '"Pick-a-Back" Plane. Two Records. Eire to Canada Non-stop.', *Sydney Morning Herald*, Saturday 23 July 1938, p. 18.
[24] The book remains in the Brisbane Grammar School archive.
[25] Don's original navigational chart, along with details of reporting codes, beacons and broadcasting stations across two pages, is in the Bennett/Vial Archive, Queensland Air Museum.
[26] Australian Associated Press correspondent, 'Atlantic Flown Twice – The Mercury Returns to England', *The West Australian*, Friday 29 July 1938, p. 24.

Chapter 17: *Mercury* and the Obsession with Records

[1] There is footage of one such launch of a Do 18 from the seaplane tender SS *Schwabenland*. See 'The S.S. Schwabenland catapults a 10 ton Dornier Do-18 flying boat', *Critical Past*, at www.youtube.com/watch?v=ReEX3fkJxfo.
[2] Different accounts offer different mileages for this particular flight – 5214, 5215, 5241 and 5278. 'The Do 18 Dornier's Whale Calf', *Air International*, Vol 18, No 4, April 1980, p. 186, offers considerable detail on D-ANHR's specifications and mentions the record-breaking flight, offers no source for the latter. Unlike the absolute distance record, this flight does not appear to have been recognised by the *Fédération Aéronautique Internationale*, the world governing body responsible for verifying record-breaking flights.
[3] Australian Associated Press London correspondent, 'Another Flight Across the Atlantic – German Catapult Aeroplane', *The Courier Mail*, 25 July 1938, p. 7.
[4] For a good summary of the aircraft and the flight, see Bryan Swopes, '*10–11 August 1938*', *This Day in Aviation*, at www.thisdayinaviation.com/10-11-august-1938/.
[5] Sims, *Adventurous Empires*, chapter 5.
[6] Ibid., p. 246.
[7] Bennett, *Pathfinder*, p. 74.
[8] Ibid., p. 77. All the original maps for this flight are in the Bennett/Vial Archive, Queensland Air Museum.
[9] Bennett, *Pathfinder*, p. 77). Don recalled in his memoirs that he crossed the coast near Bône (now Annaba), but it was west of there.
[10] Bennett, *Pathfinder*, p. 76. Don indicated the code was sent to Imperial's headquarters with each hourly signal, but the actual document – 'Code for Reporting

State of Flight' – is clear these signals were only to be sent every four hours, commencing at 4:00 pm GMT. Bennett/Vial Archive, Queensland Air Museum.
[11] Don Bennett, 'Code for Reporting State of Flight', Bennett/Vial Archive, Queensland Air Museum.
[12] Bennett, *Pathfinder*, p. 76.
[13] Ibid., p. 78.
[14] Ibid.
[15] Don recorded in pencil on his onboard map for South West Africa 'daylight' at a point just to the north-west of Maltahöhe, South West Africa. Bennett/Vial Archive, Queensland Air Museum.
[16] Original in the Bennett/Vial Archive, Queensland Air Museum.
[17] Bennett, *Pathfinder*, p. 78.
[18] From Don's French copy of *Fédération Aéronautique Internationale, Records Officiels*, Paris, 1 January 1964, Bennett/Vial Archive, Queensland Air Museum.
[19] Bennett, *Pathfinder*, p. 81. Sims (*Adventurous Empires*) says the official long-distance seaplane record was held by the Italians. The author has not been able to establish the exact nature of this record. An Italian CANT Z.501 Gabbiano set a record of 2,900 miles on 16 July 1934, but it is not known if this is the flight Don claims to have beaten. If so, he had, in fact, doubled the distance.
[20] Ibid., p. 83.
[21] Barnes, *Shorts Aircraft Since 1900*, pp. 310–11.
[22] Don Bennett, 'Long Range Routes', 2 January 1939. Bennett/Vial Archive, Queensland Air Museum. This memo does not name Don as author; that he's the author is revealed in his follow up memo of 7 March 1939.
[23] Original planning map in the Bennett/Vial Archive, Queensland Air Museum.
[24] Don Bennett, 'Investigation into the fuel consumption of the Rapier VI engine in connection with long range operation', p. 1, Bennett/Vial Archive, Queensland Air Museum.
[25] Don Bennett, 'Proposed Route for Long Distance Test Flight', 7 March 1939, p. 1, Bennett/Vial Archive, Queensland Air Museum. After the flight to South Africa, Don had updated the 5-letter code listings and these showed the bases he was considering in South America: Porto Alegre, Brazil; Montevideo, Uruguay; Buenos Aires, Bahia Blanca, Bahia San Blas, the Rio Negro and San Antonio, all in Argentina.
[26] Sims, *Adventurous Empires*, chapter 5.
[27] Don Bennett, 'Proposal for Long Range Test Flight in Mercury using 100 Octane Fuel', pre-24 March 1939, p. 1, Bennett/Vial Archive, Queensland Air Museum.
[28] Don Bennett, 'Proposed long distance test flight with route survey on return flight', 24 March 1939, p. 2, Bennett/Vial Archive, Queensland Air Museum.
[29] Bennett, *Pathfinder*, p. 84.
[30] The evolution of in-flight refuelling technology and techniques, and the dangers it posed, is covered in detail by Sims, *Adventurous Empires*, chapter 7.
[31] Handwritten note in Don's *Flying Logbook, Volume 4*, facsimile in Bennett/Vial Archive, Queensland Air Museum.
[32] Bennett, *Personal Notebook – Imperial Airways*, Bennett/Vial Archive, Queensland Air Museum.

33 See Getty images, World War II, 30th July 1939, 'The Cabot flying boat which inaugurated Britain's transatlantic air service on 5th August is shown here flying below the tanker plane just before refuelling via a mid-air pipe line connection', at www.gettyimages.ca/detail/news-photo/world-war-ii-30th-july-1939-the-cabot-flying-boat-which-news-photo/79667100?adppopup=true.
34 Bennett, *Pathfinder*, p. 87.
35 'President to Get Saturday's Grouse', *The Brooklyn Daily Eagle*, Monday 14 August 1939, p. 22. For a photo, see 'Brisbane Pilot Takes Gift to Mrs Roosevelt', *The Courier Mail*, Tuesday 26 September 1939, p. 6.

Chapter 18: War

1 Bennett, *Pathfinder*, p. 90.
2 Imperial War Museum, 'Bennett, Donald Clifford Tyndall', IWM Sound Archive 9378, 11 August 1986, Reel 1, 14.45 minute mark, at www.iwm.org.uk/collections/item/object/80009167.
3 From the Royal Aeronautical Society website, at www.aerosociety.com. See also Chris Male MRAeS, 'A Short History of the Royal Aeronautical Society', *Aerospace*, Journal of the Royal Aeronautical Society, January 2016, p. 48, available online at the RAeS website.
4 Tony Pilmer, Librarian & Archivist, Royal Aeronautical Society, email to author, 17 October 2020.
5 An account of the Cadman report and subsequent government machinations can be found in Sims, *Adventurous Empires*, chapter 4.
6 Captain Dacre Watson, 'British Overseas Airways Corporation 1940–1950 and its legacy', *Journal of Aeronautical History*, Paper No 2013/03, p. 141.
7 Ibid.
8 Don's full response, *Pathfinder*, pp. 84–5.
9 Bennett, *Pathfinder*, p. 84.
10 Bramson, *Master Airman*, p. 29.
11 Sims, *Adventurous Empires*, chapter 3.
12 Ibid., p. 261. Confirmed by Paul Sheehan.
13 Part of this detailed work comes from Shri R.S. Chugh, 'Tides in Hooghly River', *Hydrological Sciences Journal*, 1961; 6(2):10–26.
14 'The Calcutta Port Rules, 1943', at www.bareactslive.com/ACA/ACT2045.HTM?AspxAutoDetectCookieSupport=1.
15 Stuart, 'This Flier Will Make Air History', p. 82.
16 Gaetano Salvemini, 'Ciano's Diaries', *The Atlantic*, March 1946, at www.theatlantic.com/magazine/archive/1946/03/ciano-diaries/656062.
17 Sims records *Cassiopeia* flew the scheduled route four days earlier. The writing was evidently on the wall because its crew, under Captain Taylor, had said their farewells to the staff at both Brindisi and Rome, expecting it to be the last flight through Italy. This indicates BOAC's Italian bases had been expecting war for several days and were able to do at least some planning for such an eventuality. It is not beyond the bounds of possibility the manager in Rome actually had a plan to put to Bennett on his arrival into Brindisi.

[18] Bennett, *Pathfinder*, p. 91. *Mercury* was handed over to No 320 (Netherlands) Squadron.

Chapter 19: Rescuing the Polish General Staff

[1] Bennett, *Pathfinder*, pp. 91–4.
[2] Joseph Retinger, *Memoirs of an Eminence Grise*, ed. John Pomian, Butler & Tanner, Great Britain, 1972, p. 84.
[3] Imperial War Museum, photographic collection, at www.iwm.org.uk/collections/item/object/205208467.
[4] Retinger, *Memoirs of an Eminence Grise*, p. 90.
[5] Ibid.
[6] Sikorski, 'Record of Daily Events', entry for Wednesday, 19 June 1940, Polish Institute and Sikorski Museum, Archive Ref KOL 1/1 DCZ.NW, London.
[7] Józef Gula, *The Roman Catholic Church in the History of the Polish Exiled Community in Great Britain*, University of London, Great Britain, 1993, p. 57.
[8] Interview with Sir Peter Wilkinson, 'Sikorski's journey to England, June 1940', in Keith Sword, *Sikorski: Soldier and Statesman – A Collection of Essays*, London, 1990, pp. 158–66.
[9] Watson, 'British Overseas Airways Corporation 1940–1950 and its legacy', p. 141.
[10] Ibid., p. 142.
[11] Ibid.
[12] Bennett, *Pathfinder*, p. 92.
[13] Coverdale, *Celebrating Bennett the Man*, p. 91.
[14] It must not be forgotten BOAC was still performing commercial flights and Wilcockson's options were likely limited.
[15] Sims, *Adventurous Empires*, chapter 11.
[16] Sir Peter Wilkinson, in Sword, *Sikorski*, pp. 158–66.
[17] Ibid., p. 159.
[18] Ibid., p. 161.
[19] Air Ministry, *Merchant Airmen: The Air Ministry Account of British Civil Aviation, 1939–1940*, His Majesty's Stationery Office, UK, 1946, p. 28.
[20] In his interview with the Imperial War Museum, Don identifies this person as a colonel in the Polish Air Force but this is an error of memory. Imperial War Museum, 'Bennett, Donald Clifford Tyndall', IWM Sound Archive 9378, 11 August 1986, Reel 1, 16.00 minute mark, at www.iwm.org.uk/collections/item/object/80009167.
[21] Lt. Col. John Hixson & Dr Benjamin Franklin Cooling, 'Combined Operations in Peace and War', Appendix B: Anglo-French Liaison, 1939–1940, US Army Military History Institute Publication, Pennsylvania, Revised edition 1982.
[22] Peter Wilkinson was surprised to see Rozoy and didn't know him by name. 'There was one Frenchman, who was presumably the Frenchman who knew about Ultra.' Sir Peter Wilkinson, in Sword, *Sikorski*, p. 165.
[23] Ibid., p. 162.
[24] Bennett, *Pathfinder*, p. 92.
[25] Wilkinson, in Sword, *Sikorski*, p. 163.

[26] The phraseology in his memoirs appears to suggest the Biscarrosse seaplane station was some distance away. Rather, it is better to read this as Don calling the Met office from *Cathay*. Given the potential threat to *Cathay*, it is unlikely he set foot on land. Indeed, all the evidence points to him remaining on board until the following morning when they rowed ashore to see where Sikorski was.
[27] Bennett, *Pathfinder*, p. 92.
[28] Jane Anson, 'World War II Bordeaux: Life Under Occupation', at http://janeanson.com/world-war-ii-bordeaux-life-under-occupation.
[29] Les Hourtiquets Base logbook entry, 20 June 1940, supplied by Alice Vivancos, Musée de l'Hydraviation, Biscarrosse, France. Confirmed in Don's *Flying Logbook, Volume 4*.
[30] Bennett, *Pathfinder*, pp. 92–3.
[31] The assertion in the Air Ministry account (*Merchant Airmen*, p. 28) that 'air bombardment of the area began with some severity' is untrue. Marie-Paule Vié-Klaze, the French flying boat historian, also maintains *Cathay* was attacked by a German fighter but avoided the bombs by 'hydroplaning on the lake: the bombs fell in the forest', but this is also doubtful. Mme Vié-Klaze, 'En Juin 1940, Les Hourtiquets Sont le Theatre D'un Fait Historique', supplied by Alice Vivancos, Musée de l'Hydraviation, Biscarrosse, France. The document 'Rescue of Polish Cabinet from France, June 1940', provided by Musée de l'Hydraviation, Biscarrosse, France, without identifying its original source, states, 'Taxying to avoid damage from dive bomber', but this does not indicate the attack actually took place.
[32] Wilkinson, in Sword, *Sikorski*, p. 165.
[33] Sikorski, 'Record of Daily Events', entry for Wednesday, 19 June 1940.
[34] 'Rescue of Polish Cabinet from France, June 1940'. The passenger manifest, attributed to Tommy Farnsworth, *Cathay*'s first officer, was Sikorski, General Sosnkowski, Colonels Klimecki, Lunkiewicz, Demel, Noel and Wasilewski, Lt. Colonels Krubski, Iranek-Osmecki and Protasewicz, Majors Jawicz and Dzikwanowski, Lieutenant Tyszkiewicz, Tomaszewska, Rozoy and Wilkinson. For the claim that Sikorski's daughter was aboard, see Air Ministry, *Merchant Airmen*, p. 29. Bines says 'twelve high-ranking Polish staff officers', p. 16.
[35] The closest the author has come to identifying this person is a letter from one Miss Lucyna Tomaszkewska to Winston Churchill, sent from 173 Park West, Edgeware Road, London on 26 July 1943 following Sikorski's death. In this letter, she begs Churchill to take action on the latest news the Germans are committing mass murder in Poland.
[36] Maynard, *Bennett and the Pathfinders*, p. 38.
[37] Air Ministry, *Merchant Airmen*, p. 29.
[38] A.S. Jackson, *Pathfinder Bennett*, Terence Dalton Limited, Suffolk, 1992, p. 32.
[39] Mme Vié-Klaze, 'En Juin 1940'. Vié-Klaze herself casts doubt on this because the Germans didn't arrive in Biscarrosse until 24 June.
[40] Bennett, *Pathfinder*, p. 93.
[41] Ibid., p. 94.
[42] 'The Portuguese World Exhibition of 1940', at LisbonLisboaPortugal.com.

Chapter 20: Fast Tracking the Ferry

[1] From Winston S. Churchill, *Their Finest Hour, The Second World War, Volume 2*, Boston, 1949, p. 325, quoted in Christie, *Ocean Bridge*, p. 27.

[2] Watt, *I'll Take the High Road*, p. 9.

[3] F.J. Hatch, 'ATFERO and Ferry Command', RCAF Volume Two, Norman Hillmer and W.A.B. Douglas collection (89/97), Directorate of History and Heritage, Department of National Defence, Government of Canada, Box 3, File 13, p. 2. Hatch provides a breakdown of the types of aircraft and associated numbers ordered, the vast majority of which were light bombers and flying boats.

[4] Bennett, *Pathfinder*, p. 121. For Bowhill's views, see Christie, *Ocean Bridge*, p. 92. Given what transpired with the takeover of ATFERO by the RAF, and Bowhill being assigned to run it, Don, unsurprisingly, did not form a complementary view of Bowhill.

[5] Christie, *Ocean Bridge*, p. 29.

[6] Ibid. Christie cites Griffith 'Taffy' Powell as saying a director of Hunt & Holditch Ltd, a London-based exporter, had written to Beaverbrook saying he had contact with airline pilots who were positing the idea of a convoy system of Hudsons shepherded by a flying boat. Ibid., p. 29. This strongly suggests it was the pilots from Imperial Airways.

[7] Ibid., p. 28.

[8] F.J. Hatch, 'Interview with Air Vice-Marshal D.C.T. Bennett C.B., C.B.E., D.S.O. regarding the inauguration of the North Atlantic Ferry Organisation and the formation of the Pathfinder Force in Bomber Command', Biographical File: Bennett, D.C.T., Directorate of History and Heritage, Department of National Defence, Government of Canada, 1976, p. 1.

[9] Letterhead of the flying time records Don submitted on a monthly basis gives the singular 'Air Service Department'. Bennett/Vial Archive, Queensland Air Museum.

[10] A summary of the provisions of the original contract can be found in Watt, *I'll Take the High Road*, pp. 21–3.

[11] The development of Gander, known prior as Hattie's Camp, is a story in itself. See [John Pudney], Air Ministry, *Atlantic Bridge: The Official Account of R.A.F. Transport Command's Ocean Ferry*, HMSO, London, 1945, pp. 11–12. (Note: Pudney is not identified as the author in *Atlantic Bridge*. He is identified as the likely author by Christie, *Ocean Bridge*, p. 40.) For a short history of Gander, see Robert Pelley, 'Gander timeline', *Bob's Gander History*, at bobsganderhistory.com; and Darrell Hillier, 'On the Precipice of Change: Gander in 1940', *Bob's Gander History*, at bobsganderhistory.com. See also the website for the Gander Airport Historical Society.

[12] For an excellent article on Gilmore, see Robert G. Pelley, 'John Joseph Gilmore MBE', *Bob's Gander History*, at http://bobsganderhistory.com.

[13] Hatch, 'ATFERO and Ferry Command', p. 7. The BCATP was known as the Empire Air Training Scheme (EATS) in Australia.

[14] Bennett, *Pathfinder*, p. 100.

[15] It is worth noting the original order for Hudsons had been placed by then Air Commodore Arthur Harris as head of the RAF Purchasing Commission to the US

in 1938. Like Don would be later, he was impressed by Lockheed's agility in responding to suggested modifications. Saward, *'Bomber' Harris*, pp. 59–60.

[16] Watt, *I'll Take the High Road*, p. 24.

[17] The Model 18 was a development of the Model 14, on which the Hudson was based, with a longer fuselage.

[18] This is also confirmed by Watt, *I'll Take the High Road*, pp. 25–6.

[19] Office of the Historian, Foreign Service Institute, United States Department of State, 'The Neutrality Acts, 1930s', https://history.state.gov/milestones/1921-1936/neutrality-acts. For other information, see Clive Rippon, 'Legal Aspects', in Watt, *I'll Take the High Road*, Appendix II, p. 164.

[20] Bennett, *Pathfinder*, p. 100. A video of an aircraft being towed across the border is available at the British Pathé Archive, 'Planes Across the Border (1940), at www.britishpathe.com/asset/46881/. For a full and interesting article on the establishment of this 'work-around' approach to the Neutrality Act, see J.E. Vernon, 'Horses on the Payroll', *Royal Canadian Air Force Journal*, Vol 5, No 2, Spring 2016, pp. 80–95.

[21] They would be paid '$600 per trip for group leaders, $500 for aircraft captains, $400 for co-pilots, $300 for radio operators, with two trips per month guaranteed, plus expenses'. Christie, *Ocean Bridge*, p. 37.

[22] Wilcockson said some of the pilots were crop-dusters while others had come from fighting in Spain and China. Wilcockson, *The Reminiscences of A.S. Wilcockson*, p. 19.

[23] The extraordinarily diverse background of these pilots is in Air Ministry, *Atlantic Bridge*, pp. 8–9.

[24] Bennett, *Pathfinder*, p. 100.

[25] Christie, *Ocean Bridge*, p. 94.

[26] Ibid., p. 49. Two of the other pilots, Allan Andrews and STP Cripps, were from BOAC and, as such, had navigator qualifications.

[27] Hatch, 'ATFERO and Ferry Command', p. 11.

[28] Wing Commander D.C.T. Bennett, 'Calling Australia', Broadcast in the BBC Pacific Service at 0815 G.M.T. 14/4/42, transcript in Australian War Memorial, AWM 65, 292, Donald Clifford Tyndall Bennett (32065).

[29] E.J. Wynn, *Bombers Across*, E.P. Dutton, New York, 1944, quoted in Christie, *Ocean Bridge*, pp. 39–40.

[30] Wright Aeronautical Division & Curtiss-Wright Corporation, *Historical Engine Summary (beginning 1930)*, pp. 8–9, at www.enginehistory.org/Piston/Wright/C-WSpecsAfter1930.pdf.

[31] As per his logbook. He recalls it as five hours in his memoirs (*Pathfinder*, p. 98), but then again he gets the date incorrect as well by confusing his two trips to Burbank. It should also be noted that in his interview with Fred Hatch in 1976, he recalls the discrepancy as being between 10 and 12%.

[32] Bennett, *Pathfinder*, p. 99.

[33] Ibid.

[34] 'Torque is the most significant parameter, describing the mechanical powertrain and has useful information to understand machine dynamics and overall efficiency.' B. Zietek, P. Krot and P. Borkowski, *An overview of torque meters and new devices*

development for condition monitoring of mining machine, IOP Conference Series: Earth and Environmental Science, Vol 684 (2021).
[35] Hatch, 'Interview with Air Vice-Marshal D.C.T. Bennett C.B., C.B.E., D.S.O.', p. 5.
[36] Pelley, 'John Joseph Gilmore MBE'. Watt (*I'll Take the High Road*, p. 34) says, '… in the first seven Hudson IIIs to come to St. Hubert, eight hot-air shutters arrived cracked and eight air intakes were cracked and caused oil temperatures to rise to 75 or 85 degrees.'
[37] Christie (*Ocean Bridge*, p. 59) records these as two ground engineers and three labourers. Christie has multiple references regarding the preparation of Gander.
[38] Content of an interview with McTaggart-Cowan by Robert Banting, along with some basic biographic details, can be found in an article by Frank Tibbo, 'McTaggart-Cowan', *Gander Airport Historical Society*, at http://www.ganderairporthistoricalsociety.org/_html_war/McTaggert.Cowan.htm.
[39] Memo on 14 September 1972 from Charles R. Pearse to S.F. Wise, suggesting including 'the Atlantic Air Ferry Command' in the official history of the RCAF, Norman Hillmer and W.A.B. Douglas collection (89/97), Directorate of History and Heritage, Department of National Defence, Government of Canada, Box 1, File 7, p. 174.
[40] Watt, *I'll Take the High Road*, p. 135.
[41] Air Ministry, *Atlantic Bridge*, p. 21.
[42] Hatch, 'Interview with Air Vice-Marshal D.C.T. Bennett C.B., C.B.E., D.S.O.', p. 5.
[43] Ibid., p. 34.
[44] They developed not just a close working relationship but a friendship. On 25 February 1942, McTaggart-Cowan, now back in Canada after spending time in England and meeting up with Don, wrote a letter to Don to answer his questions about coefficients of viscosity, rates of evaporation of ice, etc., signing off with, 'Yours … with pink elephants, McT-C'. Bennett/Vial Archive, Queensland Air Museum.
[45] Morley Thomas, 'Oral History Project: An Interview with Dr Patrick D. McTaggart-Cowan', quoted in Christie, *Ocean Bridge*, p. 50.
[46] Ibid.
[47] Aircrew member quoted in Air Ministry, *Atlantic Bridge*, p. 13.
[48] 'Eye Witness', 'I Did the Transatlantic 'Hop' with American Bombers', *The War Illustrated*, No 89, Vol 4, May 16th 1941, p. 501. Don remarked (*Pathfinder*, p. 105) that having got T9422 in the air and settled, he handed over to Clausewitz, his second pilot, and went to the navigation table. On looking up, he was surprised to see Clausewitz was 'wearing Texas cowboy boots with fancy leather-work and high heels'. A full list of aircrew for each of the seven aircraft is in Gander Airport's Watch Log, obtained by the author from Sandra Seaward, Executive Director, North Atlantic Aviation Museum, Gander.
[49] Don Bennett, *Flying Logbook, Volume 4*. This accords with Gander airport's watch logs which record 2233 GMT, that is, all off by three minutes past 7:00 pm.
[50] Bennett, *Pathfinder*, p. 105.
[51] Air Ministry, *Atlantic Bridge*, p. 13.

[52] Ibid., p. 14.
[53] Ibid.
[54] 'Eye Witness', 'I Did the Transatlantic 'Hop' with American Bombers', p. 501.
[55] Smith, quoted in Christie, *Ocean Bridge*, p. 56.
[56] Watt, *I'll Take the High Road*, p. 42.
[57] Ibid., p. 45.
[58] Ibid., p. 98.
[59] Bennett, *Pathfinder*, p. 107.
[60] Passenger manifest for RMS *Nova Scotia*, sailing from Liverpool to Halifax on 19 November, included Don and James Giles of T9422, Sidney Cripps, Kenneth Garden and George Mullett of T9424, and Allan Andrews of T9421.

Chapter 21: Beaten by Bottlenecks and Bowhill

[1] Bennett, *Pathfinder*, p. 108. The placement in the memoirs suggests this occurred after the first formation flight on 10/11 November 1940, but this could not have been the case since the Catalinas did not begin arriving at Boucherville, Montreal, until the beginning of December. This, then, places the meeting with Beaverbrook after the fourth formation flight on 28/29 December 1940.
[2] Michael Prophet, 'The PBY Catalina (the early history)', at www.michaelprophet.com/News_articles/PBY_CAT.html. He says the first three were delivered in November; Christie says December (*Ocean Bridge*, p. 74). Prophet identifies the three as AM264, AM265 and W8405, though Hendrie identifies the first seven as AM264–270. Given the first of the W-serial Catalinas were not delivered to the RAF until March 1941, W8405 can be excluded. Andrew Hendrie, *Flying Catalinas*, Appendix B, Pen & Sword, Great Britain, 1988.
[3] Christie, *Ocean Bridge*, pp. 74–5, citing S.A. Dismore, 'Atlantic Ferrying Organisation (ATFERO): The Past', 22 May 1945, 1, Public Record Office [now National Archives UK], AIR 8/474, AIR 2/7508 and AIR 20/6090.
[4] Bennett, *Flying Logbook, Volume 4*, entry for 6 December 1940.
[5] Watt, I'll Take the High Road, p. 50.
[6] AM264 did not depart for Britain until January 1941, with AM265 leaving Halifax on 2 February. See RAF Commands forum 'Mystery over Catalina AM265 crash', at www.rafcommands.com/forum/showthread.php?29545-Mystery-over-Catalina-AM265-crash&p=176781#post176781. It is further reinforced by no entry in Don's flying logbook for departing Halifax in AM265.
[7] Bennett, *Pathfinder*, p. 108.
[8] Christie, *Ocean Bridge*, p. 76.
[9] F.J. Hatch, 'Ferry Command', in Norman Hillmer, editor-in-chief, *The Creation of the National Air Force: The Official History of the Royal Canadian Air Force, Volume II*, Appendix D, University of Toronto Press, 1986, p. 645.
[10] Ibid.
[11] Hatch, 'ATFERO and Ferry Command', p. 20.
[12] Ibid., pp. 20–1.
[13] Hatch (ibid., pp. 23–6) describes a period of vitriol in January and February 1941 when Beaverbrook and Sinclair exchanged blows about the Air Ministry's

unwillingness to support ATFERO by finding pilots, ended by Churchill with a letter on 1 March 1941 which stated: 'MAP will continue to be responsible for Atlantic Ferry Service. The Air Ministry is to supply the desired personnel or their equivalent.'

[14] He has been described as 'chief executive officer' (Christie, *Ocean Bridge*, p. 76) and 'Executive Assistant to Mr Wilson' (Watt, *I'll Take the High Road*, p. 103).

[15] Watt, *I'll Take the High Road*, p. 105.

[16] Bennett, *Pathfinder*, p. 118.

[17] Ibid., p. 108.

[18] For a full discussion of the ATFERO aircrew shortage and the political machinations surrounding ATFERO and its future, see Christie, *Ocean Bridge*, pp. 80–92.

[19] RAF Commands, 'Unaccounted Airmen 19-02-1941', at www.rafcommands.com/forum/showthread.php?30569-19410219-Unaccounted-Airmen-19-02-1941.

[20] Christie, *Ocean Bridge*, pp. 62–3.

[21] Photo of the aircraft at https://1000aircraftphotos.com/APS/2611.htm.

[22] Bennett, *Pathfinder*, pp. 113-14. Don said he phoned Mackey's wife because he had heard of instances where wives could tell telepathically if husbands were alive, and when she replied affirmatively, he swung into action.

[23] Don's original map for this flight is in the Bennett/Vial Archive, Queensland Air Museum.

[24] Quote from Don's report, in Air Ministry, *Atlantic Bridge*, p. 27.

[25] Bennett, *Pathfinder*, p. 116.

[26] Don formed a very positive impression of Dowding who, he believed, was not given due recognition for his contribution in leading Fighter Command during the Battle of Britain. 'Stuffy Dowding was a very dour old man but very intelligent – many RAF officers are the opposite – they are flamboyant or at least strongly dogmatic characters without the intelligence. Stuffy Dowding was very intelligent.' Hatch, 'Interview with Air Vice-Marshal D.C.T. Bennett C.B., C.B.E., D.S.O.', p. 7.

[27] 'Pilots' may also have meant aircrew generally. Christie notes the usual limit was 18 aircrew.

[28] Jeffrey Davis, 'ATFERO: The Atlantic Ferry Organisation', *Journal of Contemporary History*, Volume 20, No 1 (January 1985), p. 79.

[29] Hatch, 'ATFERO and Ferry Command', pp. 35–6. The American pilots would also benefit from gaining long-haul experience on the North Atlantic routes. The Americans established the Air Corps Ferrying Command on 29 May 1941. See also Davis, 'ATFERO: The Atlantic Ferry Organisation', p. 86.

[30] Christie, Ocean Bridge, p. 92.

[31] Bennett, Pathfinder, pp. 120–21.

[32] Ian Davies, 'A Casualty of War & Fate', Aeroplane, September 2001, pp. 36–41.

[33] Hatch, 'Interview with Air Vice-Marshal D.C.T. Bennett C.B., C.B.E., D.S.O.', p. 4.

[34] Air Ministry, *Atlantic Bridge*, p. 6.

[35] Ibid., p. 7.

36 Watt, *I'll Take the High Road*, Appendix I, p. 163. This would deteriorate rapidly under RAF command. For the period from 20 July to 28 September, Watt records 241 aircraft delivered, with a loss of 12 and 68 deaths.
37 Christie, *Ocean Bridge*, p. 305, and Appendix A, p. 309, has a full list of aircraft types.
38 Hatch, 'Ferry Command', p. 647.
39 Christie, *Ocean Bridge*, p. 94.

Chapter 22: A Very Short Hiatus

1 Bennett, *Pathfinder*, p. 124.
2 Harris, *Bomber Offensive*, p. 129. There is some doubt about his assertion a few sentences earlier that he also got Don the job at Imperial Airways, so it is difficult to accept this without corroboration. It is plausible Harris intervened when Don appealed being offered the acting rank of squadron leader.
3 Both letters are in the Bennett/Vial Archive, Queensland Air Museum.
4 Starting date confirmed in letters from S/L E.A. Grant to Officer Commanding, RAF Leeming, Reference No 1007/P.2, 18 December 1941. Bennett/Vial Archive, Queensland Air Museum.
5 Bennett, Personal Notebook – 1 EAOS to Formation of the Path Finder Force, Bennett/Vial Archive, Queensland Air Museum.
6 Letter to the Commanding Officer, No. 1 E.A.O.S, Eastbourne, 5 November 1941, requesting transfer to an operational squadron. Bennett/Vial Archive, Queensland Air Museum.
7 Indeed, he says members of the Directorate of Bomber Operations had suggested he '… try to get into a bomber squadron … as a preliminary measure'. Bennett, *Pathfinder*, p. 132.
8 Intriguingly, he includes the Sunderland on that list despite there being no record in his logbooks of having flown it.
9 Bennett, *Pathfinder*, p. 127.
10 See UK National Archives, *Operations Record Book*, 77 Squadron, December 1941, Summary of Events, AIR 27/655, p. 3. He wrote in his logbook 'Took Command of No 77 Squadron' after the entry on 8 December, having arrived from taking the course at 10 OTU Abingdon.

Chapter 23: 77 Squadron – Enter the 'Hot' War

1 Navigation was essentially by dead reckoning. A course was set by combining two vectors: (i) the air position vector – the course and speed of the aircraft based on zero wind and (ii) the wind vector – the direction and speed of the wind. These produced a third vector – the track and ground speed of the aircraft. With variable weather, navigators needed to check location frequently by 'taking a fix' then adjusting course as necessary. This was done terrestrially by comparing what you could see on the ground with a map (with accuracy affected, among other things, by aircraft height), or celestially by use of a sextant. The other alternative was radio-direction via communication back to the UK. All aids – eyesight, sextant and radio-

direction – suffered in one way or another from changeable European weather, and unassisted by inaccurate meteorology forecasts. For a detailed analysis, see Paul Freer, *Circumventing the Law that Humans cannot see in the Dark: an Assessment of the Development of Target Marking Techniques to the Prosecution of the Bombing Offensive during the Second World War*, Doctor of Philosophy in History, University of Exeter, 2017, pp. 40ff. See also C. Blanchett, *From Hull, Hell and Halifax: An Illustrated History of No. 4 Group 1937–1948*, Midland Counties Publications, 1992, pp. 20–3, 25.
[2] Blanchett, *From Hull, Hell and Halifax*, p. 23.
[3] Sean Feast, *Pathfinders: The Definitive Story*, Key Books, Great Britain, 2022, pp. 13–4.
[4] Summary information on the Butt Report taken from M. Middlebrook & C. Everitt, *The Bomber Command War Diaries: An Operational Reference Book 1939 – 1945*, Pen & Sword Aviation, Great Britain, 1985, e-book, pp. 309–10.
[5] D. Bensusan-Butt, *Night Photographs in June – July 1941: A Statistical Analysis*, AIR 14/1218, p. 1. He undertook the exercise with the assistance of the Photographic Interpretation Section of the Air Ministry. The report can be viewed online at https://etherwave.files.wordpress.com/2014/01/butt-report-transcription-tna-pro-air-14-12182.pdf.
[6] Ibid., 'Summary' page.
[7] This led to the formation of the Operational Research Section in Bomber Command (ORSBC) to undertake extensive ongoing analysis of the impact of the bomber offensive.
[8] Bensusan-Butt, *Night Photographs in June – July 1941*, pp. 10–1.
[9] Middlebrook and Everitt, *The Bomber Command War Diaries*, p. 310.
[10] Don Bennett, *Personal Notebook – 1 EAOS to Formation of the Path Finder Force*. Bennett/Vial Archive, Queensland Air Museum.
[11] Saward, *'Bomber' Harris*, pp. 72–3.
[12] Ibid., p. 75.
[13] Richard Mead, *Dambuster-in-Chief: The Life of Air Chief Marshal Sir Ralph Cochrane GBE, KCB, AFC*, Pen & Sword Aviation, Great Britain, 2020, p. 95.
[14] Ibid.
[15] Saward, *'Bomber' Harris*, p. 81.
[16] Bomber Command's Order of Battle for 9 January 1942. See '77 Squadron History 1937 – 1945', athttps://77squadron.org.uk/.
[17] See '77 Squadron History 1937 – 1945', at https://77squadron.org.uk/.
[18] Ibid.
[19] Squadron Leader George Seymour-Price DFC would become the CO of No 462 Squadron in September 1942, ultimately achieving the rank of air commodore.
[20] Acting F/L D.F.E.C. 'Dixie' Dean DSO DFC would become a Pathfinder stalwart. He was CO of No 35 Squadron twice (1 May 1943 – 17 November 1943; 25 July 1944 – 25 February 1945). For the intervening period he was the CO of the PFF Navigation Training Unit, rising to the rank of group captain.
[21] Staton had been the CO of No 10 Squadron earlier in the war. A veteran of the First World War, he relished flying operationally. Described as 'burly', he was '... known to his crews as "King Kong" … Absolutely fearless, this larger than life

character thought that the war was tremendously exciting and had been put on for his benefit.' Blanchett, *From Hull, Hell and Halifax*, p. 28.

22 UK National Archives, 77 Squadron, *Operations Record Book*, Detail of Work Carried Out, December 1941, AIR 27/655. Available from the 77 Squadron Association website at https://77Squadron.org.uk.

23 Middlebrook and Everitt, *The Bomber Command War Diaries*, p. 318.

24 Bennett, *Pathfinder*, p. 128.

25 It should be noted the ORB entry for Z9231 for this particular operation differs from Don's account. It records Don being in command and that they 'attacked Wesermünde from 13,300 feet, dropping his bombs in one stick', an indicator they had overshot the original target. UK National Archives, 77 Squadron, *Operations Record Book*, Detail of Work Carried Out, December 1941, AIR 27/655. Available from the 77 Squadron Association website at https://77Squadron.org.uk.

26 Bennett, *Pathfinder*, pp. 128–29.

27 Both the telegram to Don and his pencilled response are in the Bennett/Vial Archive, Queensland Air Museum.

28 Bennett, *Personal Notebook – 1 EAOS to Formation of the Path Finder Force*, Bennett/Vial Archive, Queensland Air Museum.

29 Bennett, *Pathfinder*, p. 129.

30 Ibid., p. 130.

31 Don notes this as 'Christiansund' in *Pathfinder* (p. 133). That place is Kristiansund, an entirely different place to Kristiansand.

32 Bennett, *Pathfinder*, p. 134.

33 'Flight Sergeant R J LEWIS (R/56879) of the Royal Canadian Air Force', at www.rafcommands.com/database/wardead/details.php?qnum=89824. Given the graphic nature of Don's account, not all of which appears in this book, I wrestled with whether to mention F/Sgt Richard Lewis's name. I decided Lewis's descendants should know that in spite of his injuries, he showed an extraordinary bravery despite being gravely injured. They may also find some solace in knowing it was his commanding officer who took all possible steps to save his life, even risking his own life in doing so.

34 Bennett, *Pathfinder*, p. 131.

35 Middlebrook and Everitt, *The Bomber Command War Diaries*, p. 347.

36 Ibid.

37 Bennett, *Pathfinder*, p. 131.

38 Middlebrook and Everitt, *The Bomber Command War Diaries*, pp. 182–83.

39 Ibid., p. 339.

40 Bennett, *Pathfinder*, p. 133.

41 Ibid., p. 141.

Chapter 24: Bureaucracy Sows the Seeds of Bitterness

1 Letter from S/L E.A. Grant, 4 Group HQ, to RAF Station Leeming, Reference No 1007/P.2, 17 December 1941. Bennett/Vial Archive, Queensland Air Museum.

² Letter from S/L E.A. Grant, 4 Group HQ, to Officer Commanding, RAF Station Leeming, Reference No 1007/P.2, 18 December 1941. Bennett/Vial Archive, Queensland Air Museum.
³ Letter from S/L Evans, 4 Group HQ, to Officer Commanding, RAF Station Leeming, Reference No 1001/1384/P.2, 24 March 1942. Bennett/Vial Archive, Queensland Air Museum.
⁴ Letter to the Under Secretary of State, Air Ministry, Stroud, Glos., Reference No A157375/41/Accts.2(a), 7 June 1942. Bennett/Vial Archive, Queensland Air Museum.
⁵ Bennett, *Pathfinder*, pp. 137, 140.
⁶ Wing Commander D.C.T. Bennett, 'Calling Australia', Broadcast in the BBC Pacific Service at 0815 G.M.T. 14/4/42, transcript in Australian War Memorial, AWM 65, 292, Donald Clifford Tyndall Bennett (32065).
⁷ Held in Bennett/Vial Archive, Queensland Air Museum.

Chapter 25: Shot Down

¹ Bennett, *Pathfinder*, p. 142.
² Armament of *Bismarck* and *Tirpitz*: 8 x 15-inch, 12 x 5.9-inch, 16 x 4.1-inch and multitudinous anti-aircraft guns.
³ The convoys were coded 'PQ', e.g., PQ8, for Russian-bound sailings, and 'QP' for the return journey.
⁴ This had been the catalyst for the 'Channel Dash' by *Scharnhorst*, *Gneisenau* and *Prinz Eugen* in February 1942, which had featured in Don's criticism.
⁵ Nigel Smith, *Tirpitz: The Halifax Raids*, Air Research Publications, Great Britain, 1994, p. 31–2.
⁶ PRO CAB 65-25 WM (4) 11a Conclusions, quoted in Smith, *Tirpitz*, p. 32.
⁷ Don said there were five. *Pathfinder*, p. 142.
⁸ Smith, *Tirpitz*, p. 118.
⁹ Ibid., p. 100.
¹⁰ Ibid., p. 130.
¹¹ Ibid., p. 148.
¹² Ibid., p. 150, quoting from Jack Watt's unpublished account.
¹³ Ibid., p. 160, quoting Herbert 'Mick' How, the rear gunner.
¹⁴ Statement by W/Cdr D.C.T. Bennett and Sgt H. Walmsley, No. 10 Squadron – No 4 Group, regarding 27 April 1942 attack on the *Tirpitz* at Trondheim, taken by Wing Commander S.D. Felkin, AIR 40/258, A.I.1. (k) No. E62/1942, 29 May 1942.
¹⁵ Bennett, *Pathfinder*, p. 143.
¹⁶ Smith, *Tirpitz*, p. 160.
¹⁷ Bennett, *Pathfinder*, p. 143.
¹⁸ Ibid.
¹⁹ Smith, *Tirpitz*, p. 162.
²⁰ Smith gives his name as 'Phil', but the official records show him as Thomas Henry Albert Eyles.
²¹ Smith, *Tirpitz*, p. 236. This was recounted to him by Ly Bennett, who had met Olaf Horten during a trip to Norway in 1989.

[21] This is the time Don subsequently entered into his logbook.

Chapter 26: Escape

[1] Bennett, *Pathfinder*, p. 144.
[2] Smith, *Tirpitz*, p. 247.
[3] For a full account of Colgan and Walmsley's escape, see Smith, *Tirpitz*, pp. 249–56.
[4] Bennett, *Pathfinder*, p. 145.
[5] Don recalled this as Meråker, but it was Gudå. *Pathfinder*, p. 146.
[6] Smith concluded this was probably the house of Agnes and Helmer Hagensen.
[7] Bennett, *Pathfinder*, p. 150.
[8] Ly Bennett, 'My Nostalgic and Sentimental Journey to Norway, July 1989', unpublished account, Bennett/Vial Archive, Queensland Air Museum, p. 5. Ly recorded the trip she made to meet those who helped Don escape to Sweden. John Dalamo and his wife had died in the 1970s, but Ly spent considerable time with their daughter, Mary Stenmo. Don recalled John had also lived in South Australia, though Ly did not verify this.
[9] Ibid.
[10] Bennett, *Pathfinder*, p. 153.
[11] Bennett, 'My Nostalgic and Sentimental Journey to Norway, July 1989', p. 10.
[12] Letter from John Dalamo to Don Bennett, 26 September 1945, photo in Bennett, *Pathfinder*.
[13] Smith, *Tirpitz*, p. 268.
[14] Sheet of paper from Don's time in Sweden. Bennett/Vial Archive, Queensland Air Museum.
[15] Bennett, *Pathfinder*, p. 155
[16] Some records have his rank as air commodore.
[17] It is very difficult to read this on Don's sheet of paper because it is in pencil and crosses a much-used crease.
[18] For an excellent article, see Arne Kaijser, 'Driving on wood: the Swedish transition to wood gas during World War Two', Taylor & Francis Online, 31 March 2022, at www.tandfonline.com/doi/full/10.1080/07341512.2022.2033387.
[19] David Gray, University of Dalarna, Sweden, email to author, 2 March 2023.
[20] See David Gray, 'Constance Malleson's journey in Dalarna', Dalarna Women's History Association, online article as www.dalarnas-kvinnohistoriska.se/constance-mallesons-resa-i-dalarna/.
[21] Letter from The Lady Constance Malleson to Bertrand Russell, 12 May 1942, supplied by David Gray, University of Dalarna, Sweden.
[22] Statement by W/Cdr D.C.T. Bennett and Sgt H. Walmsley, No. 10 Squadron – No 4 Group, regarding 27 April 1942 attack on the *Tirpitz*, taken by W/Cdr S.D. Felkin, 29 May 1942, A.I.1. (k) No E62/1942, AIR 40/258.
[23] Smith, *Tirpitz*, p. 271.
[24] Bennett, *Pathfinder*, p. 158.
[25] Nowlan, J.G., Honorary Secretary, Brisbane Grammar School Old Boys' Association, letter to Don Bennett on the awarding of his DSO, 16 June 1942. Held in Bennett/Vial Archive, Queensland Air Museum.

Index

Note: The rank for any person listed in the index is that applicable to their final mention in this volume of the biography. Many would go on to higher ranks after July 1942.

PEOPLE

A

Adams, Ralph 219
Aitken, Max, First Baron Beaverbrook 206-8, 212, 217-8, 221-2, 223-4, 226, 229-31
Alger, Captain Howard 129-30, 184
Anderson, Wing Commander William 'Andy' 73
Armitage, Radio Officer Jimmy 200
Arnold, General HH 'Hap' 230

B

Bailey, Captain Frank 124, 126, 129
Banting, Sir Frederick 227-8
Batten, Jean 147
Beatty, Sir Edward 207-8, 224-6
Beaverbrook, First Baron *see* Aitken, Max, First Baron Beaverbrook
Beauman, Wing Commander EP 266
Bell, Graham 189
Bennett, Arnold (brother) 2, 4, 7, 9, 16, 19-21, 26, 31-2, 49
Bennett, Aubrey (brother) 2, 7, 9, 16, 19, 25-6, 49, 117
Bennett, Celia (mother) 8-13, 19-21, 25-7, 29, 32, 49-50, 56, 114, 116-7, 130, 145-6, 229
Bennett, Clyde (brother) 2, 7, 9
Bennett, Fred (uncle) 7, 19
Bennett, George H (grandfather) 8

Bennett, George T (father) 1, 2, 7-9, 12, 19-23, 25-30, 32, 49-50, 56, 114, 116-7, 130, 145, 289
Bennett, Isaac (great-grandfather) 8
Bennett, Ly nee Gubler 107-8, 111-2, 115-7, 120, 128, 136, 141, 147, *157*, 187, 231, 277, 279, 281, 283, 286, 290
Bennett, Noreen (daughter) 128, 231, 286, 290
Bennett, Sarah nee Makepeace (grandmother) 8
Bennett, Torix (son) 141, 147, 231, 286, 290
Bensenitz, Lena 283
Bensusan-Butt, David 242
Berg, HR 32
Bernadotte, Count Folke 285, 287
Biard, Henry C 94-5
Bickell, Jack 208
Biddle, Flight Lieutenant William 198
Birdwood, Field Marshal WR 205
Blankenberg, Kapitän Joachim 171
Bowhill, Air Chief Marshal Sir Frederick 207, 223, 230-1, 234
Bowman, A McD 32, 49
Boyce, Flying Officer CDC 70, 75
Brackley, Major Herbert G 119-20, 125-6, 129-30, 134, 190
Brain, Lester 49
Breakey, Squadron Leader John 110
Brearley, Major Norman 96, 98
Brooke-Popham, Air Chief Marshal Sir Robert 125, 195
Bruce, Stanley M 146
Burgess, Captain John 125
Burchall, Colonel Harold 119, 190, 209, 224-6

C

Callaway, Group Captain William 107-8, 112, 115
Carey, Flying Officer Alban 'Crackers' 62, 79, 90, *153*
Carr, Air Vice-Marshal CR 'Roddy' 238, 262-3, 273
Chadwick, Flight Lieutenant Reg 113
Chamberlain, Neville 173, 187
Chatfield, Admiral AEM, 1st Baron Chatfield 205
Cheshire, Leonard 112
Chichester, Francis 147
Churchill, Winston 197-9, 206, 230-1, 243, 268, 270, 273

Cleland, Flying Officer Ralph 57
Clift, Flight Lieutenant 'Laddie' 61-4, 71, 78, 252
Cobham, Alan 80, 137, 184 *see also* Flight Refuelling Limited
Cochrane, Air Vice-Marshal Ralph 15, 68, 244-6
Colgan, Flight Sergeant John 276, 278-9, 281-2, 286-7
Coningham, Air Commodore Arthur 241
Coster, Radio Officer AJ 'Jimmy' 141, *160*, *162-3*, 173
Cunliffe-Lister, Sir Philip, Lord Swinton 127
Curtiss, Glenn 60

D

Dalamo, John 280, 282
Dalanes, Trygve 280, 282
Damant, Flying Officer FK 50-1
Dand, Group Captain 235-6, 238
Dawes, Group Captain H 'Daddy' 238
De Gaulle, General Charles 201
Dean, Flight Lieutenant DFEC 'Dixie' *168*, 248, 253
Dickens, CH 'Punch' 226
Dowding, Air Chief Marshal Sir Hugh 'Stuffy' 229
Draper, AJ 32
Draper, Susan (3rd wife of Dr TP Lucas) 11
Drew, AC 32, 48
Drury, Flight Lieutenant FA 248, 253
Duke, Flight Lieutenant AL 55-6, 79

E

Egglesfield, Captain Lawrence 128-9
Ellington, Sir Edward 127
Embling, Wing Commander John *168*
Evans, Charles 234
Eyles, Sergeant Tom 276, 277

F

Farnsworth, First Officer TH 'Tommy' 200, 204
Fleet, Squadron Leader LDH 283
Florman, Carl 285
Forbes, Sergeant Clive 276, 279-81, 282-3

Fysh, Hudson 25, 49

G

Garrod, Air Marshal Sir Guy 234
Geddes, Sir Eric 122
Gentry, Dana 219
George V, HM King 102, 113
George VI, HM King 205, 273
Gill, EA 9
Gillett, James 114
Gilmore, John 'Joe' 209-10, 215-6, 219
Glen, JG 32, 49
Godfrey, Mr (Rolls-Royce) 249
Goodwin, Sir John 28
Gordon, Major-General George 133
Gordon, WL 91
Gouge, Arthur 122, 126
Grace, AD 32, 49
Grace, Sergeant 249
Graham, Group Captain Strang 256, 272, 281
Grant, Squadron Leader EA 262
Gubler, Charles 107-8

H

Hall, Adjutant 248
Hansson, Löjtnant NO 286
Hargrave, Lawrence 189
Harrington, Captain JC 125
Harris, Air Marshal Sir Arthur 68-74, 76, 88, 234, 244-6, 257, 259-60, 287-8
Harvey, Ian *165*, 173-4, 176-8
Heber, Reginald 229
Henry VIII 59
Hinkler, Bert 25, 27-9, 84, 147, *149*, 289
Hitler, Adolf 171-3, 194, 205, 268
Horten, Olaf 276, 282
Hoskins, Dr Richard 35
How, Flight Lieutenant HG 'Mick' *169*, 274-6, 277-8
Howe, Clarence 143, 208
Howse, Captain Hudson 'Hoddy' 127

I

Ismay, Major-General HL 268

J

Johnson, Amy 31, 84, 289
Jones, Arthur W 5-6, 7
Jones, Flight Sergeant 248
Jones, Squadron Leader George 44-5
Juliussen, Herlof 278

K

Kelly-Rogers, Captain John 'Jack' 125, 184-5, 195
Kewell, Les 31
Kidston, Lieutenant Commander Glen 98, 100
Kingsford Smith, Charles 29, 34, 84, 147, 289
Kipling, Rudyard 14
Klose, Lieutenant EKH 32
Kringberg family 277

L

Lampson, Sir Miles 125
Lang, Squadron Leader AF 'Frank' 75-7
Lawrence of Arabia 133
Lawson, Flight Lieutenant 248
Leśniowska, Zofia 204
Lewis, Flight Sergeant Richard 254
Lindbergh, Charles 28, 238
Lindemann, Professor Frederick, Lord Cherwell 242
Liston, Adjutant 248
Littlejohn, Flying Officer Barry 32, 48-9, 111-2
Long, Harold 226, 231
Longfellow, Henry Wadsworth 73
Lucas, Mary (1st wife of Dr TP Lucas; grandmother) 10
Lucas, Mary (2nd wife of Dr TP Lucas) 11
Lucas, Samuel (great-grandfather) 10
Lucas, Susan 11
Lucas, Dr Thomas P (grandfather) 2, 8, 10-12
Ludlow-Hewitt, Air Chief Marshal Edgar 245

M

MacNeese Foster, Air Commodore WF 245
McFarlane, Squadron Leader J 114
McGinnis, Paul 25
McTaggart-Cowan, Dr Patrick 215, 217-20, 232
Mackenzie King, William L 230
Mackey, Joseph 227-8
Mahaddie, Group Captain TG 'Hamish' 248
Maitland, AJG 32
Makepeace, Sarah (grandmother) 8
Malleson, Lady Constance *170*, 283-4
Mawdesley, Flight Lieutenant Fred 75-6
Maxim, Hiram 189
Maycock, Group Captain R 283
Mayo, Major Robert H 119, 122, 137-8, 140, 142, *160*, 172, 175, 179-80, 182-3, 190
Menzies, Pilot Officer 65-6
Merer, Flight Lieutenant John 55-7, 79
Miles, JC 32, 48
Miller, Horace C 'Horrie' 97-8, 100, 104
Miller, Flight Lieutenant 273
Mitchell, RJ 60, 90
Mollison, James 'Jim' 97, 147
Moore, Steward W 'Dinty' 200
Morgan, Purser W 'Bill' 200
Mountbatten, Lord Louis 90-1
Murphy, Sergeant William A 249-50
Murray, Sergeant John 276, 277-8
Mussolini, Benito 194-5

N

Nicholl, Group Captain HR 79
Nielsen family 278
Noble, Flight Lieutenant WV 266
Nowlan, JG 288

O

Oakes, Flying Officer 248

O'Neill, Squadron Leader Henry 'Paddy' 53-4, 56
O'Rourke, Acting Flight Lieutenant 248
Øyan, Johan 278-9

P

Page, Captain Humphrey 209-10, 212, 221
Paget, JR 32, 48
Parker, John Lankester 184
Parkin, Squadron Leader LHW 248, 254
Patch, Steve 8
Peirse, Air Marshal Sir Richard 241, 243
Pétain, Marshal Philippe 197
Pickles, Flying Officer Ken TF 'Percy' 65, 79, 109-10
Pilcher, Percy 189
Pirie, Air Commodore George 207
Portal, Air Chief Marshal Sir Charles 225-6, 230, 243, 245-6
Powell, Captain Griffith 'Taffy' 129-30, 209, 215
Preston, Sergeant 44-5
Prince Edward, HRH Prince of Wales 66, 77, 82, 102
Prince George, HRH Duke of Kent 205
Prien, Günther 188

Q

Queen Mary, HM, wife of George V 102

R

Raczyński, Count EB 197, 201
Rae, RAR 32, 48
Raeder, Admiral Erich 268
Reith, Sir John 147, 189-90
Retinger, Dr Józef 197-8
Reynaud, Paul 197
Robertson, David 97, 103
Robertson, Sir Macpherson 83, 97
Robinson, George A 115, 117
Robinson, PD 86
Roosevelt, Eleanor 185
Roosevelt, Franklin D 185, 230-1

Ross, Captain Ian 209, 212
Rozoy, Colonel 201, 203-4
Russell, Bertrand 284

S

Salmond, Air Marshal Sir John 35
Sandford, Sergeant 248
Sassoon, Sir Philip 122, 126, 137, 182
Saundby, Flight Lieutenant Robert 68
Saye, Flight Lieutenant GIL 'Gill' 79, 84, 90, 108, *153*
Scherger, Flight Lieutenant Fred 40-1, 44
Schiefloe, Johan 278
Self, Sir Henry 225
Seymour-Price, Squadron Leader George *168*, 248, 253
Shackleton, Sir Ernest 238
Sikorski, General Władysław *167*, 197-204
Sinclair, Sir Archibald 226
Skoogh, Captain 281
Slessor, Air Vice-Marshal John 225
Smith, Clary 32, 48
Smith, H Gengoult 83
Smith, Ross & Keith 28
Smith, V Edward 220-1
Sosnkowski, General K 204
Stalin, Josef 268
Staton, Group Captain WE 'Bill' 249
Stone, AB 'Wizard' 5
Store, Gordon 184, 221
Stowe, AB 32
Stuart, Frank S 94-5, 193
Studd, Wing Commander TQ 82
Sundell, Åke 283, 285
Sundell, Stella 283, 285

T

Tennyson, Alfred Lord 26
Thurgood, Bill 91
Tomaszkewska, Mlle L 204
Tomlinson, Flying Officer Peter 72

Trenchard, Marshal of the RAF Sir Hugh 273
Tripp, CM 'Curly' 219-21
Tuck, Wing Commander JAH 262
Tucker, Dr WS 76
Tyszkiewicz, Count Michał 198, 200-4

U

Ulm, Charles 29, 34, 289

V

Virtue, Keith 115, 117

W

Walmsley, Sergeant Harry 276, 278-9, 281, 285-8
Watts, Flight Lieutenant Jack V 274
Wells, Geoff 173
Westbrook, Trevor 208
Widows, Pilot Officer 54-5, 57
Wikner, Geoffrey 31, 37
Wilcockson, Captain Arthur 125, 129, 139, 141-2, 147, *160, 163, 165,* 175, 199, 209, 212
Wilkinson, Major Peter 200-3
Williams, Air Commodore Richard 34-5
Williamson, Squadron Leader James J 51
Wilson, Morris 208, 226, 230
Winton, K 95
Wood, Sir Kingsley 172, 184
Woodhouse, Captain Hugh 126
Woods, James 'Jimmy' 96-103, 105, *158*
Woods Humphery, George E 190, 207-9, 224-6
Wright brothers 4

Y

Youell, Captain Alan 230
Young, Tom 117

PLACES

A

Aberdeen, UK 96
Adelaide, Australia 181
Albuquerque, USA 210
Aldergrove, Belfast, UK 219, 221, 229
Aleppo, Syria 103-4, 106, 126, *156*, *158*
Alexander Bay, *see* Orange River
Alexandria (Alex), Egypt 105, 119-27, 129-30, 132, 134, *159*, 179-81, 184, 194
Archangel, Russia 268
Ardnamurchan, UK 66
Athens, Greece 103, 123, 129, 132, 194
Atlas Mountains 174
Auchenflower (including Kellett St), Brisbane, Australia 12, 19, 21, 145, 289
Auckland, New Zealand 96
Azores *see* Horta, Azores

B

Bafia, Cameroon 174
Baghdad, Iraq 126
Bahrain 126
Basra, Iraq 126, *158*
Beachy Head, UK 64
Beira, Mozambique 132-4
Beirut 105
Belle Isle, Canada 143
Belem, Lisbon, Portugal 205
Berlin, Germany 9, 172, 247, 278, 284
Bermuda 136, 146, 223-5
Biscarrosse *see* Lake Biscarrosse
Bjornås Farm, Norway 280
Blackbushe Airport 198
Bognor Regis, UK 64
Borås Farm, Norway 275
Bordeaux, France 196-201, 203
Borneo, East Indies 181

351

Botwood, Newfoundland 136, 141, 143, 145-6, 185, 209
Boucherville, Montreal, Canada 141, 143, 209, 223
Bougie (now Béjaïa), Algeria 174
Boulogne, France 254
Boulogne-Billancourt, Paris, France 255-6
Bremen, Germany 247
Brest, France 196-7, 201, 204, 260
Brindisi, Italy 121, 123-6, 128, 131-2, 194
Brisbane, Queensland, Australia 1-2, 8-14, 19-20, 25, 27-8, 31, 33, 49, 50, 67, 114-7, 126, 144, 146, *149*, 181, 229, 288
Bundaberg, Queensland, Australia 27
Burbank, USA *see* Lockheed Aircraft Plant

C

Cairo, Egypt 125, 132-3
Calcutta, India *166*, 192
Calf of Man, UK 65
Calshot, *see* RAF Calshot
Cape Bauld, Newfoundland 143, 145
Cape Town, South Africa 126, 172, 174-80
Caravelas, Brazil 171
Camooweal, Queensland, Australia 25, 49
Carnarvon, Australia 96
Channel Ports, UK 247
Charleville, Queensland, Australia 25
Cherbourg, France 197
Cloncurry, Queensland, Australia 25
Condamine, Queensland, Australia 1, 7-8
Cowes, Isle of Wight, UK 59, 115
Crinan Canal, UK 66
Croydon, UK 28, 88, 119-20, 125
Cypress 103

D

Dar es Salaam, Tanzania 132-5
Darling Downs 7
Dartmouth, Nova Scotia *see* Halifax, Nova Scotia
Darwin, Australia 28, 31, 181
Dayboro, Queensland, Australia 31

Deniliquin, Australia 47
Douala, Cameroon 176
Douglas, UK 80
Dundee, UK *165*, 173, 178
Dunkirk, France 194, 197
Durban, South Africa 128, 132-5, 139, 141, 179-80
Durlston, UK 64
Düsseldorf, Germany 255

E

Eastbourne, UK 64 *see* No 1 Elementary Air Observer School
Eastleigh, UK 90-1, 95, 111
Emerson, Canada 211
Entebbe, Uganda 126
Exmouth, UK 64

F

Fættenfjord, Norway *169*, 267, 269-70
'Fairthorpe', Toowoomba, Queensland, Australia 7, 289
Falun, Sweden 279, 281, 283-4, 286
Feilding, New Zealand 238
Felixstowe 110, 140, 195, 199 *see also* Marine Aircraft Experimental Establishment
Flensburg, Germany 248
Flornes, Norway 276, 277
Flornesvollen, Norway 276, 277
Folkestone, UK 54
Foynes, UK 141-6, *163*, 185, 187-8
Fremantle, Australia 181

G

Gander, Newfoundland *166-7*, 209, 215-7, 227-9, 231, 254
Gaza, Palestine 126
Gennevilliers, France 253, 255
Genoa, Italy 255
Geraldton, Australia 96
Gold Coast, Queensland, Australia 20-1
Greenland 215-6, 228-9, 291

Greenock, UK 223
Gudå, Norway 278-9, 287
Guernsey, UK 64, 184
Gwadar 126

H

Halifax, Nova Scotia, Canada 223-5
Hamble, UK 90, 111
Hamburg, Germany 247
Hanworth, UK 98, 99, 105, 111
Hendon *see* RAF Hendon
Hendra, Brisbane, Queensland, Australia 31
Herøya, Norway 255
Heston, UK 91-2, 99, 101, 104
Hooghly River, Calcutta, India *166*, 192-3
Horta, Azores, Portugal 136, 146, 171
Hythe, UK 114, 128, 141-2, 173, 184, 188, 195

I

Iceland 215-6, 228
Inhambane, Mozambique 132
Iona, UK 66
Isle of Wight, UK 59, 64 *see also* Cowes

J

Jersey, UK 91-2, *155*
Juba, Sudan 125

K

Kangaroo Point, Brisbane, Queensland, Australia 8
'Kanimbla', Queensland, Australia 1-2, 7-8, 20, 23, 25, 114
Kano, Nigeria 174
Karachi, India 126, 136, 180, 191
Karima, Sudan 125
Kelvin Grove, Brisbane, Queensland, Australia 29-30
Kerguelen Islands 181
Khartoum, Sudan 132
Kisumu, Kenya 120, 126, 128-9, 132-3

Koepang, Indonesia 181
Kosti, Sudan 125
Kristiansand, Norway 253
Kuwait City 126

L

Lake Biscarrosse, France 199, 202-4
Lake Bracciano, Italy 194-5
Land's End, UK 66, 109
Laurenço Marques (now Maputo), Mozambique 132
Les Hourtiquets *see* Lake Biscarrosse, France
Libourne, France 198, 203
Libreville, Gabon
Lindi, Tanzania 132, 134-5
Lisbon, Portugal 146, 205
Liverpool, UK 221
Loch Alsh, UK 66
Loch Lomond, UK 66
Lofjord, Norway 271, 273
London, UK 10, 49, 50-51, 54, 57, 83, 88, 91, 98, 101, 104, 112, 128, 132, 144, 146, 175, 185, 188, 197-8, 200, 203, 207, 221-2, 231, 263, 268, 286-7
Longreach, Queensland, Australia 25, 49
Luderitz Bay, South West Africa 177
Luxor, Egypt 125, 132
Lympne, UK 80

M

Malakal, Sudan 125, 132
Mannheim, Germany 255, 257
Mannsæterbakken, Norway 280
Marignane, *see* Marseilles, France
Marseilles, France 103, 130-2, 195
Melbourne, Australia 11, 30, 32-5, 44, 47-8, 83, 97, 117, 131, 181
Meråker, Norway 278-9
Mildenhall, UK 101-2
Mirabella, Crete 123, 125
Mombasa, Kenya 133
Montreal, Canada 141, 143-6, 185, 208-11, 215, 223-4, 228, 230-1 *see also* Boucherville

Mount Batten, UK 65
Mount Isa, Queensland, Australia 25
Mount Kilimanjaro, Tanzania 133
Murmansk, Russia 268

N

Naples, Italy 94, 105
Narsarsuaq, Greenland 228
Newfoundland 136, 141-2, 145-6, 184, 230-1 *see also* Botwood and Gander
New York, USA 138, 143-4, 146, 171-2, 179, 185, 207-8 *see also* Port Washington
North Weald, UK 54 *see also* No 29 (Fighter) Squadron, North Weald

O

Oban, UK 66
Odda, Norway 255
Omaha, USA 210
Orange River, South Africa 177-8
Orleans 244
Oslo, Norway 253, 255, 278, 285

P

Paris, France 91, 119, 194-5, 196-7, 200, 244, 255-6 *see also* Boulogne-Billancourt
Pembina, USA 211
Pembroke Dock 66-75, 77, 78, 109, 111, 116, 199, 235 *see also* No 210 (Flying Boat) Squadron
Penzance, UK 65
Perth, Australia 96-7
Philippines 181
Point Cook *see* No 1 Flying Training School, RAAF Point Cook
Poissy, France 255
Poole, UK 188, 191, 195, 199-201, 205
Port Bell, Kampala, Uganda 132
Port-Gentil, Gabon 176
Port Nolloth, South Africa 177-8
Port Washington, New York, USA 141, 224
Portsmouth, UK 91-2

Prestwick, UK 229, 231

Q

Quelimane, Mozambique 132

R

Rio de Janeiro, Brazil 171, 181
Rhodesia 69
Rochester *see* Short Bros
Rome, Italy 103, 119, 132, 194-5 *see also* Lake Bracciano
Rotterdam, Netherlands 253, 255
Ruhr, Germany 242, 247

S

Saint-Hubert, Montreal, Canada 209, 211, 213, 216, 227, 229, 231
Saint-Malo, France 197
Saint-Nazaire, France 195, 255
St Aubin, Jersey, UK 92, *155*
St Catherine's Lighthouse, UK 64
St David's Head, UK 65
St Helier, Jersey, UK 92
Saldanha Bay, South Africa 177
San Antonio, Argentina 181
Scilly Isles, UK 75
Selsey, UK 64
Sherston, UK 94
Singapore 70, 122, 191-2
Southampton, UK 59, 80, 91-2, 107, 111-2, 114-5, 128, 130-2, 134, 139, 142, 146, 172, 179-82, 185, 187, 194, 199 *see also* Hythe
Squires Gate, Blackpool, UK 114, *154*, 230
Speke, UK 221
Start Point, Devon, UK 171
Stavanger, Norway 255
Stockholm, Sweden 283-4, 287
Storlien, Norway 279, 281
Storvallen, Norway 281
Stranraer, UK 65, 79
Stuttgart, Germany 257

Swanage, UK 64
Sydney, Australia 9, 19, 31, 34, 115, 117, 127, 145, 181

T

Tangmere 54
Toowoomba, Queensland, Australia 1, 4-9, 19, *149*
Trondheim, Norway 267-9, 287

U

Upavon, UK 54

V

Vancouver, Canada 181

W

Waalhaven, Netherlands 99
Wadi Halfa, Sudan 125, 132
Warwick, Queensland, Australia 5
West Indies 183
Weymouth, UK 111
Whitchurch, UK 80
Wilhelmshaven, Germany 247, 249, 253, 255, 267-8
Winchester, UK 115
Winton, Queensland, Australia 25
Windsor Street Railway Station, Montreal, Canada 209, 211

Z

Zurich, Switzerland 107, 115

AIR FORCE

Royal Air Force (RAF)

Groups, Squadrons, Units & Schools

No 1 Elementary Air Observer School 235-6, 261, 291
No 3 Group, Bomber Command 270
No 4 Group, Bomber Command 236, 238, 241, 246, 261-3, 265, 270
No 5 Flying Training School, RAF Sealand 50, 289
No 5 Group, Bomber Command 72, 244-6, 272
No 6 Group, Bomber Command 244-5
No 7 (Operational Training) Group, Bomber Command 245
No 7 Squadron, RAF Oakington 248
No 10 Operational Training Unit (OTU), RAF Abingdon 235, 244-6, 250
No 10 Squadron, RAF Leeming 264-5, 267, 270-4, 288, 291
No 15 Squadron 270
No 29 (Fighter) Squadron, RAF North Weald 46, 51, 53, 58, 110, *152*, 289
No 35 Squadron 272, 274
No 44 Squadron 68
No 45 Squadron 68, 245
No 58 Squadron 68
No 76 Squadron 270-2
No 77 Squadron, RAF Leeming *168*, 238, 240, 244, 246-56, 258, 261, 264, 292
No 99 Squadron 55
No 149 Squadron 270
No 191 Squadron 68
No 201 Squadron 59, 71-2
No 205 (Flying Boat) Squadron 70
No 206 Squadron 198
No 207 Squadron 55
No 210 (Flying Boat) Squadron 67, 68, 70-5, 77, 78, 85, 111, 259, 289
No 480 (Coastal Reconnaissance) Flight 60, 70
No 617 Squadron 270
Path Finder Force, Pathfinders 38, 70, 72-3, 76, 192, 204, 248, 255, 288
School of Naval Co-operation and Aerial Navigation, RAF Calshot 59, 64, 78, 90, *152*, 289
Seaplane Training Squadron, RAF Calshot 59, 79, 81, 90, 114

Bases

RAF Abingdon *see* No 10 Operational Training Unit (OTU)
RAF Calshot 57-8, 59-60, 64, 66, 68-71, 74-76, 78-82, 84-86, 89, 90, 93, 99, 105-6, 107-111, 114-116, 119, *152-3*, 235, 252, 289 *see also* No 201 Squadron, Seaplane Training Squadron, School of Naval Co-operation and Aerial Navigation, and Flying Boat Pilot's Course
RAF Cranwell 36, 53
RAF Hendon 5, 82, 112 *see also* Hendon Air Display
RAF Henlow 57-8
RAF Kinloss 248, 273
RAF Leeming 238, 244, 250, 254, 261-2, 271, 281, 286, 291 *see also* No 10 Squadron and No 77 Squadron
RAF Leuchars 285-6
RAF Linton-on-Ouse *see* No 4 Group, Bomber Command
RAF Lossiemouth 272-3
RAF North Weald *see* No 29 (Fighter) Squadron
RAF Pembroke Dock *see* No 210 (Flying Boat) Squadron
RAF Sealand *see* No 5 Flying Training School
RAF Sutton Bridge 52
RAF Tain 273
RAF Uxbridge 50
RAF Wittering 74-5
RAF Worthy Down 68

General

Air Navigator's Certificate, 1st and 2nd Class 85
'Area bombing' 257-9
Bomber Command 42, 68, 70, 72, 112, 236-8, 241-8, 250, 256-9, 262, 266, 269, 287-8
Butt Report 242-244, 248
Coastal Command 206-7, 223, 237, 272
Ferry Command 230, 232
Fighter Command 229
Flying Boat Pilot's Course, RAF Calshot 57, 61-7, 68-9, 70, 78-80, 82, 88, 90, 109, 111
Heavy Conversion Units (HCUs) 244, 246
Hendon Air Display 112-3

360

Long Navigation Course 'Long N' 90-1
'Nickel' raids 244, 247, 253, 255
Operational Training Units (OTUs) 244 *see also* No 10 Operational Training Unit (OTU)
Operations Record Books (ORBs) 241, 249, 253, 255
Photograph Reconnaissance Unit (PRU) 241-2
Royal Flying Corps (RFC) 33, 53, 69, 96
Royal Naval Air Service (RNAS) 27, 33

Royal Australian Air Force (RAAF)

Australian Flying Corps (AFC) 33
Central Flying School *see* RAAF Point Cook
No 1 Flying Training School *see* RAAF Point Cook
Pilot (wings) training *see* RAAF Point Cook
RAAF Point Cook 30-2, 33-6, 39, 41, 44, 46-49, 51-2, 58, 111, *150*, 236, 258, 288, 289
RAAF recruitment 29-30
Short-service commission 33

Royal Canadian Air Force (RCAF)

Eastern Air Command 209
No 13 Service Flying Training School 209
Nova Scotia 223-4

AIRCRAFT

A

Armstrong Ensign 180
Armstrong Whitworth F.K.8 97
Armstrong Whitworth Siskin III and IIIA 50-1, 53-7, 63, 81, *152*, 292
Armstrong Whitworth Whitley Mk.III and Mk.V 235, 238, 240, 244, 246-50, 253-6, 292
Avro 504K 25
Avro 504N 74, 292
Avro 618 Ten trimotor 115, 117, 292
Avro Anson, 292
Avro Avian 28, *149*
Avro Baby 238
Avro Lancaster 272
Avro Manchester 236, 240
Avro Tutor 292

B

Bleriot monoplane 5
Boeing 314 Clipper 186
Boeing B-17 Flying Fortress 225
Boulton & Paul P71A 119, 292
Bristol Blenheim 240
Bristol Bulldog Mk.II 57, 292

C

Calcutta *see* Short Bros S.8 Calcutta
Catalina *see* Consolidated PBY-5 Catalina
Caudron G 5, 7
Consolidated LB-30A Liberator B.Mk.I *168*, 227-8, 230, 236-7, 292
Consolidated PBY-5 Catalina 223-5, 237, 292
Curtiss Models E and F 60

D

de Havilland DH.6 96
de Havilland DH.50 25, 96
de Havilland DH.53 Humming Bird 80-1, *154-5*, 289, 292

de Havilland DH.60G Gipsy Moth 22, 31, 44, 97
de Havilland DH.60G III Gipsy Moth 117, 292
de Havilland DH.60M Metal Moth 50, 292
de Havilland DH.60X Cirrus Moth 22, 44-5, 49, *150*, 292
de Havilland DH.60 (unidentified, Qantas-owned) 49
de Havilland DH.61 Giant Moth 97
de Havilland DH.80A Puss Moth 98
de Havilland DH.?? Moth 94-5, 193
de Havilland DH.84 Dragon 92-3, 110, *155*, 292
de Havilland DH.86 Express 119, 292
de Havilland DH.88 Comet 95
Dornier Do 18 171, 178
Douglas DC-3 285

E

Empire *see* Short Bros S.23/S.30 Empire

F

Fairey IIIF 64, 292
Fairey Swordfish 267
Farman Sport 22, 31-2, 37,
Felixstowe F Series 60
Focke-Wulf Fw 200 Condor 172
Fokker FVIIB trimotor 29

H

Halifax *see* Handley Page Halifax Mk.II
Handley Page H.P.42 119, 125, *158*, 292
Handley Page Halifax Mk.II 58, *169*, 236, 240, 270-1, 274-6, 281-2, 292
Handley Page Hampden 240
Handley Page Harrow *165*, 184
Hawker Hart 95, 292
Hawker Horsley 238
Hudson *see* Lockheed Hudson Mk.I, II, III and V

J

Junkers Ju 52/3 285

K

Kent *see* Short Bros S.17 Kent
Klemm Kl32-V 111-2, 292

L

Liberator *see* Consolidated LB-30A Liberator B.Mk.I
Lockheed 14 Super Electra 285
Lockheed Hudson Mk.I, II, III and V *166*, 198, 206-7, 210-7, 219, 222, 224-5, 227, 230-2, 237, 254, 285, 291, 292
Lockheed Model 18 Lodestar 210, 292
Lockheed Vega DL-1A 96, 98-105, 110-1, 126, *156*, *158*, 292

M

Maia see Short Bros S.21 *Maia*
Mayo Composite 60, 136-42, 144, 147, *160-5*, 179-80, 182, 292 *see also* Short Bros S.20 *Mercury* and Short Bros S.21 *Maia*
Mercury see Short Bros S.20 *Mercury*
Miles M.2 Hawk 95, 292

S

Saunders-Roe A.27 London 109, 292
Saunders-Roe A.29 Cloud 81-2, 109, 111, 113, 292
Short Bros L.17 Scylla 119, 292
Short Bros S.8 Calcutta 110, 120, 123, 125, *159*, 292
Short Bros S.8/8 Rangoon 110, 123, 292
Short Bros S.14 Sarafand 110, 292
Short Bros S.17 Kent 120, 123-5, 129, *159*, 292
Short Bros S.20 *Mercury* (Mayo Composite) 23, 60, 136-47, *161-4*, 171-83, 188-9, 191, 195, 209, 214, 230, 237, 290, 292
Short Bros S.21 *Maia* (Mayo Composite) 60, 137-42, 146, *161-4*, 173, 179-80
Short Bros S.23 Empire 122-3, 126-32, 137, 139, 141, 146, *160*, 180, 183, 188, 191-5, 209, 237, 290, 292
Short Bros S.30 Empire 137, *165*, 183-6, 187, 195, 199-205, 237
Short Bros Stirling 236, 240, 270
Siskin *see* Armstrong Whitworth Siskin III and IIIA
Southampton *see* Supermarine Southampton Mk.I and II
Spartan Three Seater 91-2, 115, 292
Stinson (model not recorded) 292

Supermarine S.4 94
Supermarine Scapa 82, 109, 292
Supermarine Sea Lion II and III 94
Supermarine Southampton Mk.I and II 57, 59-66, 82, 92, 106, 109, 111, 113, 123, *152*, *154*, 292
Supermarine Spitfire 60, 90

T

Tupolev ANT-25 172

V

Vega *see* Lockheed Vega DL-1A
Vickers Vernon 68
Vickers Vimy 28, 58
Vickers Virginia 58, 59
Vickers Wellington 240

W

Wapiti *see* Westland Wapiti Mk.IA and IIA
Westland Lysander 235, 292
Westland Wapiti Mk.IA and IIA 45, 52, 236, 292
Whitley *see* Armstrong Whitworth Whitley Mk.III and Mk.V

GENERAL

A

AB Aerotransport (ABA), Sweden 285
Acoustic mirrors 'listening ears' 76, *153*
Admiral Hipper 271
Admiral Scheer 270-1
Air Defence Experimental Establishment (ADEE) 75-6, *153*
Air France 183
Air Ministry (UK) 35, 50, 85, 87, 91, 100-1, 116, 118, 137, 140, 173, 179-83, 197, 201-2, 204, 206-8, 210, 218, 225-6, 230, 232, 234, 237, 241, 247, 250-1, 257, 262-4, 266
Airlines of Australia Ltd, *see* New England Airways Ltd
Anti-submarine mine Mk.XIX 271-2, 275
Archerfield Aerodrome, Brisbane 49, 117
Arctic convoys 268, 270
Atlantic Ferry *166*, *168*, 206-33, 234-5, 237, 239, 242, 252, 265, 291
      Air Services Department, *see* Canadian Pacific Railways (CPR)
      Atlantic Ferry Organisation (ATFERO) 209, 226-7, 230, 232, 238
      Canadian Pacific Railways (CPR) 207-9, 211, 223, 226, 230, 232
      Ferry Command 230, 232
      Return Ferry Service *168*, 229-30
Australian Army 33-4, 36, 41-3,
      Citizen Forces of the Australian Army 29-30, 289
Australian Navy 33, 43

B

Bismarck 267-8
Blackbushe Airport, UK 198
Brisbane Grammar School (BGS) 19, 22-3, 78, 145-6, 288, 289
British Airways Ltd 189-91, 285
British Broadcasting Corporation (BBC) 189-90, 213, 265
British Commonwealth Air Training Plan (BCATP) 209-10, 222
British Overseas Airways Corporation (BOAC) 185, 189-91, 199-200, 213, 221-2, 231, 234, 250-1, 285 *see also* Imperial Airways and British Airways Ltd
British War Cabinet 208, 257
British War Office 42, 200-1

Bubonic plague 11

C

Canadian Airways 208, 226
Canadian Pacific Railways (CPR) 207-8, 226
Centenary Air Race 83-4, 89, 94-106, *156*, *158*, 290
Central Technical College, Brisbane, Australia 30
Challenge International de Tourisme 1932 111
Church Army 13
Companion of the Bath (CB) 20
Consolidated Aircraft Factory, San Diego, USA 227
Course Setting Bomb Sight Mk.IIB 39
Croydon Aerodrome, UK 28, 88, 119-20, 125

D

de Havilland Aircraft Company Ltd 80-1

E

Eagle Farm Aerodrome, Brisbane, Australia 25, 28-9, 31, 49
Empire Air Mail Scheme (EAMS) 98, 122, 126, 128

F

Fédération Aéronautique Internationale (FAI) 100, 172, 178
First Class Air Navigator's Licence (civil) 85-91, 94, 105, 116, 118, 129, 237, 290
Fleet Air Arm 269
Flight Refuelling Limited 137, *165*, 184 *see also* In-flight refuelling and Cobham, Alan
Ford Model A Rumble Seat Coupe 116
Ford Model T 20

G

Gneisenau 260
Ground Engineer's Licences 'A', 'C' and 'X' 111, 129, 237, 290
G.T. Bennett and Co. 7, 19, 28, *149*

H

Hampshire Aeroplane Club 90, 95

Handbook of Instructions and General Information (Imperial Airways) 119, 130-1, 134-5
HMS *Hood* 267
HMS *Iron Duke* 65, 70
HMS *Prince of Wales* 231, 268
HMS *Royal Oak* 188
HMS *Valiant* 65
Honourable Company of Air Pilots 147

I

Imperial Airways 60, 88, 110, 116-7, 119-135, 136-41, 146-7, *158-166*, 171-175, 179-186, 187-91, 205, 207, 209, 235, 281, 285, 290 *see also* British Overseas Airways Corporation (BOAC)
In-flight refuelling 137-8, 141, *165*, 182, 184-5, 187, 290 *see also* Flight Refuelling Ltd

J

Jersey Airways 91-3, 95, 110-1, *155*, 290
Johnston Memorial Trophy 147, 189, 290

K

King George V Silver Jubilee Review 113
KLM, Waalhaven 99

L

Larkin Aircraft Supply Company 117
Lockheed Aircraft Plant, Burbank, USA 98, 100, 210, 214
Lucas' Papaw Ointment 10-11
Luft Hansa 171, 183, 205

M

MacRobertson Air Race *see* Centenary Air Race
MacRobertson International Air Races *see* Centenary Air Race
MacRobertson-Miller Aviation Company Ltd (MMA) 97
Manual of Air Navigation (AP1234) 116
Marconi A.D.6m transmitter and receiver 113
Marine Aircraft Experimental Establishment (MAEE) 110, 140-1
Methodist/Methodism 8, 10, 12-14, 17-18, 115

Ministry of Aircraft Production (MAP) 206, 208, 211, 225-6, 230 *see also* Aitken, Max, First Baron Beaverbrook
MV *Ulster Prince* 66

N

Neutrality Act (USA) 210-1
New England Airways Ltd 115-7, 290

O

Operation *Barbarossa* 268
Oswald Watt Gold Medal 147, 189, 290
Overland (motor car) 21

P

Pan American Airways (Pan Am) 171-2, 183, 186
Phoney War 189, 240, 247
Pilot's 'B' Licence (civil) 86, 91-2, 237, 289
Polish General Staff 196, 200, 204, 290
Portuguese World Exhibition 205
Prince Eugen 271

Q

Qantas Airways Ltd 25, 29, 31, 49, 120, 127, 130, 191
Queensland Aero Club 49,

R

Real Estate Institute of Queensland (REIQ) 19, 145
RMS *Mauretania* 65
Royal Aero Club 100
Royal Aeronautical Society (RaeS) 189, 239, 290
Royal Federation of Aero Clubs of Australia 147

S

Scharnhorst 260
Schneider Trophy 94
Second Class Air Navigator's Licence (civil) 85-6, 237, 289
Short Brothers (Shorts), Rochester, UK 110, 122-3, 126-7, 136-7, 139, 141, *161-2*, 173, 179, 182-4

Special Operations Executive (SOE) *see* Wilkinson, Major Peter
SS *Athenia* 187, 290
SS *Bremen* 82
SS *Essex* 10
SS *Hobson's Bay* 115-6
SS *Narkunda* 49
SS *Westfalen* 171
SS *Worthing* 65
Supreme War Council 201

T

Talbot (motor car) 112
Telecommunications Research Establishment (TRE) 75
The Air Mariner, by DCT Bennett 124, 290
The Automobile Association 100
The Complete Air Navigator, by DCT Bennett 116, 118, 124, 145, 290
Tirpitz 169, 267-76, 291
Toowoomba Grammar School 9
Trans-Canada Airlines 210

U

U-47 188
U-boats 188, 206, 268
University of Queensland, Brisbane 29
USS *Tuscaloosa* 185

V

Vickers-Armstrongs Ltd 95, 208

W

War Office 42, 200-1
West Australian Airways Ltd (WAA) 96, 98
Willingdon Bridge, Calcutta, India *166*, 192-3
Windsor State School, Brisbane 19
Windsor Street Railway Station, Montreal 209, 211
Wireless Telegraphy W/T Air Operator's Licence 111, 113, 129, 237, 290

www.ingramcontent.com/pod-product-compliance
Lightning Source LLC
Chambersburg PA
CBHW041303110526
44590CB00028B/4233